MULTISCALE MODELING

FROM ATOMS TO DEVICES

MULTISCALE MODELING

FROM ATOMS TO DEVICES

EDITED BY
PEDRO DEROSA
TAHIR CAGIN

CRC Press
Taylor & Francis Group
Boca Raton London New York

CRC Press is an imprint of the
Taylor & Francis Group, an **informa** business

CRC Press
Taylor & Francis Group
6000 Broken Sound Parkway NW, Suite 300
Boca Raton, FL 33487-2742

First issued in paperback 2017

© 2011 by Taylor and Francis Group, LLC
CRC Press is an imprint of Taylor & Francis Group, an Informa business

No claim to original U.S. Government works

ISBN 13: 978-1-138-11824-9 (pbk)
ISBN 13: 978-1-4398-1039-2 (hbk)

Library of Congress Cataloging-in-Publication Data

Multiscale modeling : from atoms to devices / editors, Pedro Derosa and Tahir Cagin.
    p. cm.
  "A CRC title."
  Includes bibliographical references and index.
  ISBN 978-1-4398-1039-2 (alk. paper)
  1. Nanostructured materials--Computer simulation. 2. Nanotechnology--Data processing. 3. Multiscale modeling. I. Derosa, Pedro. II. Cagin, Tahir. III. Title.

TA418.9.N35M85 2010
620'.50285--dc22
                                           2010028762

Visit the Taylor & Francis Web site at
http://www.taylorandfrancis.com

and the CRC Press Web site at
http://www.crcpress.com

. . . To Nicholas and Camila, to Daniela, to Ana and Pedro, PD

. . . To my children, Kerem, Elif, and wife Gul, TC

Contents

Preface

Nanoscience, or nanotechnology, has become an omnipresent keyword in most scientific and technological advances. It has been shown to hold the promise of a large impact on a number of applications, especially on novel devices. Computational methods such as *ab initio*, density functional theory (based on quantum theories of electronic structure), molecular mechanics, molecular dynamics, and Monte Carlo methods (based on classical mechanics and statistical mechanics of many-body systems) are well-established and used extensively in studying the nanoscale phenomena. All of these methods are well-developed and have captivated the interest of researchers, especially within the third quarter of the twentieth century. The increasingly widespread use of computational methods in the rational design of novel materials applications and devices has led to an increasing effort to employ these methods, based on fundamental theories of physics, in problems involving phenomena coupling different length and time scales within the same application. The simplest path to take has been to force the existing methods and models to make predictions on systems and phenomena for which they simply have not been designed. To tackle problems where multiple length and time scales are coupled, a new paradigm needs to be employed: multiscale modeling and simulation. Indeed, a common property of nanosystems is that they are multiscale systems. Nanosystems are systems in which the relevant features and properties depend on nanoscopic details, yet their performance resides in the macroscopic world and, thus, all scales are relevant. This requires the use of theories and methods that are accurate enough for the nanoscale but also able to be scaled up in length and time in a consistent manner, either through a hierarchical method of successive coarsening or by devising methods to handle the coupling of scales concurrently within a well-defined framework.

There is a tendency to make a one-to-one association between coarse-graining and multiscale modeling. Clearly coarse-graining is a key multicale method, but multiscale is much more than just coarse-graining and that is what this book tries to highlight. The reader should keep in mind, however, that the key word here is "atomic." The atomic scale is indeed where things actually happen and controlling this scale is key to successfully tailoring nanosystem properties at will. Thus, being able to model the connection between the large and the atomic scale is fundamental for predictive models to be able to assist in the design of novel nanomaterials and systems.

With this in mind, in this book we put together a number of approaches—carefully described by the contributors—for which more than one scale is explicitly considered. Throughout the book, the reader will be guided to a number of alternatives, from coarse-graining sampling of the atomic and mesoscale to Monte Carlo– and thermodynamic-based models that allow sampling of increasingly large scales up to multiscale models able to describe entire devices.

In Chapter 1, Morrow describes four of the most common techniques for coarse-graining, namely, rigorous matching correlation, force-matching, and empirical coarse-graining approaches. Their theoretical background, advantages, and limitations are the focus of that chapter. Chapter 1 describes the implementation of these common techniques and addresses their limitations, particularly pointing out limited portability.

In Chapter 2, the system-specific nature of coarse-graining techniques is captured by Cranford and Buehler and they emphasize the application-dependent parameterization characteristics of coarse-graining. The "finer-trains-coarser" multiscale approach is described as a procedure for parameterization, with actual details depending on the particular application. This method stresses the inherently atomistic origin of mesoscopic processes and describes how a connection can be established such that coarse potentials inherit the atomic nature of the process. Chapter 3, by the same authors, describes three carefully selected examples in which the parameterization, although based on the same principles, depends on the actual application. The first application illustrates how parameterization should be conducted to capture the structure–property relationship and its dependence on a system. The second application describes an approach to study processes at time scales that classical molecular dynamics (MD) cannot access. Finally, the purpose of the third application is to illustrate cases where the objective is to minimize the number of degrees of freedom.

Chapter 4 by Maroudas, Amat, and Kevrekidis is a transition between coarse-graining multiscale and other alternatives. This chapter describes an alternative use of coarse-graining, an elegant strategy to inherently study slow phase transitions that does not require the explicit coarse-graining of the dynamics. This minimizes or eliminates some of the most common problems of coarse-graining, some of which are described in Chapters 1 and 2. The coarse molecular-dynamics method instead consist of the selection of an appropriate coarse variable to describe the thermodynamic state of the system and a mapping between the scales, i.e., between the coarse and atomic variables. The method is beautifully exemplified by applying it to three different processes: melting, stress-induced solid–solid transitions, and order–disorder transitions of physisorbed layers.

The following two chapters describe the use of *ab initio* and density functional theory to obtain parameters needed for larger scale models, leading to a hierarchical parameterization. In Chapter 5, Dandala, Jansen, and Mainardi describe the alternative use of density functional theory parameters in a Monte Carlo method and illustrate the approach modeling an enzymatic reaction. Biological systems are multiscale in nature and reactions occur in very small domains (active sites) of gigantic macromolecules and both the electronic interaction at the atomic level and the effect at the macroscopic level are of relevance. Kinetic Monte Carlo is used to model methanol dehydrogenase oxidation by methanol dehydrogenase (MDH). A clear description of how a collection of atomic-resolution techniques is used to parameterize the coarse calculation of the process is offered.

The use of *ab initio* and density functional theory as the atomistic technique underlying the calculation of thermodynamics properties of alloy phase stability is explored by van de Walle in Chapter 6. A set of strategies to circumvent the need to model a large number of configurations necessary to accurately build the partition function for the systems is described with an application in phase diagram modeling for alloys. This chapter also bridges the use of multiscale modeling for large-scale materials science applications (continued in the next chapter), exploring the modeling of nanocomposites and offering the transition to multiscale models for systems at scales larger than the mesoscopic.

In Chapter 7, Elsbernd and Spanos describe their approach to nancomposite modeling. Modeling nanocomposites is one of the most challenging tasks for multiscale modelers. These materials owe their macroscopic properties to the interaction between the matrix and the nanoinsert and, thus, for a model to have any hope of appropriately predicting these materials' properties, both scales have to be explicitly included and properly connected. The chapter presents the embedded fiber finite element method (EFFEM). This method was designed for carbon nanotubes polymer nanocomposites. Within the framework of this method, nanotube morphology is explicitly considered such that the properties of the representative volume element greatly depend on that morphology. The implementation of the model and its virtues and limitations in predicting elastic and thermal properties of nanocomposites is presented.

The last two chapters describe multiscale models reaching all the way to devices. In Chapter 8, Sevik describes the ensemble Monte Carlo method as applied to high field-charge transport problems in electronic devices and demonstrates the use of electronic structure data obtained by the *ab initio* method in the solution of non-linear integro-differential equations describing charge transport processes in these devices.

Chapter 9 by Siddiki describes modeling of two-dimensional charge devices and the application of modern many-body quantum theories in detail. Here, solutions to Schrödinger and Poisson equations are developed through the use of finite element methods and applied to gating, quantum dots, and the development of solutions to these equations in the presence of a magnetic field, quantized Hall effects, and related phenomena.

Contributors

Miguel A. Amat
Department of Chemical Engineering
and Program in Applied and
Computational Mathematics
Princeton University
Princeton, New Jersey

Markus J. Buehler
Laboratory for Atomistic and Molecular
Mechanics
Department of Civil and Environmental
Engineering
Massachusetts Institute of Technology
Cambridge, Massachusetts

Steve Cranford
Laboratory for Atomistic and Molecular
Mechanics
Department of Civil and Environmental
Engineering
Massachusetts Institute of Technology
Cambridge, Massachusetts

Nirmal Kumar Reddy Dandala
Chemical Engineering Program
Institute for Micromanufacturing
Louisiana Tech University
Ruston, Louisiana

Paul Elsbernd
Department of Mechanical Engineering
Rice University
Houston, Texas

A. P. J. Jansen
Schuit Institute of Catalysis
Eindhoven University of Technology
Eindhoven, the Netherlands

Ioannis G. Kevrekidis
Department of Chemical Engineering
and Program in Applied and
Computational Mathematics
Princeton University
Princeton, New Jersey

Daniela Silvia Mainardi
Chemical Engineering Program
Institute for Micromanufacturing
Louisiana Tech University
Ruston, Louisiana

Dimitrios Maroudas
Department of Chemical Engineering
University of Massachusetts–Amherst
Amherst, Massachusetts

Timothy Morrow
Institute for Micromanufacturing
and Department of Chemical
Engineering
Louisiana Tech University
Ruston, Louisiana

Cem Sevik
Artie McFerrin Department of
Chemical Engineering
Texas A&M University
College Station, Texas

Afif Siddiki
Physics Department
Faculty of Sciences
Istanbul University
Istanbul, Turkey

Pol Spanos
Department of Mechanical Engineering
Rice University
Houston, Texas

Axel van de Walle
Engineering and Applied Science
　Division
California Institute of Technology
Pasadena, California

1 Overcoming Large Time- and Length-Scale Challenges in Molecular Modeling: A Review of Atomistic to Mesoscale Coarse-Graining Methods

Timothy Morrow

CONTENTS

1.1 INTRODUCTION

In the quest for knowledge, it is not uncommon for researchers to push the limits of simulation techniques to the point where they have to be adapted or totally new techniques or approaches become necessary. True multiscale modeling techniques are becoming increasingly necessary given the growing interest in materials and processes on which large-scale properties are dependent or that can be tuned by their low-scale properties. An example would be nanocomposites, where embedded nanostructures completely change the matrix properties due to effects occurring at the atomic level. Complex physical systems like nanocomposites, fuel cell membranes, electrolyte systems, and polymer systems owe some of their properties to processes occurring at the atomic-length scale and femto-nanosecond timescale. Controlling these small-scale properties can be the key to tuning the properties of these materials and opens up myriad potential applications. Unfortunately, it can be impractical and sometimes impossible to study these systems with molecular modeling methods using fully atomistic descriptions of the system due to the large time and length scales that must be accessed (i.e., timescales > 1 ns, length scales > 20 nm, system sizes > 1,000,000 atoms). To access such large time and length scales, the molecular models used to represent the physical system must be simplified or coarse-grained in such a way as to preserve only the interesting degrees of freedom; however, to account for the influence of the underlying atomistic system responsible for some of the macroscopic properties, a careful protocol must be followed. That protocol consists of defining the right parameters (selection or development of the coarse-graining method), systematically determining the values for those parameters from atomistic simulations, and clearly specifying the range of validity of those methods and parameters.

Multiscale modeling is very much under development; among other problems, the connection between scales is not fully resolved for many applications and parameters are not general and can only be expected to work for the conditions and systems for which they were developed, assuming the parameters exist at all. The current state of multiscale modeling consists of a set of robust methods for a limited set of applications in well-defined conditions. This review will not discuss every coarse-graining technique that has been developed, but will instead present the theoretical and computational background behind some of the main techniques and discuss their advantages and disadvantages. This review is divided into four parts: (1) discussion of rigorous coarse-graining techniques, (2) discussion of matching correlation function techniques, (3) discussion of force-matching techniques, and (4) discussion of empirical coarse-graining techniques.

1.2 RIGOROUS COARSE-GRAINING METHOD

Of all the coarse-graining methods that have been proposed to date, the method developed by Dijkstra et al. [1] and the similar method of Bolhuis et al. [2] stand alone as the only methods that rigorously map the partition function of the fully atomistic system onto the partition function of the coarse-grained system, which ensures that the coarse-grained system reproduces all the equilibrium thermodynamic properties of the fully atomistic system. Following the earlier works of McMillan and Mayer [3], Dijkstra et al. obtained the expression for the effective potential through rigorously coarse-graining uncharged systems of spherical particles in the semigrand canonical ensemble. In this rigorous method, the semigrand ensemble Hamiltonian of a two-component system is mapped onto an effective one-component canonical ensemble Hamiltonian whose expression is composed of a series of onebody, two-body, threebody, and larger terms that depend upon the temperature of the system but that are independent of the density of the canonically treated component. The effective potential terms are given by particle-insertion formulas. To derive the expressions for the coarse-grained potentials, one starts by requiring that the partition function of the degrees of freedom in the coarse-grained system be identical to the partition function of the same degrees of freedom in the original atomistic system. This condition yields an equation for the effective potential in the coarse-grained system that preserves the thermodynamic and structural properties of the original atomistic system. In a two-component semigrand ensemble, the number of particles of one of the components and the chemical potential of the other component are fixed, along with the volume and temperature (N_A, μ_B, V, T), where N_A is the number of molecules of the component to be preserved and μ_B is the chemical potential of the component to be coarse-grained out of the system. Dijkstra et al. showed that, in this ensemble, the effective potential naturally splits into a sum of a volume term along with onebody, twobody, threebody, and larger N_A-body interactions. The effective Hamiltonian obtained for the coarse-grained system is given by

$$H^{\mathrm{eff}} = H_{AA} + \Omega(\mu_B), \tag{1.1}$$

where $H_{AA} = \sum_{ij} \phi_{AA}\left(r_{ij}\right)$ is the interaction between A particles and Ω is the effective interaction. Ω is given by

$$\Omega = \sum_{n=0}^{N_A} \Omega_n = -\frac{\ln \Xi_0}{kT} + N_A \omega_1 + \sum_{i,j} \omega_2(r_{ij}) + \sum_{i,j,k} \omega_3(\mathbf{r}_i, \mathbf{r}_j, \mathbf{r}_k) + \cdots \tag{1.2}$$

The first term on the far right side of Equation 1.2 can be identified as the grand canonical potential of a pure fluid of species B (where Ξ_0 is its grand canonical partition function). The rest of the terms are the onebody (ω_1), twobody (ω_2), threebody (ω_3), and larger terms. As noted above, a very important property of the expressions for ω_1 in Equation 1.2 is that they do not depend on the total number of particles of species A, N_A. Thus, the individual potentials ω_1 are species A density-transferable (but still depend on T and μ_B). On the other hand, Ω in Equation 1.2 contains terms for all numbers of particles of species A up to N_A. Thus the density transferability is only an advantage if the series of effective interactions in Equation 1.2 converges after the first few terms. Following the work of Dijkstra et al., Chennamsetty et al. [4] devised a technique to compute the onebody, twobody, and larger N_A-body potentials using Widom's particle insertion method. They also showed that the twobody term in Equation 1.2 can be computed from the potential of mean force (PMF) of a system of two species A particles in a grand-canonical sea of species-B particles. This PMF route avoids the difficulties of using Widom's particle insertion method at high densities [5].

Using an effective Hamiltonian truncated at the twobody term, Dijkstra et al. computed the phase diagrams of size—asymmetric binary hard sphere mixtures with size ratios ranging from 5:1 to 30:1. They found that the one-component effective Hamiltonian accurately reproduced the two-component phase diagram for all size ratios studied, including the small size ratio of 5:1, where they could not justify the twobody approximation from geometrical arguments. They were also able to use coarse-grained simulations to predict the phase diagram for mixtures with size ratios of 20:1 and 30:1, where ergodicity problems made simulations of the two-component system intractable. Chennamsetty et al. [4] have used Dijkstra's method to coarse-grain a binary mixture of argon and krypton into an effective pure-krypton system. They found that truncation of the effective Hamiltonian at the twobody term only provided an accurate representation of the two-component system for krypton compositions up to 20%. While Dijkstra's method is desirable as a rigorous coarse-graining method, calculation of the terms in the effective Hamiltonian is computationally expensive and the method is not practical if the series does not converge rapidly. The method cannot be used to coarse-grain ionic systems unless approximations to the coulomb potential are made that make the series convergent [6]. Lastly, the method has yet to be extended to the coarse-graining of structured molecules.

The primary advantage of Dijkstra's method over other, nonrigorous, coarse-graining methods is that the coarse-grained potentials computed from the nonrigorous methods are strictly valid only for the state point (i.e., temperature and density) from which they were generated. Such potentials are often computed from fully atomistic simulations at a relatively low density and then used to perform mesoscale simulations at other, usually higher, density state points, but there is no underlying mathematical proof that the coarse-grained potentials will provide accurate mesoscale results at higher density state points. In Dijkstra's method, it can be proven that the coarse-grained potentials are independent of the density of the effective one-component system, so, in principle, the coarse-grained potentials computed using this method would provide accurate mesoscale results at any density. A secondary advantage of Dijkstra's method is that the thermodynamic properties of the original two-component system (pressure, enthalpy, chemical potential, etc.) are preserved in the one-component coarse-grained system, while this information is lost when using a nonrigorous coarse-graining method. An important consequence of this is that phase equilibrium properties of two-component mixtures can be predicted using coarse-grained one-component simulations with Dijkstra's method, but not with any of the nonrigorous methods.

As noted above, Dijkstra's work demonstrated that effective one-component potentials truncated at the twobody term provide accurate mesoscale results for binary mixtures of hard spheres in which the size ratio between the components is ≥ 5.1. However, mixtures of more realistic systems will possess not only a size ratio between the components, but also an energy ratio, which describes the relative strengths of the components' attractive intermolecular forces. To date, no systematic study of the range of size and energy ratios over which twobody coarse-grained potentials can provide accurate results has been undertaken.

1.2.1 Coarse-Graining by Matching Correlation Functions: Potential of Mean Force and Integral Equation Approaches

The potential of mean force between two particles is defined in terms of the pair radial distribution function (RDF) by

$$u_{ij} = -kT \ln[g(r_{ij})], \tag{1.3}$$

where $g(r_{ij})$ is the RDF between particles i and j in the fully atomistic system. It can be shown that for a two-component mixture, which is infinitely dilute in one of the components, the PMF of the dilute component becomes equal to the twobody potential term in Equation 1.2. Thus, the PMF can be thought of as an effective one-component coarse-grained potential that is rigorously correct in the infinitely dilute limit, but becomes less accurate as the density of the preserved species (species A in the context of Section 1.2) is increased. This method has been used extensively in the literature to calculate coarse-grained potentials. In a Langevin dynamics study of the surfactant n-decyltrimethylammonium chloride in water, the water was coarse-grained out of the system using an effective potential approximated as the potential of mean force between different groups of atoms, designated as "head" or "tail" (shown in Figure 1.1) [7]. Using the resulting coarse-grained potential, the authors observed the onset of surfactant self-assembly over a simulation period of 12 ns. Bolhuis et al. [8], however, showed that the PMF method was unable to reproduce the structures of polymer systems at high densities, where threebody and higher potentials of mean force must be included in the coarse-grained potential. The contributions from higher-body PMFs can be approximately accounted for in a two-body coarse-grained potential using integral equations. Integral equations relate the

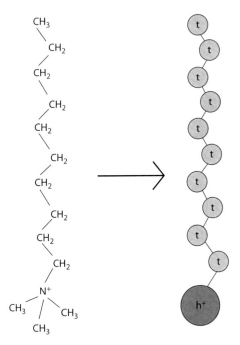

FIGURE 1.1   Coarse-graining of the atoms of the surfactant cation n-decyltrimethylammonium into nonpolar tail (t) groups and a positively charged, polar head group (h$^+$).

RDF to the direct correlation function $c(r)$. An example is the Ornstein-Zernike [9] integral equation:

$$h\left(r_{AA}\right) = c\left(r_{AA}\right) + \rho_A \int h\left(r_{AA}\right) c\left(r_{AA}\right) dr_{AA}, \tag{1.4}$$

where $h(r_{AA}) = g(r_{AA}) - 1$, r_{AA} is the distance between two A-type particles in the fully atomistic simulation, $c(r_{AA})$ is the direct correlation function between two A-particles in the fully atomistic simulation, and ρ_A is the density of species A. The twobody coarse-grained potential can be related to the direct correlation function $c(r_{AA})$ by using a closure relation like the Percus-Yevick [10] equation:

$$u\left(r_{AA}\right) = kT \ln\left(1 - \frac{c\left(r_{AA}\right)}{g\left(r_{AA}\right)} \right), \tag{1.5}$$

where $u(r_{AA})$ is the effective one-component (species A) pair potential that approximately accounts for the effects of threebody and higher interactions. By performing a fully atomistic simulation of a binary mixture of species A and B and calculating $g(r_{AA})$, Equations 1.4 and 1.5 can then be used to calculate the coarse-grained potential $u(r_{AA})$. Silbermann et al. [11] have used the integral equation procedure to coarse-grain out water from ethanol/water mixtures and simultaneously coarse-grain the intramolecular degrees of freedom of ethanol. They observed that the integral equation procedure satisfactorily reproduced the fully atomistic RDF in the coarse-grained simulations up to concentrations of 50% (by weight) ethanol in water, but showed significant deviations at higher concentrations. It is also important to note that coarse-grained potentials determined using the integral equation approach are not guaranteed to accurately reproduce all of the thermodynamic properties of the fully atomistic system, even when the RDF of interest is accurately reproduced. Only coarse-grained potentials calculated by equating the fully atomistic- and coarse-grained system partition functions, as is done in Dijkstra's method, are capable of reproducing all thermodynamic properties of the fully atomistic system.

1.2.2 COARSE-GRAINING BY MATCHING CORRELATION FUNCTIONS

The coarse-graining methods of Section 1.2.1 ensure that the coarse-grained system reproduces a set of correlation functions of the atomistic system, either a set of PMFs or a set of RDFs. An alternative method to using an integral equation approach to calculate a twobody coarse-grained potential that will reproduce a set of RDFs is to iterate the coarse-grained potential until the desired set of atomistic-system correlation functions are satisfactorily reproduced in the coarse-grained system. This method will approximately account for contributions to the coarse-grained potential from threebody and higher interactions, but it will not necessarily ensure that all of the equilibrium thermodynamic properties of the atomistic system will be reproduced by the coarse-grained system, and the coarse grained potentials obtained will strictly be valid only for the specific state point for which they were calculated from the fully atomistic system (i.e., the potentials will be temperature- and density-dependent). This is in contrast to coarse-grained potentials obtained using Dijkstra's method, which are temperature-dependent but not density-dependent. That is not to say that coarse-grained potentials calculated from matching correlation function approaches are not useful: they can be used to study the state point of interest; may be transferable over a certain range of state points; and are very useful for studying systems with long-range electrostatic forces or complex intramolecular degrees of freedom since these systems are very difficult to coarse-grain using rigorous methods.

One of the most popular matching correlation function coarse-graining methods is the well-known iterative Boltzmann inversion method (IBI) [12]. This involves iteratively fitting the effective twobody potentials based upon the differences between the site–site RDFs in the coarse-grained system and those of the fully atomistic system. The coarse-grained potentials are converged in the IBI procedure using the following equation:

$$u_{k+1}\left(r_{ij}\right) = u_k\left(r_{ij}\right) + kT \frac{g_k\left(r_{ij}\right)}{g_{ref}\left(r_{ij}\right)}, \qquad (1.6)$$

where $u_k(r_{ij})$ is the coarse-grained interaction potential between sites (or beads) i and j calculated at iteration k, $g_k(r_{ij})$ is the RDF between beads i and j in the coarse-grained simulation at iteration k, and $g_{ref}(r_{ij})$ is the RDF between beads i and j in the fully atomistic simulation. The PMF from Equation 1.3 usually gives a good starting value for $u_1(r_{ij})$, which is used in a coarse-grained simulation to calculate $g_1(r_{ij})$. The difference between $g_1(r_{ij})$ and $g_{ref}(r_{ij})$ determines the value of $u_2(r_{ij})$, which is then used to determine $g_2(r_{ij})$ from another coarse-grained simulation. This procedure is repeated until $g_k(r_{ij}) = g_{ref}(r_{ij})$ to within some acceptable tolerance. The IBI method has been used to coarse-grain pure alkane systems with chain lengths ranging from 16 to 96 [13], lipid bilayers [14], polymeric systems [15], and has been used to study surfactant self-assembly on solid surfaces [16]. The effective potentials obtained from this method are strictly valid only for the temperature and density at which they were computed, and to obtain the most accurate results the IBI procedure should be performed at every state point of interest. Silbermann et al. [11] used the IBI procedure to coarse-grain an ethanol/water system. They observed that the IBI procedure satisfactorily reproduced the atomistic RDF at all ethanol concentrations they examined (up to 70% ethanol by weight), and they observed that the coarse-grained potentials calculated using the IBI procedure provided better results than those obtained using an integral equation procedure with the hypernetted chain closure relation.

Lyubartsev and Laaksonen developed a matching correlation function coarsegraining method called inverse Monte Carlo (IMC) [17]. This approach is similar to the IBI method, but the two methods differ in the way the coarse-grained potentials are updated at each iteration step. As with the IBI method, the IMC procedure begins by choosing an initial estimate, usually from the PMF, for the coarse-grained potential $u_1(r_{ij})$, and a coarse-grained simulation is performed to calculate the RDF, $g_1(r_{ij})$. The differences between atomistic RDF and coarse-grained RDF, $\Delta g_1 = g_{ref}(r_{ij}) - g_1(r_{ij})$

are calculated, and $\Delta g_1(r_{ij})$ is used in a linear equation to calculate a correction to the coarse-grained potential, $\Delta u_1(r_{ij})$. The resulting coarse-grained potential is used in a simulation to calculate $g_2(r_{ij})$, and the iteration procedure is repeated until $g_k(r_{ij}) = g_{ref}(r_{ij})$ to within some acceptable tolerance. The IMC procedure has been used to calculate coarse-grained potentials for ionic salts in water [18], to study the interactions of different alkali ions with DNA [19], and in a study of the behavior of cholesterol/phospholipid bilayers [20], the authors reported a computational speedup of approximately eight orders of magnitude using the coarse-grained model.

1.3   COARSE-GRAINING BY MATCHING FORCES

Izvekov and Voth [21] have recently proposed an alternative to the matching correlation function approach based upon a force-matching (FM) approach originally proposed by Ercolessi and Adams [22] and referred to as the "multiscale coarse-graining" method. In this method, the forces acting between sites on the coarse-grained model are fitted to the forces on those sites obtained from a fully atomistic molecular dynamics simulation.

The FM procedure is performed as follows: First, a fully atomistic simulation is performed, and the average force acting on a predefined set of coarse-grained sites (such as the center of mass of a functional group) is calculated. The forces on the sites in the coarse-grained model are then set equal to the average forces on the sites from the atomistic simulation. The force on a coarse-grained site is assumed to be the sum of a short-range force and a coulombic force [23]:

$$\mathbf{f}_{ij}^{CG}\left(\mathbf{r}_{ij}\right) = -\left[f_{ij}\left(r_{ij}\right) + \frac{q_i q_j}{r_{ij}^2}\right]\frac{\mathbf{r}_{ij}}{r_{ij}}, \tag{1.7}$$

where $\mathbf{f}_{ij}^{CG}(\mathbf{r}_{ij})$ is the force acting on site i due to site j, $f_{ij}(r_{ij})$ is the short-range component of the force represented by a third-order polynomial (such as a cubic spline) which is to be fitted, q_i and q_j are the charges on sites i and j which are also to be fitted, $\mathbf{r}_{ij}$ is the distance vector between the two sites, and r_{ij} is the magnitude of this vector. Constraints are added to the fitting procedure to ensure that the proper charge on each coarse-grained molecule is obtained, and that the virial of the coarse-grained system matches that of the atomistic simulation, which helps improve the coarse-grained model's prediction of thermodynamic properties such as pressure and density [23].

As with the matching correlation function methods, the effective potentials obtained from the FM approach are temperature- and density-dependent. The FM approach has been used to perform coarse-grained simulations of lipid bilayers [24], nanoparticles [25], and ionic liquids [26]. Recent efforts have been made to develop a formal statistical mechanical framework for the FM method, including: a proof [27] that coarse-grained potentials computed using the FM method will be consistent with their underlying atomistic models in both momentum and configurational space; a numerical procedure for computing force-matched coarse-grained potentials using a set of basis functions [28]; an investigation of the range of binary

Lennard–Jones mixture concentrations over which effective one-component simulations using twobody force-matched coarse-grained potentials can accurately reproduce the properties of the underlying two-component mixture [29]; and a method for using atomistic simulation data at one temperature to construct force-matched coarse-grained potentials that will be valid at other temperatures [30]. While the FM approach provides an advantage over the matching correlation function methods in that no iteration of coarse-grained simulations is required, no direct comparison of the two methods for the same system has yet been undertaken to determine if one method provides clearly superior coarse grained potentials.

1.4   EMPIRICAL COARSE-GRAINING TECHNIQUES

Coarse-grained potentials can be determined empirically by choosing a mathematical form for the coarse grained potentials (such as the Lennard–Jones equation) and fitting the adjustable parameters to fully atomistic simulation data (such as densities, vapor pressures, etc.). Smit et al. [31] developed coarse-grained potentials for aqueous surfactant systems by fitting Lennard–Jones parameters for the coarse-grained surfactant sites. Using the empirically coarse-grained potentials, they were able to observe the formation and breakdown of micelles. A similar empirical coarse-graining approach was used by Marrink and Mark [32] to study the behavior of lipid bilayers. In this study, Lennard–Jones potentials for the coarse-grained lipid molecules were fitted to pure liquid density data and oil/water mutual solubility data. The coarse-grained simulations showed spontaneous formation of the lipid molecules into bilayers, and the structural properties of the bilayer closely matched experimental data. Burov et al. [33] used an empirical coarse-graining approach to study micelle formation in aqueous solutions of ionic surfactants, and Suter et al. [34] reviewed the use of molecular dynamics simulations with empirically coarse-grained models to study the properties of clay mineral systems.

While empirically fit coarse-grained potentials can be successful in reproducing the properties to which they were fit, there is no guarantee that the model will satisfactorily reproduce any other properties.

1.5   SUMMARY

Many interesting but complex systems such as membranes, polymers, biomolecules, and surfactant solutions can be impractical or impossible to study using molecular modeling methods with fully atomistic descriptions of the system due to the extremely large time and length scales that are necessary. To access such large time and length scales, the molecular models used to represent the physical system must be coarse-grained in such a way as to preserve only the interesting degrees of freedom but still account for the influence of the underlying atomistic system. This review discusses four popular methods for coarse-graining complex physical systems into models that can be simulated over large time and length scales: a rigorous coarse-graining technique; matching correlation function techniques; force-matching (FM) techniques; and empirical coarse-graining techniques.

In the rigorous method, the partition function of the fully atomistic system is set equal to the partition function of the coarse-grained system. The resulting expression for the coarse-grained potential is comprised of a series of onebody, twobody, threebody, and larger terms that depend upon the temperature of the system but are independent of the density of the component in the coarse-grained simulation. The main advantage of this method is that the coarse-grained potential is guaranteed to reproduce all the thermodynamic and structural properties of the original atomistic system, provided that the coarse-grained potential series converges at the twobody term. However, a procedure for rigorously coarse-graining systems with complex intramolecular degrees of freedom or long-ranged forces (i.e., ionic or polar compounds) has not yet been worked out. Such systems are usually coarse-grained by matching a specific structural or thermodynamic property between the atomistic and coarse-grained simulations. Such nonrigorous techniques are not guaranteed to reproduce all the thermodynamic and structural properties of the original atomistic system, and the coarse-grained potentials generated from nonrigorous techniques are valid only in the vicinity of the state point from which they were computed. The nonrigorous techniques can be categorized into matching correlation function, FM, and empirical coarse-graining techniques.

One of the simplest matching correlation function techniques is the use of the potential of mean force (PMF), which is a rigorous coarse-grained potential in the infinite dilution limit. While coarse-grained potentials obtained from PMFs can provide satisfactory results for dilute systems, the results become unsatisfactory at moderate and high densities due to the neglect of the contributions of many-body interactions in the coarse-grained potential. The effects of these interactions can be approximately accounted for in the coarse-grained potential in a fairly simple way by using integral equations and a closure relation for the radial distribution functions (RDFs) that are to be matched.

A more effective but computationally expensive method for approximating the effects of many-body interactions in matching correlation function coarse-grained potentials is to iteratively adjust the coarse-grained potential until the RDFs of interest from a fully atomistic simulation and those from coarse-grained simulations agree to within a specified tolerance. The iterative Boltzmann inversion (IBI) and inverse Monte Carlo (IMC) methods are examples of such techniques. Coarse-grained potentials computed using iterative matching correlation function approaches have been shown to provide satisfactory results even at high solute densities, but the potentials are density-dependent.

FM is a recently proposed alternative to the matching correlation function approach. In this method, the forces acting between sites on the coarse-grained model are fitted to the forces on those sites obtained from a fully atomistic molecular dynamics simulation. This approach has been shown to work well for a number of systems, and has an advantage over the matching correlation function methods in that no iteration of coarse-grained simulations is required.

Another alternative coarse-grained method is to choose a mathematical form for the coarse-grained potentials and then fit the potentials' parameters to data from fully atomistic simulations or experiments. While this empirical approach can be successful in reproducing the properties to which they were fit, there is no guarantee that the model will satisfactorily reproduce any other properties.

REFERENCES

1. M. Dijkstra, R. van Roij, and R. Evans. 1999. Phase diagram of highly asymmetric binary hard-sphere mixtures. *Physical Review E* 59:5744–5770.
2. P. G. Bolhuis, A. A. Louis, and J. P. Hansen. 2001. Many-body interactions and correlations in coarse-grained descriptions of polymer systems. *Physical Review E* 64:021801.
3. J. E. Mayer and M. G. Mayer. 1940. *Statistical mechanics.* New York: Wiley.
4. N. Chennamsetty, H. Bock, and K. E. Gubbins. 2005. Coarse-grained potentials from Widom's particle insertion method. *Molecular Physics* 103:3185.
5. D. Frenkel and B. Smit. 2001. *Understanding molecular simulation: From algorithms to applications*, 2nd ed. San Diego: Academic Press.
6. C. Russ, H. H. von Grünberg, M. Dijkstra, and R. van Roij. 2002. Three-body forces between charged colloidal particles. *Physical Review E* 66:011402.
7. H. Shinto, S. Morisada, M. Miyahara, and K. Higashitani. 2004. Langevin dynamics simulations of cationic surfactants in aqueous solutions using potentials of mean force. *Langmuir* 20:2017.
8. P. G. Bolhuis, A. A. Louis, and J. P. Hansen. 2001. Many-body interactions and correlations in coarse-grained descriptions of polymer systems. *Physical Review E* 64:21801.
9. L. S. Ornstein and F. Zernike. 1914. Accidental deviations of density and opalescence at the critical point of a single substance. *Proceedings of the Academy of Science of Amsterdam* 17:793.
10. J. P. Hansen and I. R. McDonald. 1986. *Theory of simple liquids*, 2nd ed. London: Academic Press.
11. J. R. Silbermann, S. H. L. Klapp, M. Schoen, N. Chennamsetty, H. Bock, and K. E. Gubbins. 2006. Mesoscale modeling of complex binary fluid mixtures: Towards an atomistic foundation of effective potentials. *Journal of Chemical Physics* 124:074105.
12. A. K. Soper. 1996. Empirical potential Monte Carlo simulation of fluid structure. *Chemical Physics* 202:295–306.
13. A. S. Ashbaugh, H. A. Patel, S. K. Kumar, and S. Garde. 2005. Mesoscale model of polymer melt structure: Self-consistent mapping of molecular correlations to coarse-grained potentials. *Journal of Chemical Physics* 122:104908.
14. M. Müller, K. Katsov, and M. Schick. 2006. Biological and synthetic membranes: What can be learned from a coarse-grained description? *Physics Report* 434:113–176.
15. Q. Sun, F. R. Pon, and R. Faller. 2007. Multiscale modeling of polystyrene in various environments. *Fluid Phase Equilibria* 261:35–40.
16. W. Shinoda, R. DeVane, and M. L. Klein. 2008. Self-assembly of surfactants in bulk phases and at interfaces using coarse-grain models. In *Coarse-graining of condensed phase and biomolecular systems*, ed. G. A. Voth. Boca Raton, FL: CRC Press.
17. A. Lyubartsev and A. Laaksonen. 1995. Calculation of effective interaction potentials from radial distribution functions: A reverse Monte Carlo approach. *Physical Review E* 52:3730.
18. A. Lyubartsev and A. Laaksonen. 1996. Concentration effects in aqueous NaCl solutions. A molecular dynamics simulation. *Journal of Physical Chemistry* 100:16410.
19. A. Lyubartsev and A. Laaksonen. 1999. Effective potentials for ion–DNA interactions. *Journal of Chemical Physics* 111:11207.

20. T. Murtola, E. Falck, M. Patra, M. Karttunen, and I. Vattulainena. 2004. Coarse-grained model for phospholipid/cholesterol bilayer. *Journal of Chemical Physics* 121:9156.
21. S. Izvekov and G. A. Voth. 2005. A multiscale coarse-graining method for biomolecular systems. *Journal of Physical Chemistry B* 109:2469–2473.
22. F. Ercolessi and J. B. Adams. 1994. Interatomic potentials from first-principles calculations: The force-matching method. *Europhysics Letters* 26:583.
23. S. Izvekov and G. A. Voth. 2005. Multiscale coarse graining of liquid-state systems. *Journal of Chemical Physics* 123:134105.
24. J. W. Chu, S. Izvekov, and G. A. Voth. 2006. The multiscale challenge for biomolecular systems: Coarse-grained modeling. *Molecular Simulation* 32:211–218.
25. S. Izvekov, A. Violi, and G. A. Voth. 2005. Systematic coarse-graining of nanoparticle interactions in molecular dynamics simulation. *Journal of Physical Chemistry B* 109:17019–17024.
26. Y. Wang, S. Izvekov, T. Yan, and G. A. Voth. 2006. Multiscale coarse-graining of ionic liquids. *Journal of Physical Chemistry B* 110:3564–3575.
27. W. G. Noid, J.-W. Chu, G. S. Ayton, V. Krishna, A. Izvekov, G. A. Voth, A. Das, and H. C. Andersen. 2008. The multiscale coarse-graining method. I. A rigorous bridge between atomistic and coarse-grained models. *Journal of Chemical Physics* 128:244114.
28. W. G. Noid, P. Liu, Y. Wang, J.-W. Chu, G. S. Ayton, A. Izvekov, H. C. Andersen, and G. A. Voth. 2008. The multiscale coarse-graining method. II. Numerical implementation for coarse-grained molecular models. *Journal of Chemical Physics* 128:244115.
29. A. Das and H. C. Andersen. 2009. The multiscale coarse-graining method. III. A test of pairwise additivity of the coarse-grained potential and of new basis functions for the variational calculation. *Journal of Chemical Physics* 131:034102.
30. V. Krishna, W. G. Noid, and G. A. Voth. 2009. The multiscale coarse-graining method. IV. Transferring coarse-grained potentials between temperatures. *Journal of Chemical Physics* 131:024103.
31. B. Smit, K. Esselink, P. A. J. Hilbers, N. M. van Os, L. A. M. Rupert, and I. Szleifer. 1993. Computer simulation of surfactant self-assembly. *Langmuir* 9:9.
32. S. Marrink and A. Mark. 2003. Molecular dynamics simulation of the formation, structure, and dynamics of small phospholipid vesicles. *Journal of the American Chemical Society* 125:15233.
33. S. V. Burov, N. P. Obrezkov, A. A. Vanin, and E. M. Piotrovskaya. 2008. Molecular dynamic simulation of micellar solutions: A coarse-grain model. *Colloid Journal* 70:1–5.
34. J. L. Suter, R. L. Anderson, C. Greenwell, and P. V. Coveney. 2009. Recent advances in large-scale atomistic and coarse-grained molecular dynamics simulation of clay minerals. *Journal of Materials Chemistry* 19:2482–2493.

2 Coarse-Graining Parameterization and Multiscale Simulation of Hierarchical Systems. Part I: Theory and Model Formulation

Steve Cranford and Markus J. Buehler

CONTENTS

2.1 INTRODUCTION

Coarse-grain models provide an efficient means to simulate and investigate systems in which the desired behavior, property, or response is inherently at the mesoscale—those that are both inaccessible to full atomistic representations and inapplicable to continuum theory. Granted, a developed coarse-grain model can only reflect the behavior included in their governing potentials and associated parameters, and consequently, the source of such parameters typically determines the accuracy and utility of the coarse-grain model. It is our contention that a complete theoretical foundation for any system requires synergistic multiscale transitions from atomic to mesoscale to macroscale descriptions. Hierarchical "handshaking" at each scale is crucial to predict structure–property relationships, to provide fundamental mechanistic understanding, and to enable predictive modeling and material optimization to guide synthetic design efforts. Indeed, a finer-trains-coarser approach is not limited to bridge atomistic to mesoscopic scales (which is the focus of the current discussion), but can also refer to hierarchical parameterization transcending any scale, such as mesoscopic to continuum levels. Such a multiscale modeling paradigm establishes a fundamental link between atomistic behavior and the coarse-grain representation, providing a consistent theoretical approach to develop coarse-grain models for systems of various scales, constituent materials, and intended applications.

2.1.1 MOTIVATION: HIERARCHICAL SYSTEMS AND EMPIRICAL LINKS

Many biological tissues are composed of hierarchical structures, which provide exceptional mechanical, optical, or chemical properties due to specific functional adaptation and optimization at all levels of hierarchy. Nature has shown that a material's structure—and not its composition alone—must be considered in the design of new material systems for use in high-performance applications. Fundamental structural arrangements and the mechanistic properties are inherently linked. Indeed, nature's integration of robustness, adaptability, and multifunctionality requires the merging of structure and material across a broad range of length scales, from nano to macro, and is apparent in biological materials such as bone, wood, and protein-based materials [1–3]. The analysis of such hierarchical materials is an emerging field that uses the relationships between multiscale structures, processes, and properties to probe deformation and failure phenomena at the molecular and microscopic levels [4].

For the current discussion, the term *hierarchical* is used loosely to indicate a material system with at least a single distinct differentiation between constituent material *components* and global system *structure*. For example, in Chapter 3 we discuss both coarse-grain modeling of carbon nanotube arrays and collagen fibrils. For the nanotube arrays, the components are defined by individual carbon nanotubes, and thus the array is considered a hierarchical structure. For the collagen fibrils, the components are defined as tropocollagen molecules, while the system of interest is the entire fibril. It is noted that tropocollagen fibrils are themselves composed of a hierarchical arrangement of polypeptide chains, which are also composed of constituent amino acids. Thus, the defined components need not be the fundamental building blocks of the system. In contrast, we define the components of alpha-helical proteins (see case study in Chapter 3) as a single-protein convolution, while the system is characterized by the entire protein, recognizing the hierarchical effects of single molecular conformations. The coarse-graining procedures discussed here focus on constituent materials that can be modeled by full atomistic techniques using classical molecular dynamics, while the system structure (and relevant behaviors) requires a fully informed coarse-grain representation. The atomistic model provides the underlying physics to the coarse-grain potentials, allowing the investigation of structure–property functions at the required length scale.

As an analogy, such multiscale model transitions can be compared to the parameterization of atomistic force fields by detailed quantum mechanical methods (Figure 2.1). It is possible to derive the force constant parameters used in full atomistic models from first principle quantum mechanical calculations [5]. More sophisticated techniques, such as density functional theory (DFT), are simplified in the parameters and functions of atomistic potentials. The atomic structure (bosons, fermions, etc.) can be considered as the material components, while bonded atoms (such as carbon–carbon bonds) can be thought of as the structural system. From this perspective, atomistic force fields can be considered as a thorough and systematic coarse-graining of quantum mechanics. The direct application of quantum mechanical results to large macromolecular systems (such as proteins) is not yet feasible because of the large number of atoms involved. As a logical extension, we analyze the results of full atomistic simulations and implement them in the fitting of mesoscopic coarse-grain potentials. Each transition involves an increase in accessible system size, length scale, and time scale, as well as a simplification of the governing theoretical models. As a consequence, details of interactions are lost, but the desired fundamental behavior

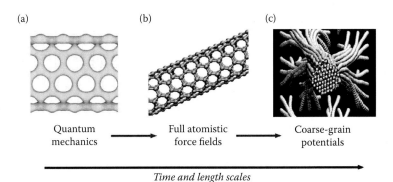

(a)   (b)   (c)

Quantum   ⟶   Full atomistic   ⟶   Coarse-grain
mechanics   force fields   potentials

Time and length scales

FIGURE 2.1 Rationale for coarse-graining approach through analogous comparison between the parameterization of atomistic force fields by detailed quantum mechanical methods to the development of coarse-grain potentials by the results of full atomistic simulations. (a) Electron charge isosurface plot of (5,5) carbon nanotube via density functional theory (DFT). (b) Segment of full atomistic model of identical carbon nanotube using carbon–carbon interactions. (c) Top view of adhered bundle of coarse-grain carbon nanotubes using potentials developed from atomistic results. Each transition involves an increase in accessible system size, length scale, and time scale, as well as a simplification of the governing theoretical models.

is maintained. It is noted that this "desired fundamental behavior" is dependent on the system application, and can include mechanistic behaviors, thermal or electrical properties, molecular interactions, equilibrium configurations, etc. To illustrate, an atomistic representation of carbon bonding can include terms for atom separation, angle, and bond order (such as CHARMM [6]- or AMBER [6,7]-type force fields), but lacks any description of electron structure, band gaps, or transition states found in quantum mechanical approaches such as DFT. However, the atomistic model maintains the accurate behavior of the individual carbon atoms and neglecting the effects of electrons is deemed a necessary simplification. Likewise, the coarse-graining of a carbon nanotube integrates the effects of multiple carbon bonds into a single potential. The exact distribution of carbon interactions is lost, but the behavior of the molecular structure is maintained.

The development of coarse-grain models allows the simulation of events on physical time and size scales, leading to a range of possible advances in nanoscale design and molecular engineering in a completely integrated bottom-up approach. A coarse-graining approach is intended to develop tools to investigate material properties and underlying mechanical behavior typically required for material design that systematically integrates characteristic chemical responses. However, it is emphasized that the intention is not to circumvent the need for full atomistic simulations—in contrast, coarse-graining requires accurate full atomistic representations to acquire the necessary potential parameters. The finer-trains-coarser procedure described here is an attempt to reconcile first principles derivations with hierarchical multiscale techniques by using atomistic theory with molecular dynamics simulations in lieu of empirical observations in a unified and systematic approach.

2.1.2 DIVERSITY OF SYSTEMS AND APPLICATIONS: NEED FOR A SYSTEM-DEPENDENT APPROACH

Atomistic force fields, which must encompass atom-atom bonds and interactions, can be developed in a general formulation due to the common molecular components. For example, the behavior of carbon–carbon bonds in a carbon nanotube is similar to the alpha-carbon backbone of a protein sequence in terms of strength and bond length. A thoroughly developed atomistic force field is capable of representing a vast assortment of molecular systems. Coarse-grained potentials, conversely, are usually developed to represent a particular system, and consequently feature unique idiosyncrasies in construction and behavior. Accordingly, differences in system complexity and goals of modeling lead to difficulties in developing a universal method for coarse-graining.

Attempts to avoid this system-dependent approach (i.e., generalized coarse-graining frameworks) can potentially result in a complex coarse-grain description to account for the multitude of molecular interactions a general description must encompass. The introduction of many potentials and parameters essentially mimic the function and form of full atomistic force fields, albeit at a coarse-grain scale. Other generalized frameworks attempt to simplify the description of molecular interactions as much as possible. For example the MARTINI force field [8] has been successful in the intent of efficiently modeling larger systems, and warrants further discussion (see Section 2.3). To capture fundamental interactions, general coarse-grain potentials developed for application to multiple systems are limited to few atom mappings (i.e., two-, four-, and six-bead models) and are beneficial and appropriate for systems where the interactions are still at the atomistic scale (i.e., proteins, polymers, etc.).

It is apparent that different material systems can be characterized by mechanical properties at the atomistic to microscale, requiring a more general coarse-graining approach to model the intended structure–property behavior. It is our proposition that a hierarchical multiscale system requires a system-dependent approach to coarse-graining. Accurate system representation is maintained at a cost of coarse-grain potential versatility. By applying a system-dependent finer-trains-coarser approach, a variety of systems can be coarse-grained and modeled for different purposes, transcending different scales and functionalities. Such derived coarse-grain potentials are not meant to be universally applicable, but provide a means to investigate a specific system under specific conditions.

2.1.3 ALTERNATIVE COARSE-GRAINING ADVANTAGES: PRAGMATIC SYSTEM SIMPLIFICATION

A commonly stated motivation and presumed primary benefit for the development of coarse-grain potentials and models is to allow the simulation of larger systems at longer time scales. Indeed, the reduction of system degrees-of-freedom and the smoother potentials implemented allow larger time-step increments (and thus time scales) and each element is typically an order of magnitude or more larger in length scale. Additionally, cheaper potential calculations (in terms of computational efficiency) can be exploited to either increase the number of coarse-grain elements (thus representing even larger systems) or simulate a relatively small system over more integration steps (further extending accessible time scales). Such benefits are inherent to any coarse-grain representation, and can serve as a *de facto* definition of the coarse-graining approach.

Nevertheless, a pragmatic approach of system simplification is found in many engineering disciplines. Complex electronic components are designed based on simplified models of circuits, with element behavior defined by such general properties as current, voltage, and resistance. Robust building structures are analyzed via notions of beam deflections and beam-column joint rotations, among other simplifying assumptions. In both cases, more detailed system representations are known and can be implemented (e.g., implementation of temperature and material effects in a transistor, or a detailed frame analysis including stress concentrations in bolts or welds). It is apparent that such additions result in a more accurate representation of the modeled system, but also serve to increase the computational expense of analysis as well as introduce a more sophisticated theoretical framework (which subsequently requires a more detailed set of material and model parameters). Rarely, however, is the use of simplified and computationally efficient models justified by inaccessible time and length scales of the more detailed description. Such models are applied for analysis in lieu of a more detailed description because they provide an accurate representation of the system-level behavior and response with confidence in the properties of the model components.

For example, a common application of system-level analysis is used in the study of structural frames. Required loads and resistances are computed at the system-level, and the structural components are assumed to behave according to a generalized theoretical framework (in terms of beam deflections, rotations, etc.). Detailed structural elements, such as steel joist girders, are presumed to behave according to their representative models and do not warrant a full analysis at the structural level. Of course, the joist girders themselves are rigorously tested at the component level to determine the necessary parameters (ultimate loads, yield stresses, etc.). We apply the same approach to molecular coarse-graining. Full atomistic simulations are used to probe and acquire the behavior of molecular components and parameterize coarse-grain potentials. These potentials, in turn, are implemented in coarse-grain representations that can facilitate the analysis of large-scale system-level phenomenon. There is an inherent assumption that the coarse-grain theoretical framework provides an accurate representation of the atomistic behavior, just as application of elastic beam theory assumes the reliability of a steel joist girder. See Figure 2.2 for a schematic comparison.

It is stressed that even if full atomistic representations are computationally possible, a coarse-grain description can still be suitable for systematic analysis of variable

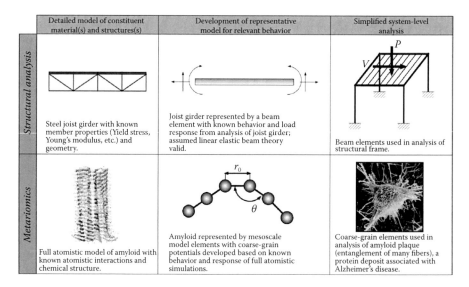

	Detailed model of constituent material(s) and structures(s)	Development of representative model for relevant behavior	Simplified system-level analysis
Structural analysis	Steel joist girder with known member properties (Yield stress, Young's modulus, etc.) and geometry.	Joist girder represented by a beam element with known behavior and load response from analysis of joist girder; assumed linear elastic beam theory valid.	Beam elements used in analysis of structural frame.
Materiomics	Full atomistic model of amyloid with known atomistic interactions and chemical structure.	Amyloid represented by mesoscale model elements with coarse-grain potentials developed based on known behavior and response of full atomistic simulations.	Coarse-grain elements used in analysis of amyloid plaque (entanglement of many fibers), a protein deposit associated with Alzheimer's disease.

FIGURE 2.2 (See color insert.) Analogous comparison of system simplification between structural analysis and materiomics. Detailed model constituents and their structural arrangement are analyzed to parameterize a coarse-grain representative model, maintaining relevant behaviors and implemented for simplified analysis. Here, we see the transition of steel joist girders to beam elements to structural frame analysis, paralleled by a full atomistic model of an amyloid, a coarse-grain mesoscopic representation, and the system-level amyloid plaque (aggregation of thousands of amyloids). (Image of amyloid plaque reprinted with permission from Macmillan Publishers Ltd., Pilcher, H.R., *Nature News*, 2003. Copyright 2003.)

system configurations, which requires a large number of simulations. We offer an alternative motivation for a coarse-graining approach complementary to extension of accessible time and length scales, where the catalyst for coarse-grain potential development is not the extension of traditional molecular dynamics, but to provide an accurate and reliable method for system-level analysis and probe the mechanical response and structure–property relation for hierarchical systems. Furthermore, mesoscopic models provide simple and efficient modeling techniques for experimentalists, allowing a more direct comparison between simulation and a vast variety of experimental techniques, without requiring specialized molecular dynamics tools such as specialized computer clusters with complex software.

2.1.4 WHEN TO COARSE-GRAIN: APPROPRIATE SYSTEMS AND CONSIDERATIONS

It is emphasized that not all systems will benefit from a coarse-grain representation and prudent consideration must be given regarding the system characterization and intent of the simulations. For some systems, complex behaviors may require full atomistic representation, or material inhomogeneities may not be able to be described by coarse-grain elements. Such systems may benefit and indeed require full atomistic representations. Typical motivations for a coarse-graining approach include:

1. Inaccessible time scale for phenomenon or behavior via full atomistic representation
2. Inaccessible length scale for phenomenon or behavior via full atomistic representation
3. Focus on global system properties and/or mechanical behavior rather than on molecular structure and/or chemical interactions
4. Desire for a direct simplified analysis of simulation results and system behavior

In addition to these motivating factors, some systems are more conducive to coarse-graining due to chemical composure and/or structural arrangement. Coarse-grain models are easily adapted for atomistically homogenous systems, consisting of repetitive structures such as carbon nanotubes (with a uniform cylindrical nanostructure). By extension, atomistically heterogeneous materials, such as protein-based materials, are applicable to coarse-graining if they are mesoscopically homogeneous, where the local effects of distinct amino acids (or other molecular inhomogeneities) produce similar global behaviors and are deemed negligible. Other system properties to consider include the discretization or material units and/or mechanical behavior, with a logical correlation to coarse-grain elements, and any hierarchical structures in which the intended structure–property relation is to be investigated.

2.2 EXAMPLES OF COARSE-GRAINING METHODS

In the past decade, various simple models have been used to describe the large-scale motions of complex molecular structures where more detailed classical phenomenological potentials [6,7] involving all atoms cannot be used because of the restrictions on the amount of time that can be covered in computer simulations. The reader is referred to recent reviews for a more thorough discussion of techniques and applications [9–12].

Single-bead models are the most direct approach taken for studying macromolecules. The term *single bead* derives from the idea of using single beads, that is, point masses, for describing a functional group, such as single amino acids, in a macromolecular structure. The elastic network model (ENM) [13], Gaussian network model [14], and GO model [15] are well-known examples that are based on such bead model approximations. These models treat each amino acid as a single bead located at the C_α position, with mass equal to the mass of the amino acid. The beads are connected via harmonic bonding potentials, which represent the covalently bonded protein backbone. Elastic network models have been used to study the properties of coarse-grained models of proteins and larger biomolecular complexes, focusing on the structural fluctuations about a prescribed equilibrium configuration, such as normal vibrational modes, and are discussed further in Section 2.2.1.

In GO-like models, an additional Lennard–Jones-based term is included in the potential to describe short-range nonbonded interactions between atoms within a finite cutoff separation. Despite their simplicity, these models have been extremely successful in explaining thermal fluctuations of proteins [11] and have also been implemented to model the unfolding problem to elucidate atomic-level details of deformation and rupture that complement experimental results [16–18]. A more recent direction is coupling of ENM models with a finite element-type framework for mechanistic studies of protein structures and assemblies [19].

Using more than one bead per amino acid provides a more sophisticated description of protein molecules. In the simplest case, the addition of another bead can be used to describe specific side-chain interactions in proteins [20]. Higher-level models (e.g., four- to six-bead descriptions) capture more details by explicit or united atom description for backbone carbon atoms, side chains, and carboxyl and amino groups of amino acids [21,22]. Coarser-level multiscale modeling methods similar to those presented here have been reported more recently, applied to model biomolecular systems at larger time and length scales. These models typically employ super-atom descriptions that treat clusters of amino acids as beads. In such models, the elasticity of a polypeptide chain is captured by simple harmonic or anharmonic (nonlinear) bond and angle terms. These methods are computationally quite efficient and capture shape-dependent mechanical phenomena in large biomolecular structures [23], and can also be applied to collagen fibrils in connective tissue [24] as well as mineralized composites such as nascent bone [25]. The development of the coarse-grain model for collagen will be discussed further as a case study in Chapter 3.

We proceed to provide a brief discussion of some of the aforementioned approaches implemented for coarse-grain potentials as an overview of other methods. The following discussion is not intended to include all the intricacies and details of the development and application of each method, but merely to provide examples of various types of coarse-graining procedures and the progression of adding more complexity to the coarse-grain representation.

2.2.1 ELASTIC NETWORK MODELS

In the simplest form, a coarse-grain model can be defined by a single potential for all beads. For example, the aforementioned ENMs can be thought of as a single pair potential between neighbor atoms, such that:

$$E_{CG} = E_{ENM} = \sum \phi_{ENM}(r), \tag{2.1}$$

where

$$\phi_{ENM}(r) = \frac{1}{2} K_r \left(r - r_0 \right)^2 \tag{2.2}$$

Here, each pair of atoms is assigned a harmonic spring bond potential with stiffness, K_r, about an initial equilibrium spacing, r_0 (see Figure 2.3 for an example).

The elastic constant, K_r, can be defined as a constant [13] or, as another example, as a function of r_0:

$$K_r \left(r_0 \right) = \bar{K} \left(\frac{\bar{r}}{r_0} \right) \tag{2.3}$$

(a) Full atomistic representation   (b) Elastic network model

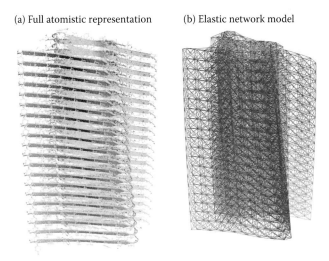

FIGURE 2.3 (**See color insert.**) (a) Full atomistic representation and (b) elastic spring network representation of an amyloid fibril. Here, springs are connected to all neighbor atoms within a cutoff of 10 Å, and spring stiffnesses are assigned according to an exponentially decaying function (Equation 2.4). Model was implemented to determine normal vibration modes and structure stiffness. (Model images courtesy of Dr. Z. Xu, Massachusetts Institute of Technology.)

$$K_r(r_0) = \bar{K} \exp\left[-\left(\frac{r_0}{r}\right)^2\right] \tag{2.4}$$

Equation 2.3 scales the mean value of the elastic constant, $\bar{K}$, with the initial bond length, r_0, such that $K_r(\bar{r}) = \bar{K}$. If the ENM is representative of a solid or continuous media, this method treats all elastic elements as if they had the same cross-sectional area, A, and a constant Young's modulus, E, such that the quantity (EA/r_0) is the same for all elements in the network [26]. The second method subjects the elastic constant to an exponentially decaying function with respect to a mean bond distance ($\bar{r}$). Such a formulation of K_r can represent weak interactions of atoms at a distance (such as van der Waals interactions) and provide a more complex description of interactions [27,28].

Such ENMs fall into two broad classes, those with Hookean springs that describe the rigidity of the macromolecule [13], and those describing the connectivity of the macromolecule [29]. For illustrative purposes, we limit our discussion to scalar spring constants and forces along the vector between two atoms, but also note that more complex model representations introducing directionality and anisotropy can be formulated [30]. As such, even a single potential description can become increasingly sophisticated as applications attempt to probe complex molecular deformations such as protein residue fluctuations [31] and equilibrium state transitions [32,33].

2.2.2 Two Potential Freely Jointed Chain Polymer Models

Although useful in the application of normal mode analysis of single protein macromolecules, ENM techniques lack the description necessary for intermolecular interactions and large deformation from equilibrium conditions. The subsequent step in coarse-grain model development is the combination of two simple potentials to represent the intermolecular and intramolecular interactions of individual macromolecules. Polymer systems are frequently represented by two such coarse-grain potentials, which encompass intramolecular (bonded) interactions and intermolecular (nonbonded) interactions, respectively, in a freely jointed chain (FJC) representation (no angular constraints) or:

$$E_{CG} = E_{bonded} + E_{nonbonded} = \sum \phi_{FENE}(r) + \sum \phi_{WCA}(r) \tag{2.5}$$

For bonded beads, a finitely extensible nonlinear elastic (FENE) potential [34–36] is implemented to maintain distance between connected beads and prevent polymer chains from crossing each other:

$$\phi_{FENE}(r) = -\frac{1}{2}kr_0^2 \ln\left[1-\left(\frac{r}{r_0}\right)^2\right] \quad \text{for } r < r_0, \tag{2.6}$$

$$\phi_{FENE}(r) = \infty \quad \text{for } r \geq r_0$$

The Weeks–Chandler–Anderson (WCA) potential [37,38] is the Lennard–Jones 12:6 potential truncated at the position of the minimum and shifted to eliminate discontinuity:

$$\phi_{WCA}(r) = 4\varepsilon \left[\left(\frac{\sigma}{r} \right)^{12} - \left(\frac{\sigma}{r} \right)^{6} \right] \quad \text{for} \quad \frac{r}{\sigma} < \sqrt[6]{2}, \qquad (2.7)$$

$$\phi_{WCA}(r) = 0 \quad \text{for} \quad \frac{r}{\sigma} \geq \sqrt[6]{2}$$

The WCA potential results in a purely repulsive potential. The combination of the FENE attractive potential with the WCA repulsive interaction creates a potential well for the flexible bonds that can maintain the topology of the molecule [39]. Such models can successfully represent stretching, orientation, and deformation of polymer chains and simple biomolecules.

Due to the inherent flexibility of the freely jointed chains, this coarse-graining approach is particularly suited for systems defined by long-chain polymers with relatively short persistence lengths, or systems that are entropically driven. As such, example applications include the investigation of viscoelastic behavior of polymer melts [40], stretching of polymers in flow [41], and other rheological properties [42,43]. More complex formulations use the combination of FENE and WCA potentials to define the coarse-grain model, and add additional potentials for a more robust description of the system, such as the introduction of angular terms (for stiffness) and Coulombic terms (for electrostatic interactions) to model complex macromolecules such as DNA [44].

2.2.3 GENERALIZATION OF INTERACTIONS: THE MARTINI FORCE FIELD

The lack of intermolecular interaction characterization in the previous coarse-graining approaches required a more complex formulation—one that maintained a definitive structure of the macromolecule, as well as integrated the interactions between macromolecules. The development of the MARTINI force field attempted to provide a general coarse-grain model that could be efficiently adapted for a multitude of biological systems by taking advantage of the fact that the majority of biological molecules (such as protein structures and lipids) are composed of the same categories of functional groups at the atomistic level [8,45,46]. Such approaches were previously implemented with various degrees of complexity for application to specific systems, integrating multiple beads per functional group, or more complex parameter formulations (see Marrink et al. [45] for a discussion and Shelley et al. [47,48] as examples). Essentially, by coarse-graining the functional groups, molecules with different architectures can be easily built and simulated. The aim of the MARTINI force field was to retain the chemical nature of the molecular components while defining as few bead types as possible. The applied philosophy was to avoid a focus on the reproduction of structural details for a particular system, but rather aim for a broad range of applications without the need to reparameterize the model each time. As a result, there is a slight trade-off between accuracy and applicability, with many different biological systems able to be modeled. Only four interaction types are considered—polar, nonpolar, apolar, and charged—and subtypes of these interactions are used to define hydrogen-bonding characteristics (see Figure 2.4 for coarse-grain representations of amino acids).

The MARTINI force field can be represented by

$$E_{\text{MARTINI}} = E_{\text{bond}} + E_{\text{angle}} + E_{\text{nonbonded}} = \sum \phi_T(r) + \sum \phi_\theta(\theta) + \sum \phi_{\text{LJ}}(r), \quad (2.8)$$

where harmonic potentials are implemented for the bonded and angle potentials, or:

$$\phi_T(r) = \frac{1}{2} k_T \left(r - r_0 \right)^2 \tag{2.9}$$

$$\phi_\theta(r) = \frac{1}{2} k_\theta \left(\cos\theta - \cos\theta_0 \right)^2 \tag{2.10}$$

The harmonic potentials are parameterized by constants (k_T, r_0, k_θ, and θ_0) common to all bead-types. Such an approach increases the general applicability and versatility of the force field. Nonbonded interactions between interaction sites i and j are described by the Lennard–Jones 12:6 potential:

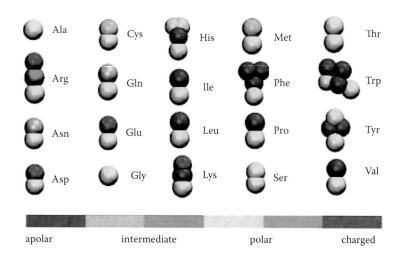

apolar   intermediate   polar   charged

FIGURE 2.4   (See color insert.) Coarse-grain representations of all amino acid types for MARTINI force field formulation for proteins. Different colors represent different particle types consisting of four main types of interaction sites: polar, nonpolar, apolar, and charged. (Reprinted with permission from Monticelli, L., et al., *J. Chem. Theory Comput.*, 4, 819–834, 2008. Copyright 2008 American Chemical Society.)

$$\phi_{LJ}(r) = 4\varepsilon_{ij} \left[\left(\frac{\sigma_{ij}}{r} \right)^{12} - \left(\frac{\sigma_{ij}}{r} \right)^{6} \right], \tag{2.11}$$

with σ_{ij} representing the effective minimum distance between two particles, and ε_{ij} the strength of their interaction. The uniqueness of the MARTINI force field parameterization is the arrangement of both σ_{ij} and ε_{ij} into distinct subsets of interaction groups.

The major advantage of the MARTINI force field is the simplicity and versatility of the parameterization—many proteins, lipids, and molecules can be built using the same set of coarse-grain building blocks. However, these MARTINI-type formulations restrict the coarse-graining to distinct functional groups to accurately maintain atomistic interactions while increasing computational efficiency. The obvious disadvantage of such approaches lies in the restriction of the coarse-graining scale. For example, on average, each bead of the MARTINI model has the volume of four water molecules. As a result, the time and length scales of MARTINI and MARTINI-type simulations are still limited, albeit much more efficient than full atomistic approaches. Again, prudence is required in selecting the level of detail of the coarse-grain representation.

For appropriate systems such as biological membranes consisting of lipid bilayers (Figure 2.5) and lipid-protein interactions, such approaches provide a powerful simulation framework (see Venturoli et al. [49] and Nielson et al. [50] for more thorough reviews and discussion of such methods).

FIGURE 2.5 **(See color insert.)** Simulation snapshot of DPPC/cholesterol bilayer structure with MARTINI coarse-grain model representation. Cholesterol molecules are displayed in green, with red hydroxyl groups. The DPPC lipid tails are shown in silver. Lipid head groups are displayed in purple and blue. Such a complex system can only be simulated via a coarse-graining approach. (Reprinted with permission from Marrink, S.J., de Vries, A.H., and Mark, A.E., *J. Phys. Chem. B*, 108, 750–760, 2004. Copyright 2007 American Chemical Society.)

2.2.4 Universal Framework, Diverse Applications

The aforementioned coarse-graining techniques, ENM models, FJC representations, and MARTINI-type formulations have been successful in their own applications. ENMs reflect the atomistic structure, FJC representations can accurately reproduce polymer flow and rheological properties, while MARTINI-type models accurately capture the interactions of functional groups for applications in assembly. However, such techniques focus on molecular geometry and interactions rather than reflect accurate mechanical properties and response. We predict that the investigation of the mechanical behavior of hierarchical systems requires a system-dependent parameterization of coarse-grain potentials, with a focus on maintaining both molecular interactions and molecular mechanics. Thus, we introduce a universal framework through a finer-trains-coarser multiscale paradigm, which effectively defines coarse-grain potentials via the response of full atomistic simulations, while introducing relevant mechanical properties at the mesoscopic scale. The approach is unique in its ability to transcend multiple scales, enabling the investigation of a broad regime of systems from protein-based materials to synthetic composite structures, while emphasizing a theoretical foundation on full atomistic investigations and asserting energetic equivalence and consistent mechanical behavior between all levels of modeling. The result is a set of problem-specific coarse-grain representations with diverse applications. The discussion in Section 2.3 presents guiding principles behind model formulation.

2.3 MODEL FORMULATION

From the previous discussion, it is apparent that a universal, stepwise procedure for coarse-graining a system is limited by the inherent simplifications and intent of the coarse-grain representation. A model developed to investigate arrays of carbon nanotubes (see Chapter 3, Case Study I) will differ from a model developed for the simulation of alpha-helix unfolding (see Chapter 3, Case Study II), even though the coarse-grain elements and potentials are similar. The coarse-graining of a particular system can be characterized by trial-and-error and subjective omissions or inclusions of pertinent behaviors, and as such considered just as much an art as a science. Careful consideration must be taken for the intended application and purpose of the coarse-grain model. Indeed, such a judicious approach is the motivation for a finer-trains-coarser approach. A deliberate focus on the atomistic behavior to characterize the coarse-grain model not only provides a complete "bottom-up" theoretical basis for a material system, but also assists in the formulation of the coarse-grain model by delineating relevant behavior and interactions. Nevertheless, although we contend that a step-by-step coarse-graining "recipe" is problematic, we can outline the "ingredients" in a general, coarse-graining framework:

1. Potentials required to define and characterize the mesoscale system
2. Full atomistic "test suite" applied to determine required coarse-grain potential parameters

3. Fitting of atomistic results to coarse-grain potential parameters via energy equivalence and consistent mechanical behavior
4. Validation of coarse-grain model with full atomistic and/or empirical results

We proceed to discuss each in detail.

2.3.1 CHARACTERIZE THE SYSTEM: COARSE-GRAIN POTENTIAL TYPE AND QUANTITY

The first step in the development of a coarse-grain model is to determine the necessary potentials to characterize the system. In general, we define the total energy of a coarse-grain system as the sum of these potentials:

$$E_{\text{System}} = E_{\text{CG}} = \sum \phi_{\text{CG}}, \qquad (2.12)$$

where ϕ_{CG} is the defined coarse-grain potentials. Although a simple statement, it is not trivial in implementation. Indeed, the type of developed potentials can serve to both broaden and restrict the applications of the model. Specifically, the coarse-grain potentials must be complex enough to represent the molecule in the intended mesoscopic simulation. One must consider whether provisions are necessary for fracture, intermolecular interactions, plasticity, formation of secondary structures, or a host of other molecular behaviors depending on the intent of the investigation. It is again stressed that a coarse-grain model for a system is not universal. The goal is to utilize the fewest and simplest potentials as possible that represent the system structure(s), mechanical properties, and interactions. Such an approach can be facilitated by the utilization of harmonic spring potentials to reflect mechanical response, where

$$\phi_i(\psi) = \frac{1}{2} k_i \left(\psi - \psi_0 \right)^2 \qquad (2.13)$$

Here, k_i refers to a harmonic spring stiffness while ψ typically refers to either an interatomic distance, r, or an angle, θ, for stretching and bending deformations, respectively. The harmonic spring potential results in a linear relationship between force and deformation, allowing efficient computation of interactions. Indeed, more complex nonlinear behavior can be approximated by combinations of linear functions (such as bilinear or trilinear formulations) to maintain computational efficiency in lieu of the introduction of more complex potentials (such an approach is discussed further in the ensuing chapter). A possible deficiency of harmonic potentials is the continuous increase in force with deformation. Systems subject to large deformations will subsequently deviate from true behavior, and thus, without provisions such as a potential cutoff, the use of the harmonic potential should be limited to small deformation. Selection of the harmonic spring potential can also be affected by the atomistic response by which it is parameterized. For example, a linear elastic strain response can be sufficiently modeled by harmonic spring potentials, albeit limited to small deformation.

2.3.2 FULL ATOMISTIC TEST SUITE

The finer-trains-coarser approach necessitates the parameterization of coarse-grain potentials from full atomistic results. An atomistic test suite is thus developed to obtain the necessary mechanical response and molecular interactions to be integrated into coarse-grain potentials. Typically, a relatively simple atomistic simulation is applied to isolate a single molecular behavior, mimicking simple material specimen tests. For example, uniaxial stretching can be applied to obtain the force-displacement or stress-strain response of a macromolecule, allowing the calculation of Young's modulus, and the parameterization of coarse-grain bond strength. Further, a three-point bending test can be utilized to determine the bending stiffness of a molecule, thereby allowing the parameterization of the coarse-grain bending potential. For molecular interactions, pairs of molecules are simulated to determine relative adhesion. In essence, a single atomistic investigation is implemented to characterize a single coarse-grain potential, ensuring accurate representation of each behavior. The number of required tests depends on the number of coarse-grain potentials (and associated parameters) implemented to describe the system.

2.3.3 FITTING COARSE-GRAIN POTENTIALS

The fundamental principle underlying the parameterization of coarse-grain potentials is the assertion of energy conservation. Energy equivalence is imposed between potential energy functions and relevant atomistic energy results. The basis of the formulation of the energy functions, either through elastic strain energy, deformation energy, adhesion energy, or other techniques, essentially defines the accuracy and behavior of the representative coarse-grain system. Two such approaches of energy conservation warrant separate discussions: (1) the application of direct energy equivalence between atomistic and coarse-grain representations and (2) the assertion of energy conservation via consistent mechanical behavior.

2.3.4 DIRECT ENERGY EQUIVALENCE

Nonbonded interactions representing macromolecular adhesion, attraction, or repulsion innately have no mechanical property analogue for parameterization. For such energetic potentials as the Lennard–Jones 12:6 function, or Coulombic interactions, the coarse-grain equivalent is extracted directly from the energetic results of a full atomistic representation. Potential energy as a function of separation can be derived between individual macromolecules. The coarse-grain potential can then be expressed as

$$\phi_{CG}(R) = \sum_{atoms} \omega_{ij} \phi_{ij}\left(r_{ij}\right),$$

(2.14)

where ϕ_{CG} is the coarse-grain potential, R is the macromolecular separation, and $\sum_{\text{atoms}} \omega_{ij}\phi_{ij}(r_{ij})$ is a weighted summation of the full atomistic potentials, ϕ_{ij}. This is essentially how coarse-grain potentials are developed for a small number of atoms per superatom (such as MARTINI-type force fields), where the superposition of multibody effects is apparent in the mapping. The relation becomes more complex as the number of incorporated/mapped atoms increase, requiring explicit atomistic simulation of the desired macromolecules. The above relation should not be thought of as a simple summation of potentials, or:

$$\phi_{CG\text{-}LJ}(R) = 4\varepsilon\left[\left(\frac{\sigma}{R_{ij}}\right)^{12} - \left(\frac{\sigma}{R_{ij}}\right)^{6}\right] \neq \sum_{\text{atoms}}\left(4\varepsilon_{ij}\left[\left(\frac{\sigma_{ij}}{r_{ij}}\right)^{12} - \left(\frac{\sigma_{ij}}{r_{ij}}\right)^{6}\right]\right). \quad (2.15)$$

The coarse-grain potential incorporates effects of the entire represented system (such as hydrophobic and other solvent effects, electrostatic interactions, hydrogen bonding between molecules, and/or entropic effects). We thereby introduce unknown weighting coefficients, ω_{ij}, in the summation of Equation 2.14. The complexity necessitates a full atomistic investigation to derive the effective coarse-grain potential parameters.

2.3.4.1 Consistent Mechanical Behavior

Bonded and angle potentials effectively represent molecular mechanical behavior. As there can be many contributions to even simple mechanical processes (such as the breaking of hydrogen bonds or solvent friction during molecular stretching or bending), we introduce conservation of energy for the relevant mechanical response (as opposed to the potential energy of direct energy equivalence) resulting in consistent mechanical behavior between full atomistic and coarse-grain representations. The approach is to apply energy equivalence between the coarse-grain potential and the observed strain or deformation energy of the full atomistic system:

$$\phi_{CG}(\psi) = U(\psi), \quad (2.16)$$

where $\phi_{CG}(\psi)$ is again the coarse-grain potential, and $U(\psi)$ is the representative energy function for the mechanical response. It is an underlying assumption of the model that the response of the full atomistic system is accurately described by the energy function, $U(\psi)$. The choice of the strain or deformation energy function is dependent on the system to be coarse-grained, and assumptions of elastic/plastic behavior, material isotropy, system failure such as yielding or fracture, etc. affect the definition (and interpretation) of atomistic results. As an example, for an elastic-isotropic material, we can define strain energy as

$$U(\varepsilon) = \frac{1}{2}\int_{V}[\sigma][\varepsilon]\,d\overline{V}. \quad (2.17)$$

To equate with $\phi_{CG}(\psi)$, we must formulate U as a function of ψ. For axial stretching, this requires the formulation of strain as a function of bond length, $\varepsilon(r)$, while for shearing this requires the formulation of strain as a function of shearing angle, $\varepsilon(\theta)$.

Individual parameterization of coarse-grain potentials via consistent mechanical behavior has an inherent limitation by the decoupling of mechanical behavior. To illustrate, as bond and angle coarse-grain potentials are independent under simulation, there is no coupling between stretching and bending of the molecule and effects such as twisting or torque are completely neglected. However, under full atomistic representations, such coupling can serve to increase stresses and induce failure. Ultimately, the model is tailored to the desired application, and systems with obvious or significant coupling effects should be avoided.

2.3.5 VALIDATION

Once the coarse-grain potentials are developed, validation of the coarse-grain model is a necessary step to assure an accurate representation. The primary basis for validation is through a direct comparison to full atomistic simulations. Such an approach may seem counterintuitive, redundant, or self-serving as the coarse-grain potentials were developed from such simulation results. However, the test suites applied were intended to result in a single behavior/response for each coarse-grain potential. The full coarse-grain representation can possibly involve the interaction and/or coupling of individual potentials. The combination of two behaviors (bending and stretching, for example) must be explored to justify the coarse-grain model. Relatively simple validation simulations can be designed, combining multiple system behaviors, and a *one-to-one* comparison made between full atomistic and coarse-grain results.

Due to limitations of full atomistic simulations, such validation is restricted to component-level behavior in which the coarse-grain model is specifically developed to circumvent. If available, secondary support and model validation is found in experimental data. Experimental techniques such as nanoindentation or atomic or chemical force microscopy (AFM/CFM) can directly probe materials at the mesoscale and determine system level characteristics and mechanical properties. The development of coarse-grain models is intended to elucidate such system-level behavior and thus correlation with experimental results is essential. Reciprocally, an accurate coarse-grain model can serve to validate and support experimental results.

It is noted that advances in experimental techniques are continually probing smaller scales, reaching atomistic precision, and allowing single molecule investigations [such as optical tweezer methods, atomic force microscopy (AFM), etc.]. Such data can serve to reinforce both the full atomistic representation and the developed coarse-grain model, thereby increasing confidence in both representations and supporting the finer-trains-coarser approach. Figure 2.6 provides a schematic of the relationship between atomistic and coarse-grain representations with experimental data for model validation.

2.4 SUMMARY AND CONCLUSIONS

The interplay between material properties and structural arrangement of hierarchical systems provides unique and robust mechanical behavior that transcends

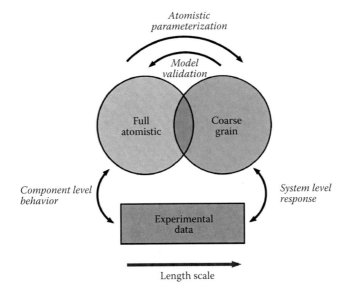

FIGURE 2.6 Validation of coarse-grain model through component behavior at atomistic scale and system characterization from experimental data. Experimental data ranges from component (molecular) to system (mesoscopic) length scales, providing a reciprocal loop of validation with both atomistic and coarse-grain models.

multiple length scales. Coarse-grain models provide an efficient means to simulate and investigate such systems in which the desired behavior, property, or response is inherently at the mesoscale. The relatively few parameters and potentials that define a coarse-grain representation must accurately represent the full atomistic behavior of the structure in both mechanical response and intermolecular interactions. Concurrently, the diverse nature and possible components of hierarchical systems makes a general formulation difficult. We thus introduce and discuss a system-dependent finer-trains-coarser approach to coarse-grain model development, requiring the parameterization of coarse-grain potentials via full atomistic results in a consistent and systematic manner.

A well-defined coarse-grain representation simultaneously allows investigation of molecular structures and behavior at multiple levels of hierarchy while intrinsically maintaining full atomistic behavior. This simplification also allows a pragmatic approach to system analysis, delineating global system behavior from constituent element properties, thereby encompassing multiscale structure–property relations. Extension of accessible time and length scales can allow direct correlations with empirical investigations, providing a novel tool for experimental design and material characterization.

Coarse-grain model formulation can be described in terms of four guiding principles: (1) selection and characterization of governing potential types and quantity; (2) application of full atomistic test suite to reflect independent behaviors and determine relevant coarse-grain potential parameters; (3) fitting of atomistic results to necessary potential parameters via direct energy equivalence of conservation of energy through consistent mechanical behavior; and (4) validation of the developed model via comparison with full atomistic results (component-level) or correlation with empirical data (system-level).

ACKNOWLEDGMENTS

This research was supported by the Army Research Office (W911NF-06-1-0291), the National Science Foundation (CAREER Grant CMMI-0642545 and MRSEC DMR-0819762), the Air Force Office of Scientific Research (FA9550-08-1-0321), the Office of Naval Research (N00014-08-1-00844), and the Defense Advanced Research Projects Agency (DARPA) (HR0011-08-1-0067). M. J. B. acknowledges support through the Esther and Harold E. Edgerton Career Development Professorship.

REFERENCES

1. Fratzl, P. and R. Weinkamer. 2007. Nature's hierarchical materials. *Progress in Materials Science*. 52:1263–1334.
2. Buehler, M. J., S. Keten, and T. Ackbarow. 2008. Theoretical and computational hierarchical nanomechanics of protein materials: Deformation and fracture. *Progress in Materials Science* 53:1101–1241.
3. Espinosa, H. D., et al. 2009. Merger of structure and material in nacre and bone—Perspectives on de novo biomimetic materials. *Progress in Materials Science* 54:1059–1100.
4. Buehler, M. J. and Y. C. Yung. 2009. Deformation and failure of protein materials in physiologically extreme conditions and disease. *Nature Materials* 8(3):175–188.
5. Car, R. and M. Parrinello. 1985. Unified approach for molecular dynamics and density-functional theory. *Physical Review Letters* 55(22):2471–2474.
6. Brooks, B. R., et al. 1983. CHARMM: A program for macromolecular energy, minimization, and dynamics calculations. *Journal of Computational Chemistry* 4(2):187–217.
7. Pearlman, D. A. et al. 1995. AMBER, a package of computer programs for applying molecular mechanics, normal mode analysis, molecular dynamics and free energy calculations to simulate the structural and energetic properties of molecules. *Computer Physics Communications* 91(1):1–41.
8. Marrink, S. J. et al. 2007. The MARTINI force field: Coarse Grained Model for Biomolecular Structures. *Journal of Physical Chemistry B* 111:7812–7824.
9. Tama, F. and C. L. Brooks, III. 2006. Symmetry, form, and shape: Guiding principles for robustness in macromolecular machines. *Annual Reviews in Biophysics and Biomolecular Structure* 35:115–133.
10. Bahar, I. and A. J. Rader. 2005. Coarse-grain normal model analysis in structural biology. *Current Opinion in Structural Biology* 15:586–592.
11. Tozzini, V. 2005. Coarse-grained models for proteins. *Current Opinion in Structural Biology* 15:144–150.
12. Sherwood, P., B. R. Brooks, and M. S. P. Sansom. 2008. Multiscale methods for macromolecular simulations. *Current Opinion in Structural Biology* 18:630–640.
13. Tirion, M. M. 1996. Large amplitude elastic motions in proteins from a single-parameter, atomic analysis. *Physical Review Letters* 77(9):1905–1908.
14. Haliloglu, T., I. Bahar, and B. Erman. 1997. Gaussian dynamcs of folded proteins. *Physical Review Letters* 79(16):3090–3093.

15. Hayward, S. and N. Go. 1995. Collective variable description of native protein dynamics. *Annual Review of Physical Chemistry* 46:223–250.
16. West, D. K. et al. 2006. Mechancial resistance of proteins explained using simple molecular models. *Biophysical Journal* 90(1):287–297.
17. Dietz, H. and M. Rief. 2008. Elastic bond network model for protein unfolding mechanics. *Physical Review Letters* 100:098101.
18. Sulkowska, J. I. and M. Cieplak. 2007. Mechanical stretching of proteins—a theoretical survey of the Protein Data Bank. *Journal of Physics: Condensed Matter* 19:283201.
19. Bathe, M. 2007. A finite element framework for computation of protein normal modes and mechanical response. *Proteins: Structure, Functions, and Bioinformatics* 70(4):1595–1609.
20. Bahar, I. and R. L. Jernigan. 1997. Inter-residue potentials in globular proteins and the dominance of highly specific hydrophilic interactions at close separation. *Journal of Molecular Biology* 266(1):195–214.
21. Nguyen, H. D. and C. K. Hall. 2004. Molecular dynamics simulations of spontaneous fibril formation by random-coil peptides. *Proceedings of the National Academy of Sciences* 101(46):16180–16185.
22. Nguyen, H. D. and C. K. Hall. Spontaneous fibril formation by polyalanines: Discontinuous molecular dynamic simulations. *Journal of the American Chemical Society* 128(6):1890–1901.
23. Arkhipov, A., et al. 2006. Coarse-grained molecular dynamics simulations of a rotating bacterial flagellum. *Biophysical Journal* 91:4589–4597.
24. Buehler, M. J. 2006. Nature designs tough collagen: Explaining the nanostructure of collagen fibrils. *Proceedings of the National Academy of Sciences* 103(33):12285–12290.
25. Buehler, M. J. 2007. Molecular nanomechanics of nascent bone: Fibrillar toughening by mineralization. *Nanotechnology* 18:295102.
26. Hansen, J. C. et al. 1996. An elastic network model based on the structure of the red blood cell membrane skeleton. *Biophysical Journal* 70:146–166.
27. Hinsen, K. 1998. Analysis of domain motions by approximate normal mode calculations. *Proteins: Structure, Functions, and Genetics* 33:417–429.
28. Hinsen, K., A. Thomas, and M. J. Field. 1999. Analysis of domain motions in large proteins. *Proteins: Structure, Functions, and Genetics* 34:369–382.
29. Bahar, I., A. R. Atilgan, and B. Erman. 1997. Direct evaluation of thermal fluctuations in proteins using a single-parameter harmonic potential. *Folding and Design* 2:173–181.
30. Atilgan, A. R. et al. 2001. Anisotropy of fluctuation dynamics of proteins with an elastic network model. *Biophysical Journal* 80:505–515.
31. Doruker, P., R. L. Jernigan, and I. Bahar. 2002. Dynamics of large proteins through hierarchical levels of coarse-grained structures. *Journal of Computational Chemistry* 23(1):119–127.
32. Navizet, I., R. Lavery, and R. L. Jernigan. 2004. Myosin flexibility: Structural domains and collective vibrations. *Proteins: Structure, Functions, and Bioinformatics* 54:384–393.
33. Zheng, W. and S. Doniach. 2003. A comparative study of motor-protein motions by using a simple elastic-network model. *Proceedings of the National Academy of Sciences* 100(23):13253–13258.
34. Warner, H. R., Jr. 1972. Kinetic theory and rheology of dilute suspensions of finitely extendible dumbells. *Industrial and Engineering Chemistry Fundamentals* 11(3):379–387.
35. Grest, G. S. and K. Kremer. 1986. Molecular dynamics simulation for polymers in the presence of a heat bath. *Physical Review A* 33(5):3628–3631.
36. Koplik, J. and J. R. Banavar. 2003. Extensional rupture of model non-Newtonian fluid filaments. *Physical Review E* 67:011502.

37. Weeks, J. D., D. Chandler, and H. C. Anderson. 1971. Role of repulsive forces in determining the equilibrium structure of simple liquids. *Journal of Chemical Physics* 54(12):5237–5247.

38. Heyes, D. M. and H. Okumura. 2006. Equation of state and structural properties of the Weeks-Chandler-Anderson fluid. *Journal of Chemical Physics* 124:164597.

39. Kremer, K. and G. S. Grest. 1990. Dynamics of entangled linear polymer melts: A molecular-dynamics simulation. *Journal of Chemical Physics* 92(8):5057–5086.

40. Cifre, J. G. H., S. Hess, and M. Kroger. 2004. Linear viscoelastic behavior of unentangled polymer melts via non-equilibrium molecular dynamics. *Macromolecular Theory and Simulations* 13:748–753.

41. Cheon, M., et al. 2002. Chain molecule deformation in a uniform flow—a computer experiment. *Europhysics Letters* 58(2):215–221.

42. Kroger, M. and S. Hess. 2000. Rheological evidence for a dynamical crossover in polymer melts via nonequilibrium molecular dynamics. *Physical Review Letters* 85(5):1128–1131.

43. Jeng, Y.-R., C.-C. Chen, and S.-H. Shyu. 2003. A molecular dynamics study of lubrication rheology of polymer fluids. *Tribology Letters* 15(3):293–299.

44. Stevens, M. J. 2001. Simple simulations of DNA condensation. *Biophysical Journal* 80:130–139.

45. Marrink, S. J., A. H. de Vries, and A. E. Mark. 2004. Coarse Grained Model for Semiquantitative Lipid Simulations. *Journal of Physical Chemistry B* 108:750–760.

46. Monticelli, L., S. K. Kandasamy, X. Periole, R. G. Larson, D. P. Tieleman, and S.-J. Marrink. 2008. The MARTINI coarse-grained force field: Extension to proteins. *Journal of Chemical Theory and Computation* 4:819–834.

47. Shelley, J. C. et al. 2001. A coarse grain model for phospholipid simulations. *Journal of Physical Chemistry B* 105:4464–4470.

48. Shelley, J. C., et al. 2001. Simulations of phospholipids using a coarse-grain model. *Journal of Physical Chemistry B* 105:9785–9792.

49. Venturoli, M. et al. 2006. Mesoscopic models of biological membranes. *Physics Reports* 437:1–54.

50. Nielson, S. O. et al. 2004. Coarse grain models and the computer simulation of soft materials. *Journal of Physics: Condensed Matter* 16:R481–R512.

51. Pilcher, H. R. 2003. Alzheimer's abnormal brain proteins glow. *Nature News* September 23, 2003.

52. Xu, Z., R. Paparcone, and M. J. Buehler. 2010. Alzheimer's Aβ(1-40) amyloid fibrils feature size dependent mechanical properties. *Biophysical Journal* 98(10):2053–2062.

3 Coarse-Graining Parameterization and Multiscale Simulation of Hierarchical Systems. Part II: Case Studies

Steve Cranford and Markus J. Buehler

CONTENTS

3.1   INTRODUCTION

To illustrate the coarse-grain parameterization and multiscale methodology, we provide a more thorough discussion regarding the development of specific mesoscopic models. The focus is on the coarse-grain potential development, to provide examples of the model formulation framework described in depth in Chapter 2. It is noted that some equations are repeated to provide a self-contained and complete description of each model formulation within each case.

The three chosen case studies are presented to exemplify both the finer-trains-coarser multiscale paradigm (that is, use of full atomistic test suites to parameterize the coarse-grain potentials), as well as a system-dependent approach (the developed coarse-grain potentials are unique to the intended application) as previously discussed. Our intent is to differentiate general coarse-graining frameworks [such as the aforementioned elastic network model (ENM) or MARTINI force field] from system-specific coarse-graining development. The merits of either approach, of course, are both subjective and problem-specific and judicious consideration of the resulting simulation design and intent is required.

For a broader perspective, each case study presented represents a fundamental advantage to all coarse-graining approaches, and thus can be considered as archetype coarse-graining problems. Sections 3.1.1 through 3.1.3 discuss characteristics of the case studies presented in this chapter.

3.1.1   Investigate the Structure–Property Relation at the Mesoscale

A primary motivation for the development of a coarse-grain representation is to provide a means to directly model system behavior at the mesoscopic scale. With the coarse-grained representation, structures approaching micrometers in scale can be efficiently modeled. Because of the use of a finer-trains-coarser approach, the atomistic behavior and intramolecular interactions are maintained, thereby providing a necessary intermediate step reconciling the gap between atomistic and continuum theory. Here, we specifically wish to investigate two materials, carbon nanotubes and collagen, that form hierarchical secondary structures at length scales beyond the capacity of full atomistic representation, but yet are dependent on intermolecular adhesion and interaction. These secondary structures (nanotube arrays and collagen fibrils) have unique structural arrangements that directly affect the mechanical properties. Thus, neither an atomistic chemical description nor continuum material properties are sufficient to describe the structure-property relation at the mesoscale. It is intended that the coarse-grain representation can be used to investigate such behavior. The coarse-grain models we present provide a method to model and investigate this class of nanostructures that fall precariously between atomistic and continuum techniques.

3.1.2 Extend Atomistic Behavior to Inaccessible Time- and Length-Scales

A fundamental limitation of full atomistic molecular dynamics simulations is the accessible (or inaccessible) time and length scales. As a consequence, it is frequently difficult to extend theoretically well-described atomistic behavior to physically relevant time and length regimes. By integrating atomistic behavior into developed coarse-grain potentials, larger systems can be simulated for longer time spans, while representing complex molecular interactions and properties. Precise definition of the intended behavior is required to develop accurate coarse-grain representation. Case Study II presents the development of a coarse-grain model to represent the unfolding behavior of alpha-helical protein domains. The unfolding behavior is initiated by the rupture of hydrogen bonds—an unquestionable atomistic response—and is supported by full atomistic simulations. Coarse-graining is introduced here to both examine the length dependencies of this unfolding response and investigate such proteins in networked systems (as found in biological cells and membranes).

3.1.3 Minimize Degrees of Freedom for Large Systems

Simulations of molecular systems in solution are often hindered by the computational overhead of calculation of the reactions of water molecules, regardless of the interaction with the relevant macromolecule. Indeed, the degrees of freedom associated with the solvent can exceed the macromolecule by an order of magnitude or more. By design, parameterization of coarse-grain potentials fully integrates the effects of water molecules, eliminating the need for either explicit water molecule representations or applied implicit water force fields. Case Study III illustrates the coarse-graining of polymer-tethered fullerenes to allow the investigation of self-association of such large nanoparticles in solution and efficient investigation of the effects of parameters such as molecular weight, polymer architecture, and particle density.

There is significant overlap for all presented case studies in the sense that each investigates the structure–property relationship, extends the accessible time and length scales, and reduces the number of degrees of freedom. However, each is differentiated by the associated advantage for coarse-graining, and the approach for each system subsequently differs based on intent and utilization of the model. Relevant applications for each model are described to provide an understanding of the intentions and benefits of the coarse-grain representation. The discussion of applications is relatively brief, to emphasize model development and intent rather than results of specific investigations. The reader is encouraged to refer to the cited literature for each case study for more details.

3.2 CASE STUDY I: CARBON NANOTUBES AND TROPOCOLLAGEN

At the atomistic scale, carbon nanotubes (CNTs) [1,2] and collagen differ in terms of structure and behavior. Carbon nanotubes consist of rolled sheets of graphene (single-layer bonded carbon), which form rigid cylindrical structures with high aspect ratios (Figure 3.1a). Carbon nanotubes are among the most widely studied

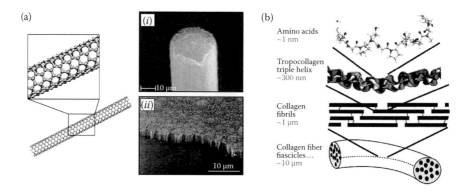

FIGURE 3.1 **(See color insert.)** Overview of carbon nanotube and tropocollagen systems. (a) Depiction of full atomistic representation of (5,5) carbon nanotube. Subplots show higher order hierarchical arrangements, including (i) SEM of single bundle of carbon nanotubes (Reprinted in part with permission from McClain, D., et al., *J. Phys. Chem. C*, 111(20), 7514–7520, 2007. Copyright 2007 American Chemistry Society.) and (ii) SEM micrograph of vertically aligned carbon nanotube array (Reprinted in part with permission from Yang, J. and Dai, L., *J. Phys. Chem. B*, 107, 12387–12390, 2003. Copyright 2003 American Chemical Society.). (b) Schematic view of some of the hierarchical features of collagen, ranging from the amino acid sequence level at nanoscale up to the scale of collagen fibers with lengths on the order of 10 μm. (From Buehler, M.J., Keten, S., and Ackbarow, T., *Prog. Mater. Sci.*, 53, 1101–1241, 2008.) The coarse-grain model development discussed here is focused on the behavior of tropocollagen molecules (component level) and their role in the mechanical behavior and properties of collagen fibrils (system level).

nanomaterials, with many potential applications that take advantage of their unique mechanical, electrical, thermal, and optical properties [6]. There are many concurrent investigations involving carbon nanotubes, ranging from experimental synthesis to atomistic and continuum modeling with a focus on a variety of properties, behaviors, and applications. The superior mechanical properties of carbon nanotubes are appealing for their potential use in novel nanomaterials. For instance, the Young's modulus of a single-walled nanotube approaches a terapascal (10^{12} Pa) [7], implying one of the strongest known synthesized materials in terms of elastic modulus and ultimate tensile strength [8]. Collagen, in contrast, is a protein-based material, composed of polypeptide chains of various constituent amino acids (Figure 3.1b). A tropocollagen molecule is composed of three polypeptide chains arrange in a helical structure, stabilized by hydrogen bonding between different residues [3–5,9]. The Young's modulus of tropocollagen is on the order of 4 to 10 GPa [3, 10–12]. Materials based on proteins hold particular promise because of their great flexibility in usage and their applications and the potential integration of technology and biology, allowing translation of nature's structural concepts into engineered materials [13].

The motivating factor for the coarse-graining of both carbon nanotubes and collagen fibrils (and fibers) is the investigation of the materials at the mesoscopic hierarchical level. For carbon nanotubes, a well-known behavior is intertube bonding due to weak van der Waals interactions, which results in the formation of bundles that contain hundreds or thousands of individual nanotubes (Figure 3.1a, inset). At the microscale, collagen fibrils consisting of staggered, crosslinked tropocollagen molecules form the basis for biological tissues such as tendon and bone. The formation of these mesoscopic structures complicates the full atomistic investigation of the mechanical properties. However, due to the homogeneous and fibrillar structure of both carbon nanotubes and tropocollagen, the formulation of their respective coarse-grain models is the same.

3.2.1 MODEL DEVELOPMENT

The coarse-grain model developed is intended to capture two essential components of both carbon nanotubes and tropocollagen: (1) the mechanical behavior of the fibrillar structure for both stretching and bending and (2) the intermolecular interactions between adjacent macromolecules. The intent is to apply a coarse-graining approach to achieve a mechanical response while maintaining atomistic interactions, an approach more apropos than equivalent continuum or elasticity techniques due to the system dependence on intermolecular interactions. We thus define the energy landscape as

$$E_{CG} = E_{bond} + E_{angle} + E_{pair} \qquad (3.1)$$

To obtain the necessary parameters for these potentials, the atomistic behavior of each must be investigated and full atomistic molecular dynamics simulations are undertaken to determine key mechanical property values.

For the current fibrillar structures, the coarse-grain bond potential, E_{bond}, is representative of axial strain. Furthermore, the intended coarse-grain application is limited to tensile stretching. Thus, a simple simulation is developed to determine the force-displacement or stress-strain relationship of the macromolecule. We apply tensile deformation by keeping one end of the molecule fixed and slowly displacing the other end in the axial direction (Figure 3.2a). In terms of mechanical properties, this relationship can be converted to Young's modulus, E. It is noted that the atomistic behavior is not meant to correspond one-to-one with continuum properties such as Young's modulus, which is typically limited to an elastic isotropic material, but such properties provide appropriate, conventional, and convenient measures for behavior such as axial stretching.

From the full atomistic results (Figure 3.3), the axial stretching behavior of both carbon nanotubes and tropocollagen consists of two regimes. For carbon nanotubes, there is nonlinear softening and plastic deformation due to the yielding of carbon bonds, whereas tropocollagen undergoes nonlinear stiffening due to the extension and unfolding of the helical arrangement and transition to direct straining of the protein backbone. Both carbon nanotubes and tropocollagen fracture at an ultimate strain. Although the nonlinear behavior is the result of complex atomic interactions, it is a trivial simplification to integrate the desired effect into the coarse-grain potential. We can determine Young's modulus either directly from stress-strain results, or indirectly via force-displacement results, depending on the output and sophistication of the full atomistic simulation, where

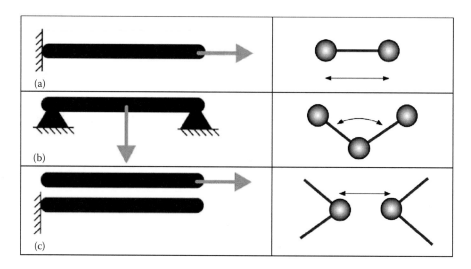

FIGURE 3.2 Atomistic "test suite" and corresponding coarse-grain potential behavior: (a) axial stretching, bond potential; (b) three-point bending, angle potential; (c) surface adhesion; pair potential.

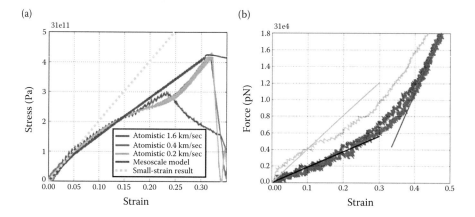

FIGURE 3.3 **(See color insert.)** Full atomistic simulation results for the axial stretching of (a) a single-walled carbon nanotube (From Buehler, M.J., *J. Mater. Res.* 21(11), 2855–2869, 2006. With permission.) and (b) a tropocollagen molecule (From Buehler, M.J., *J. Mater. Res.*, 21(8), 1947–1961, 2006. With permission.). The nanotube results depict a softening behavior as the carbon bonds yield at high strain, while the tropocollagen results depict a stiffening behavior, as the molecule undergoes extension of the helical structure before direct straining of the protein backbone.

$$E = \frac{\partial \sigma}{\partial \varepsilon} \approx \frac{\Delta \sigma}{\Delta \varepsilon} = \frac{r_0}{A_c} \frac{\Delta F}{\Delta r} \qquad (3.2)$$

where σ and ε are the stress and strain, F and r are the force and displacement, and A_c and r_0 are the cross-sectional area (assumed constant) and initial length. For small deformation (initial stretching regime), the Young's modulus for a (5,5) single-walled carbon nanotube was calculated to be approximately 2 TPa, while the modulus for tropocollagen was determined to be on the order of 8 GPa. For nonlinear behavior, the Young's modulus is calculated for each regime independently. The total bond energy of the coarse-grain system is given by the sum over all bonded interactions or:

$$E_{bond} = \sum_{bonds} \phi_t(r) \qquad (3.3)$$

For axial stretching, a simple harmonic spring is used to determine the energy between all bonded pairs of particles in the system, given by

$$\phi_t(\Delta r) = \frac{1}{2} k_t \left(r - r_0 \right)^2 = \frac{1}{2} k_t \Delta r^2, \qquad (3.4)$$

with k_t as the spring constant relating distance, r, between two particles relative to the equilibrium distance, r_0. We assume each linear regime can be approximated using the equivalent elastic strain energy,

$$U(\varepsilon) = \frac{1}{2} \int_V (\sigma \varepsilon) \, d\bar{V} = \frac{1}{2} \frac{A_c E \Delta r^2}{r_0} = U(\Delta r) \qquad (3.5)$$

For the integration over the volume, $\bar{V}$, we assume a constant cross-section, A_c, such that $\bar{V} = A_c r_0$, define strain, $\varepsilon = \Delta r / r_0$, and stress, $\sigma = E\varepsilon$. We note that we utilized the full atomistic simulations to determine Young's modulus specifically to allow this formulation of strain energy in our parameterization. Caution must be taken not to overextend the significance of the atomistic to continuum equivalence. Here, we only apply Young's modulus to characterize the work required to stretch our atomistic model and thus train the coarse-grain potential. It is not implied that either carbon nanotubes or tropocollagen can be suitably modeled by elastic formulations. Indeed, unlike the Young's modulus of elastic isotropic materials, the modulus determined by atomistic simulation can differ depending on system properties, atomistic force field, boundary conditions, and loading rates (see Buehler [3,4], for example). For equivalent energy and consistent mechanical behavior, we let $\phi_t(\Delta r) = U(\Delta r)$ and find

$$k_t = \frac{A_c E}{r_0} \tag{3.6}$$

To account for the nonlinear stress-strain behavior under tensile loading, a bilinear model that has been used successfully in previous studies [15,16] is applied where:

$$\phi_t(r) = H\left(r_{\text{fracture}} - r\right) \begin{cases} \dfrac{1}{2} k_t^0 \left(r - r_0\right)^2, & r < r_1 \\[2mm] \beta(r) + \dfrac{1}{2} k_t^1 \left(r - r_1\right)^2, & r \geq r_1 \end{cases} \tag{3.7}$$

where $H(r_{\text{fracture}} - r)$ is the Heaviside function $H(a)$, which is defined to be 0 for $a < 0$ and 1 for $a \geq 0$, k_t^0 and k_t^1 are the spring constants for the different deformation regimes, and $\beta(r)$ is obtained from continuity conditions where:

$$\beta(r) = \frac{1}{2} k_t^0 \left(r_1 - r_0\right)^2 + k_t^0 \left(r_1 - r_0\right)\left(r - r_1\right) \tag{3.8}$$

A schematic of the resulting force-displacement relationship is shown in Figure 3.4. The same technique can be extended to multiple linear regimes, maintaining a computationally inexpensive harmonic potential while incorporating nonlinear effects.

For the angle potential, E_{angle}, the bending stiffness and force-displacement behavior of each structure is required. A simple three-point bending test is simulated via full atomistic representation, with the macromolecule subjected to bending by a center point load (Figure 3.2b). From the results, we can determine the bending stiffness

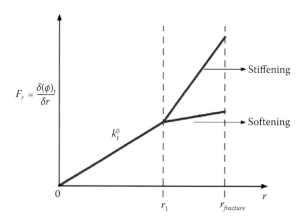

FIGURE 3.4   Plot of bilinear force model to account for nonlinear stiffening (collagen) or softening (carbon nanotubes) and fracture, as described by the potential in Equation 3.24.

of the molecule, which we label *EI*, using continuum beam theory to describe the mechanics of our system:

$$EI = \frac{L^3}{48}\left(\frac{F}{d}\right), \tag{3.9}$$

where L is the bent length of the molecule, F is the applied load at the center of the span, and d is the maximum displacement (at the load point). Application of beam-theory to atomistic simulations is a matter of judgment, as considerations must be made for deformation mechanisms (i.e., the presence of nonlinear plastic hinging or shear deformation). Furthermore, it is again stressed that the continuum interpretation of *EI*, the product of Young's modulus and area moment of inertia, is not applicable to all atomistic simulations. Here we use *EI* as a convention to characterize the bending stiffness of the molecule and assist in the formulation of the coarse-grain potential. However, for a rigid molecule such as a carbon nanotube, a continuum approximation can provide support to help validate resultant simulation values. Indeed, for the (5,5) single-walled carbon nanotube, full atomistic simulated bending results in a bending stiffness, *EI*, of 6.65×10^{-26} N-m². Using the previously determined E of 2 TPa, with a conservative approximation of I (assuming a solid cylinder with diameter 6.8 Å), we calculate a bending stiffness, $EI_{\text{theoretical}}$, of 2.1×10^{-26} N/m², which is on the same order of magnitude as the atomistic results. The bending energy is given by a sum over all triples in the system, given by

$$E_{\text{angle}} = \sum_{\text{triples}} \phi_\theta(\theta) \tag{3.10}$$

For bending, a rotational harmonic spring potential is used to determine the energy between all triples of particles in the system:

$$\phi_\theta(\theta) = \frac{1}{2}k_\theta\left(\theta - \theta_0\right)^2, \tag{3.11}$$

with k_θ as the spring constant relating bending angle, θ, between three particles relative to the equilibrium angle, $\theta_0 = 180°$. Using the equivalent elastic energy [17],

$$U(d) = \frac{48EI}{\left(2r_0\right)^3}d^2 \tag{3.12}$$

For small deformation, $\theta - \theta_0 \approx 2d/r_0$, and letting $\phi_\theta(d) = U(d)$:

$$k_\theta = \frac{3EI}{r_0} \tag{3.13}$$

We next characterize weak interactions (van der Waals interactions) between all pairs of coarse-grain elements, E_{pair}. The weak interactions represent the adhesion between adjacent macromolecules, thus a full atomistic simulation with two molecules (usually copies of the original) is simulated to determine the adhesion energy (Figure 3.2c). The energy barrier and equilibrium distance can be quickly determined by minimizing the atomistic system at two distinct states: (1) when the molecules are in contact energy minimum and (2) when the molecules are arbitrarily separated such that the interaction is negligible (the required separation is dependent on the relative adhesion strength of the simulated system). Differences in energy minima can be used to extract potential energy gain of adhesion ($E_{adhesion}$), while the geometric configuration at contact can be used to determine equilibrium distances ($D_{adhesion}$). A more sophisticated approach would be to determine the potential energy as a function of separation for a more accurate fitting of the coarse-grain potential. However, here we assume a LJ 12:6 function to represent adhesion, requiring only the potential energy well depth and equilibrium spacing for parameterization.

For the current bead-spring representation, we require the adhesion energy per unit length. From the atomistic simulation results, with an adhesion energy gain, $E_{adhesion}$, and a total molecular contact length, L, we define the adhesion energy per unit length, E_L, as

$$E_L = \frac{1}{2}\frac{E_{adhesion}}{L} \tag{3.14}$$

The total adhesion energy of the coarse-grain system is given by the sum over all pairs or

$$E_{pairs} = \sum_{pairs} \phi_{LJ}(r) \tag{3.15}$$

We use the LJ 12:6 function for each pair interaction:

$$\phi_{LJ}(r) = 4\varepsilon\left[\left(\frac{\sigma}{r}\right)^{12} - \left(\frac{\sigma}{r}\right)^{6}\right], \tag{3.16}$$

where ε describes the energy well depth at equilibrium, and σ is the distance parameter. We assume that a pair-wise interaction between different particles is sufficient to describe the adhesion between the coarse-grain elements, and that there are no multibody considerations. For both carbon nanotubes and tropocollagen, this assumption is deemed appropriate. As the coarse-grain particles are fundamentally point masses, we must assign a representative thickness to our representation via the pair potential. For the carbon nanotube, this is representative of the diameter of the tube. For a molecule such as tropocollagen, a thickness is approximated based on the molecular cross-section and assumed boundaries of atomistic interaction. We can then determine the distance parameter for the LJ function:

$$\sigma = \frac{D_{\text{adhesion}} + t}{\sqrt[6]{2}},$$ (3.17)

where D_{adhesion} is the equilibrium distance between macromolecules determined via atomistic simulation, and t is the representative thickness. To illustrate, the equilibrium distance between two (5,5) single-walled carbon nanotubes was determined to be 3.70 Å, and each tube has a diameter of approximately 6.8 Å. Using the Equation 3.17, we find $\sigma \approx 9.35$ Å. Note that the equilibrium spacing between coarse-grain particles is now approximately 10.5 Å, on the order of the bead-spring bond length. Choice of bond length, r_0, and choice of thickness, t, are critical parameters determining bead-bead interactions, and a balance may be required if intermolecular adhesion is a pertinent system behavior (Figure 3.5).

If the bond length is much greater than the equilibrium distance of the pair potential, $r_0 \gg r_{\text{LJ}}$, it is possible for coarse-grain molecules to be in close contact, or even pass through each other under certain conditions. If the bond length is relatively large, $r_0 > r_{\text{LJ}}$, the energy landscape about the equilibrium conformations is not smooth and can result in a "stick-slip" mechanism for molecules in contact as particles pass from neighbor to neighbor. Further, if the bond length is relatively short in comparison to equilibrium separation, $r_0 < r_{\text{LJ}}$, configurations may occur where a particle is in equilibrium with its second-nearest neighbors, while being repelled by the closest molecular surface. All arrangements of bond distance maintain energy equivalence between atomistic and coarse-grain potentials, but do not result in consistent mechanical behavior. In general, the bond length is chosen around the pair potential equilibrium distance, to ensure consistent mechanical behavior and a smooth energy landscape. For carbon nanotubes, we choose an equilibrium bead-spacing, r_0, of 10 Å ($r_{\text{LJ}} = 10.5$ Å), while for tropocollagen, we choose a bead-spacing of 14 Å ($r_{\text{LJ}} = 16.5$ Å).

The potential minimum, represented by the adhesion energy per unit length, E_{L}, is given by ε for the LJ 12:6 function. This parameter in the coarse-grain model is chosen such that the interaction of a single pair of beads is the same as the adhesion energy for the representative length of the full atomistic model. For nearest neighbors only, we find:

$$\varepsilon = E_{\text{L}} r_0$$ (3.18)

To account for the interactions of next-nearest neighbors in the energy contribution to the atomistic results, we note, at equilibrium:

$$E_{\text{L}} r_0 = \phi^{(1)}(r_{\text{LJ}}) + \phi^{(2)}(r_2) + \phi^{(3)}(r_3) + \ldots + \phi^{(n)}(r_n)$$ (3.19)

Or, the total adhesion energy along the coarse-grain element length is the summation of the nearest-neighbor interactions, and "n" next-nearest neighbor interactions, $\phi^{(n)}(r_n)$, at distances r_n. Thus, for more than one nearest neighbor,

$$\varepsilon = E_{\text{L}} r_0 \left[\left(1 + \pi^{(2)} + \pi^{(3)} + \ldots + \pi^{(n)} \right)^{-1} \right],$$ (3.20)

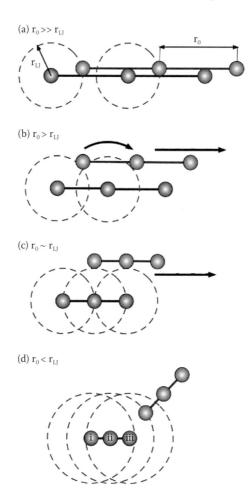

FIGURE 3.5 Energy landscape as a result of the relative magnitudes of bond distance, r_0, and pair potential equilibrium distance, r_{LJ}. (a) $r_0 \gg r_{LJ}$, potentially resulting in false equilibrium configurations, or allowing the passing of particles through bonds; (b) $r_0 > r_{LJ}$, rough energy landscape resulting in a "stick-slip" mechanism for adjacent molecules; (c) $r_0 \sim r_{LJ}$, best practice for consistent mechanical behavior and smooth energy landscape; (d) $r_0 < r_{LJ}$, potential equilibrium with next-nearest neighbor, particle (ii) while being repelled by nearest neighbor, particle (iii), yet attracted to third-nearest neighbor, particle (i), resulting in inconsistent mechanical behavior.

where $\pi^{(i)} = \phi^{(1)}(r_{LJ})/\phi^{(i)}(r_i)$. We define the term $(1 + \pi^{(2)} + \pi^{(3)} + \ldots + \pi^{(n)}) = \beta^{(n)}$, and then the above equation reduces to

$$\varepsilon = \frac{E_L r_0}{\beta^{(n)}}, \tag{3.21}$$

where $\beta^{(n)}$ is a numerical factor to account for next-nearest neighbor interactions. The calculation $\beta^{(n)}$ depends on the geometry of the coarse-grain system, as well as consideration for the cutoff of the pair potential (which dictates the extent of considered neighbors). For the current model of bead-springs for carbon nanotubes and tropocollagen, we use the first six nearest neighbors, and find $\beta^{(6)} \cong 1.1$. The factor, $\beta^{(n)}$, thus represents a reduction in the energy well depth of individual pairs, as the full atomistic representation implicitly accounts for the interaction with next-nearest neighbors.

The final consideration for the mesoscopic coarse-grain model is the assignment of mass to the particles. The mass of each bead is determined by assuming a homogeneous distribution of mass in the molecular model. Given the homogeneous structure of CNTs and tropocollagen, this is a reasonable approximation. The mass of each bead will scale with the selection of equilibrium bond distance, r_0, as each bead is representative of a larger portion of the full atomistic model. It behooves us to note that the full implication of this mass assignment approach to coarse-grain models on such thermodynamic properties as temperature effects has not been thoroughly investigated. However, the approach has proven adequate for investigations focusing on mechanical behavior at constant temperature conditions.

Finally, we can now define the mesoscopic model potentials by six parameters: k_t, r_0, k_θ, θ_0, σ, and ε. The results from the described atomistic simulations are used to determine these six parameters via equilibrium conditions (r_0, θ_0, σ) and energy conservation (k_t, k_θ, ε) by imposing energy equivalence and consistent mechanical behavior. The parameters can be extended to represent nonlinear effects, as illustrated by the bilinear function implemented for the softening or stiffening of the carbon nanotube or tropocollagen respectively. As such, the parameter k_t can be thought of as a set of parameters, depending on the complexity of the developed potential. All parameters of the coarse-grain potentials developed for a (5,5) SWCNT and a tropocollagen molecule are given in Table 3.1, derived completely from the results of full

TABLE 3.1
Summary of Coarse-Grain Parameters for the Bond, Angle, and Pair Potentials of a (5,5) SWCNT and a Tropocollagen Molecule

Parameter	SWCNT	Tropocollagen
Equilibrium bead distance, r_0 (Å)	10.00	14.00
Tensile stiffness parameter, k_t^0 (kcal/mol/Å²)	1000.00	17.13
Tensile stiffness parameter, k_t^1 (kcal/mol/Å²)	700.00	97.66
Hyperelastic parameter, r_1 (Å)	10.50	18.20
Fracture parameter, r_{fracture} (Å)	13.20	21.00
Equilibrium angle, θ_0 (degrees)	180.00	180.00
Tensile stiffness parameter, k_θ (kcal/mol/ rad²)	14300.00	15.00
LJ parameter, ε (kcal/mol)	15.10	10.6
LJ parameter, σ (Å)	9.35	14.72

Source: Buehler, M. J., *J. Mater. Res.*, 21(11), 2855–2869, 2006; Buehler, M. J., *J. Mech. Behav. Biomed. Mater.*, 1(1), 59–67, 2008.

Note: Derived from atomistic modeling and corresponding to Equations 3.23, 3.30, 3.34, and 3.38, as well as Section 3.3.1 (units in brackets).

atomistic simulations and the formulation described herein, specifically Equations 3.6, 3.13, 3.17, and 3.21.

It is noted that in the case of tropocollagen, which is typically found in solution, the influence of the solvent on the behavior of the macromolecules is captured in the aforementioned model constants, such that no explicit modeling of solvent is required. Typically, such models do not require an explicit solvent nor an implicit solvent force field or frictional coefficient. The effect of solvation is captured implicitly by the derived parameters and integrated into the coarse-grain potentials.

3.2.2 MODEL APPLICATIONS

3.2.2.1 Application 1: Self-Folding of Large Aspect Ratio Carbon Nanotubes and Nanotube Bundles

Large aspect ratio CNTs are extremely flexible and can be deformed into almost arbitrary shapes with relatively small energetical effort [1]. As illustrated by the development of the coarse-grain pair potential, different adjacent CNTs attract each other via van der Waals forces. If different parts of the same tube come sufficiently close, these attractive forces can initiate the formation of self-folded structures, where adjacent tube sections align, forming a racket-like structure. Such structures have been observed in MD simulations [18] as well as experimentally [19]. The stability, self-assembly, and mechanical properties of these structures are difficult to probe experimentally, and become computationally expensive for full atomistic simulations as the length of required nanotube increases. The described coarse-grain model was implemented to investigate the stability of folded structures, as well as the variation of folded configurations as a function of adhesion strength. Further, coarse-grain nanotubes were simulated in bundled configurations (up to 100 nanotubes per bundle) to determine mechanical properties and behavior under compressive, tensile, and bending deformations (see Figure 3.6).

The simple mesoscale model developed can easily be adapted for different types of carbon nanotubes, and allows the direct simulation of hierarchical bundled structures. Such investigations can potentially be of use for the development of carbon-nanotube-reinforced nanocomposites that attempt to utilize the adhesion properties

FIGURE 3.6 Simulation snapshot of response of a CNT bundle under mechanical compressive loading depicting significantly deformed/buckled shape. The bundle consists of 81 nanotubes using coarse-grain representation. (From Buehler, M.J., *J. Mater. Res.* 21(11), 2855–2869, 2006. With permission.)

of nanotube clusters and exploit energetically favorable folded configurations and manipulate stable adhesion domains.

3.2.2.2 Application 2: Mechanical and Surface Properties of Vertically Aligned CNT Arrays

Synthesis techniques have become adept at producing arrays of carbon nanotubes consisting of thousands of aligned tubes with similar diameters, lengths, and aspect ratios [2,20]. The properties of such arrays can be exploited to produce novel materials with unique, amplified, and controlled mechanical properties. Again, it is difficult to simulate such systems via full atomistic representations due to the sheer number of required nanotubes and timescales required to mimic real physical and experimental processes.

A vertically aligned array of nanotubes was constructed using the discussed coarse-grain CNT model, and then subjected to nanoindentation simulations (Figure 3.7). The goal was to probe the global behavior and mechanical properties of the array, through variations in nanotube parameters and application of external forces (in the form of a magnetic field). We again stress the use of a coarse-graining approach to investigate the system-level response of the array as opposed to the constituent nanotubes (component-level).

The coarse-grain model inherently allows the efficient varying of array geometry (aspect ratios, array spacing, etc.) to investigate behavior dependencies and pattern formation. Combined with a representation of physical experimental techniques (nanoindentation) to derive physical properties, such models can serve to facilitate empirical investigations by providing efficient means of prediction and a theoretical basis for behavior, thereby providing a crucial link between simulation and reality.

3.2.2.3 Application 3: Mechanical Property Variation through Collagen Fibril Crosslink Density

Natural collagen-based tissues are composed of staggered arrays of ultralong tropocollagen molecules extending to several hundred nanometers [4]. Although the macroscopic properties of collagen-based tissues (such as bone and tendon) have

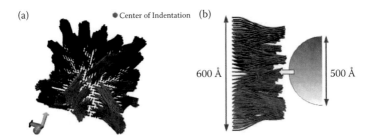

FIGURE 3.7 Depiction of nanoindentation simulation of coarse-grain nanotube array consisting of a 30 × 30 grid of nanotubes with a height of 30.0 nm [2]. (a) Top view, indenter not shown. (b) Side view with relative size of indenter depicted. The coarse-grain simulation consists of approximately 30,000 beads, whereas the equivalent full atomistic representation of the system would consist of over 4 million carbon atoms.

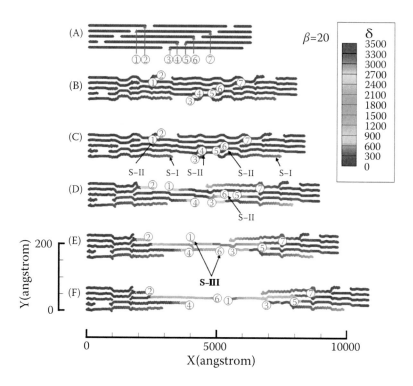

FIGURE 3.8 (See color insert.) Mesoscale model of collagen fibril, consisting of a two-dimensional array of ultralong coarse-grain tropocollagen molecules. The snapshots show the molecular structure as the fibril undergoes tensile deformation, where the color is defined by the magnitude of the slip vector [61]. A detailed analysis of the molecular deformation mechanisms suggests that intermolecular slip plays a major role in mediating large tensile strains in collagen fibrils leading up to failure, following a significant elastic regime. (Adapted from Tang, Y., Ballarini, R., Buehler, M. J., and Eppell, S. J., *J. Roy. Soc. Interface*, 7(46), 839, 2010.)

been studied extensively, less is known about the nanomechanical properties at the mesoscale—the hierarchical structure formed by the staggered tropocollagen molecules. A coarse-grain representation is uniquely suited to investigate the behavior of the structure and interaction of collagen fibrils. One such investigation probed the effect of crosslink density on the mechanical strength, deformation, yield, and fracture behavior on collagen fibrils (Figure 3.8). Crosslink-deficient collagen fibrils show a highly dissipative deformation behavior with large yield regimes, while increasing crosslink densities leads to stronger fibrils that display increasingly brittle behavior.

Collagen is such a fundamental constituent of biological materials that an improved understanding of the relevant nanomechanics can facilitate the development of novel biomimetic materials and aid in the understanding of injury and pathology processes. Indeed, the mutable collagenous tissue of echinoderms serves as an inspiration of new pharmacological agents and composite materials with biomedical applications [21]. In addition, diseases such as *osteogenesis imperfecta* [22] are caused by defects in the molecular structure of collagen, altering the intermolecular and molecular properties due to genetic mutations [23–26]. Investigations of the effects of such mutations on the subsequent mechanical behavior and properties of collagen structures can serve to elucidate the characterization and diagnosis of diseased tissues and the pathology of similar genetic diseases [27,28]. Such investigations are only possible via multiscale coarse-graining approaches that transcend the hierarchy of collagen fibrils, from the constituent polypeptides to tropocollagen molecules to collagen fibrils, penultimately leading to a deeper understanding of biological tissues such as nascent bone and associated disease states.

3.3 CASE STUDY II: FOLDING/UNFOLDING OF ALPHA-HELICAL PROTEIN DOMAINS

Proteins constitute the critical building blocks of life, providing essential mechanical functions to biological systems, and the focus of many molecular and atomistic level simulations [29,30]. In particular, alpha-helical (AH) protein domains are the key constituents in a variety of biological materials, including cells, hair, hooves, and wool. While continuum mechanical theories have been very successful coupling the atomistic and macro scales for crystalline materials, biological materials and soft condensed matter (such as polymer composites) require different approaches to describe elasticity, strength, and failure. The fundamental deformation and failure mechanisms of biological protein materials remain largely unknown due to a lack of understanding of how individual protein building blocks respond to mechanical load.

It has been determined both experimentally [31] and via simulation [32] that the mechanical response of biological materials is a combination of molecular unfolding or sliding, with a particular significance of rupture of reversible chemical bonds such as hydrogen bonds (H-bonds), covalent crosslinks, or intermolecular entanglement. The dominance of specific mechanisms can emerge at different time and length scales, chemical environments of the protein, and hierarchical arrangements/structures. As such, it is difficult to generalize fully atomistic results from nano to macro. Figure 3.9 displays an example hierarchical alpha-helical protein system.

A coarse-grain model is developed here to investigate the unfolding behavior of alpha-helical domains. The coarse-grain representation integrates parameters that define the energy landscape of the strength properties of alpha-helical protein domains, including energy barriers, unfolding and refolding distances, and the location of folded and unfolded states, and is implemented to investigate the variations of strength with respect to length and loading rate of alpha-helical protein filaments. Although unfolding of short alpha-helical segments can be modeled using full atomistic techniques, a coarse-grain representation is required to fully investigate the length dependence on mechanical response, as well as integration of alpha-helices into higher-level hierarchical arrangements. Such an approach intends to extend a known atomistic behavior to larger systems via coarse-grain potentials, which differs from the intent of the aforementioned carbon nanotube or collagen models that focused on system-level behavior and mechanical response.

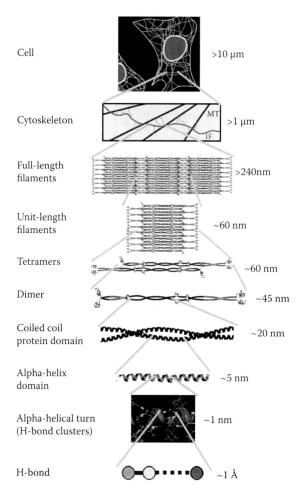

Cell		>10 μm
Cytoskeleton		>1 μm
Full-length filaments		>240nm
Unit-length filaments		~60 nm
Tetramers		~60 nm
Dimer		~45 nm
Coiled coil protein domain		~20 nm
Alpha-helix domain		~5 nm
Alpha-helical turn (H-bond clusters)		~1 nm
H-bond		~1 Å

FIGURE 3.9 **(See color insert.)** Schematic depicting hierarchical structure of alpha-helix protein-based intermediate filaments (IFs), which provide structural tensegrity to the cytoskeleton of cellular membranes. Over seven levels of hierarchy are transcended, from hydrogen bonds to alpha-helical turns, alpha-helical proteins (which are the focus of coarse-graining discussed here), dimers (coiled-coiled protein domain), tetramers, unit-length filaments, and full-length filaments to the cellular level. (Adapted from Qin, Z., Kreplak, L., and Buehler, M.J., *PLos ONE*, 4(10), e7294, 2009.)

3.3.1 MODEL DEVELOPMENT

The setup of the coarse-grain model for alpha-helical protein domains is based on the geometry of an AH, which features a linear array of turns or convolutions stabilized through the presence of H-bonds between sequential amino acid residues. During mechanical loading, any one of these convolutions can possibly rupture. As such, the coarse-grain representation is rationally discretized into bead-spring elements representing a single convolution consisting of approximately 3.6 amino acid residues. To achieve the coarse-grained description, the entire sequence of amino acids that constitute the alpha-helices is replaced by a collection of mesoscopic bead-spring elements (see Figure 3.10).

Similar to the previous linear, one-dimensional bead-spring models, we define the energy landscape of the coarse-grain system by three potentials:

$$E_{AH} = E_{bond} + E_{angle} + E_{pair} \tag{3.22}$$

Here, the bond potential must represent the structural backbone protein domain, and also the energetic features of the stabilizing H-bonds. The aim is to capture the structural and energetic features of an alpha-helical protein domain. A double-well potential is chosen to capture the existence of two equilibrium states for a convolution, folded and unfolded (see Figure 3.11). The model does not involve explicit solvent; rather, the effect of the solvent on the breaking dynamics of alpha-helical convolutions is captured by an effective double-well potential, parameterized by full atomistic simulations that implemented explicit solvent.

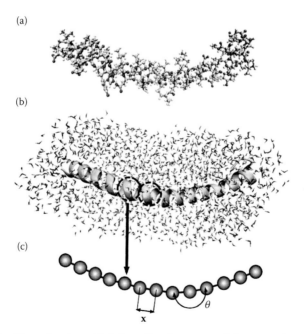

(a)

(b)

(c)

x

θ

FIGURE 3.10 **(See color insert.)** Schematic of coarse-graining procedure, in which full atomistic representation is replaced by a mesoscopic bead-spring model. A pair of beads represents one turn in the alpha-helix (also called a convolution), and thus 3.6 residues with the corresponding mass. (a) Full atomistic representation depicting all atoms and bonds; folded states of the turns are stabilized by the presence of hydrogen bonds between residues (not shown); water molecules not shown for clarity. (b) Ribbon representation of protein, illustrating alpha-helical folded conformation of backbone chain and individual convolutions; explicit solvent (water molecules) shown. (c) Developed coarse-grain representation, with a single bead per convolution; need for explicit solvent eliminated in coarse-grain model, as effects are integrated into coarse-grain potentials.

The bond potential can describe the microscopic details of the rupture mechanism of the convolution H-bonds under force, as well as the transition from a folded to unfolded state, through the prescribed energy barrier of the potential. The description is sufficiently coarse to enable significant computational speedup and efficiency compared with a full atomistic description.

Again, the total bond energy of the alpha-helical system is given by the sum over all bonded interactions or:

$$E_{bond} = \sum_{bonds} \phi_{bond}(x) \tag{3.23}$$

The double-well potential, $\phi_{bond}(x)$, is given by

$$\phi_{bond}(x) = \begin{cases} \dfrac{E_b}{x_b^4}\left(x - x_{tr}\right)^2 \left(x - x_{tr} - \sqrt{2} \cdot x_b\right)\left(x - x_{tr} + \sqrt{2} \cdot x_b\right), & x < x_{tr} \\[2ex] \dfrac{E_r}{x_r^4}\left(x - x_{tr}\right)^2 \left(x - x_{tr} - \sqrt{2} \cdot x_r\right)\left(x - x_{tr} + \sqrt{2} \cdot x_r\right), & x \geq x_{tr} \end{cases} \tag{3.24}$$

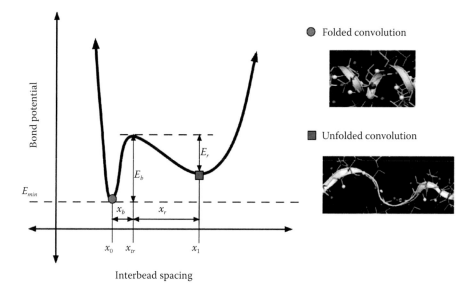

FIGURE 3.11   (See color insert.) Double-well profile of the bond-stretching potential of the coarse-grain model, representing the energy landscape associated with the unfolding of one convolution (see Equation 3.24). The values of the equilibrium states, x_0 and x_1, energy barriers, E_b and E_r, and the transition state, x_{tr}, are obtained from geometric analysis of the alpha-helix geometry, as well as the full atomistic simulations. The transition state (local energy peak) corresponds to the breaking of hydrogen bonds between convolutions of the alpha-helix. After failure of these weak bonds, the convolution unfolds to a second equilibrium state with a large interparticle distance. Under further loading, the covalent bonds begin to stretch, which leads to a second increase of the potential at large deformation.

The first equilibrium reaction coordinate, x_0 (first potential minimum), corresponds to the folded state of one alpha-helical convolution under no force applied. The transition state (energy barrier E_b), with position x_{tr} (peak of potential between the two wells), corresponds to the breaking of the H-bonds between alpha-helical convolutions. After failure of these weak bonds, the alpha-helix unfolds to a second equilibrium state. This corresponds to the second potential minimum with a larger interbead distance, x_1. Under further loading, the backbone bonds begin to be stretched, leading to a second increase in the potential (see Figure 3.11). This formulation does not include the rupture of the covalent backbone bonds. The parameters x_b and E_b represent the distance and energy barrier required to unfold one convolution, which x_r and E_r correspond to the refolding process. It is noted that the energy barrier for refolding, E_r, must be smaller than the energy barrier for unfolding, E_b, since the folded state is the most favorable state for a convolution in equilibrium [33].

The representation of two equilibrium states also requires a transition of the bending stiffness from a folded to unfolded state. In order to distinguish the bending stiffness for each state (which entails a severe structural change), we define a stiffness parameter, K_b, as a function of bead distance, x:

$$K_b(x) = K_{b,\text{fold}}\left[\alpha - \frac{(1-\alpha)}{\pi}\left(\arctan\left(100\left(x - x_{tr}\right)\right) - \frac{\pi}{2}\right)\right], \tag{3.25}$$

with

$$\alpha = \frac{K_{b,\text{unfold}}}{K_{b,\text{fold}}} \tag{3.26}$$

From full atomistic simulations, the bending stiffness of the protein, EI, is determined. We let

$$K_{b,\text{fold}} = \frac{3EI_{\text{fold}}}{x_0} \text{ and } K_{b,\text{unfold}} = \frac{3EI_{\text{unfold}}}{x_1}, \tag{3.27}$$

where EI_{fold} and EI_{unfold} are the bending stiffnesses of the folded and unfolded AH, respectively. Again, the total bending energy of the alpha-helical system is given by the sum over all bead triples (angles), or:

$$E_{\text{angle}} = \sum_{\text{triples}} \phi_{\text{angle}}(x,\theta) \tag{3.28}$$

We can then define the coarse-grain angle potential as

$$\phi_{\text{angle}}(x,\theta) = \frac{1}{2}K_b(x)\left(\theta - \theta_0\right) \tag{3.29}$$

Finally, the total intermolecular interaction energy, E_{pair}, is again represented by the sum over pairwise interactions between beads of different alpha-helical protein

TABLE 3.2

Summary of Parameters for a Coarse-Grain Alpha-Helical Protein Model

Parameter	Numerical Value
Equilibrium distance, folded state, x_0 (Å)	5.4
Equilibrium distance, unfolded state, x_1 (Å)	10.8
Distance between folded state and transition state, x_b (Å)	1.2
Energy barrier, folded state and transition state, E_b (kcal/mol)	11.1
Energy barrier, unfolded state and transition state, E_r (kcal/mol)	6.7
Bending stiffness, folded state, $K_{b,\text{fold}}$ (kcal/mol/rad^2)	21.6
Bending stiffness, unfolded state, $K_{b,\text{unfold}}$ (kcal/mol/rad^2)	0.665
Equilibrium angle, θ_0 (degrees)	180
Pair potential, LJ distance parameter, σ (Å)	10.8
Pair potential, LJ energy parameter, ε (kcal/mol)	6.815
Mass of mesoscale bead (amu)	400

Source: Bertaud, J., et al., J. Phys. Condens. Matter, 2009.

Note: Derived from atomistic modeling, corresponding to Figure 3.12 and representing the constants required for the coarse-grain potentials as discussed in Section 3.3.1 and Equations 3.22 through 3.29 (units in brackets).

domains. The adhesion potential (ϕ_{pair}) is again formulated by a LJ 12:6 potential, via an energy minimum (ε) and distance parameter (σ), in a manner discussed in Section 3.2.1. Table 3.2 lists the parameters implemented in the mesoscopic bead-spring alpha-helix model, from previous full atomistic simulations [29,32,34] depicted in Figure 3.12. The mass of each bead corresponds to the approximate average mass of each convolution (400 amu).

It is quite apparent that, although a similar double-well potential can be developed for other systems with distinct equilibrium conformations, the current coarse-grain description is uniquely developed for the alpha-helix. Specifically, it represents the atomistic rupture behavior of alpha-helical protein domains during mechanical loading under a limited range of loading rates (pulling speeds under 0.3 m/s) at a specific temperature (300 K) and particular environmental conditions (explicit waterbox) implemented in the full atomistic simulations in which the distance and energy barrier parameters were obtained [32,34]. However, the model could be easily adapted to other classes of protein filaments that feature serial arranged domains that undergo unfolding or a transition to distinct equilibrium states under mechanical loading and strain.

3.3.2 MODEL APPLICATIONS

3.3.2.1 Application 1: Time Scale Extension

The developed coarse-grain model for the alpha-helix was implemented to investigate the length and rate dependence of the Bell model, a theoretical strength model that can be applied to describe the mechanical behavior of molecules with reversible

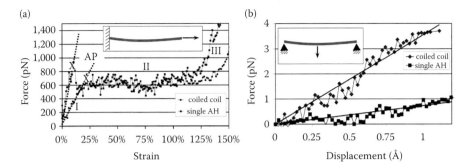

FIGURE 3.12 **(See color insert.)** Full atomistic tests of alpha-helical protein (single-helix and coiled-coiled conformations) used to parameterize the coarse-grain model. (a) Force-strain results of direct tension simulations. The first regime (I) consists of a linear increase in force, until a strain of approximately 13% for the single alpha-helix, noted as the angular point (AP), which corresponds to the rupture and unfolding of an alpha-helix convolution. The second regime (II) represents the unfolding of the helix under approximately constant force. The third regime represents a nonlinear increase in strain due to backbone stretching of the protein. (b) Force-displacement results of three-point bending simulations. The slope of the curve is proportional to the bending stiffness. Only the single alpha-helix values were used in the development of the coarse-grain representation. (Adapted from Ackbarow, T., and Buehler, M.J., *J. Mater. Sci.*, 42(21), 8771–8787, 2007.)

bonds (see Bell [35], Evans [36], Evans and Ritchie [37], and Walton, Lee, and Van Vliet [38], in addition to the primary references, Bertaud et al. [29,30], for details). Essentially, the Bell model presents a logarithmic relationship between reversible bond strength and loading rate (or molecule pulling speed). Full atomistic simulations are limited to time scales on the order of nanoseconds, limiting the pulling of alpha-helix stretching to approximately 0.01 m/s. Such a relatively high loading rate prevents a one-to-one correspondence with experimental results. However, the use of the coarse-grain potential extends the accessible time-scales to an order of microseconds, allowing pulling speeds on the order of 0.0001 m/s, representing a 100-fold increase in time scale. Experimental results of stretching and breaking single AH domains [39,40] report forces corresponding to the force level predictions at ultraslow pulling speeds of the coarse-grain model. Additionally, the coarse-grain representation can still be implemented at time scales on the order of full atomistic studies, allowing the validation of mesoscopic and full atomistic results.

3.3.2.2 Application 2: Length Dependence

The model was further implemented to investigate the length dependence on the rupture strength of proteins to extend the Bell model to ultralong protein regimes. Both experimental and full atomistic investigations are limited to an alpha-helix length on the order of 10 nm. This limitation only represents approximately 20 bead-spring elements of the coarse-grain representation, and thus can be easily surmounted using the coarse-grain model. Extending the length of AH protein domains to approximately 50 nm, coarse-grain simulations resulted in a logarithmic decrease in rupture strength as the protein length increased or

$$f(L) = a \ln(L/L_0) + b \qquad (3.30)$$

Due to the logarithmic dependence on length, this relation can only be investigated by simulating proteins with lengths extending several orders of magnitude. Such an expanse of scales is not accessible to full atomistic representations. This weakening behavior can be attributed to an increase in potential H-bond locations as the number of convolutions increases, each of which can break with the same probability. Since failure of one convolution is sufficient to initiate failure of the entire system, we expect longer molecules to be weaker, as observed in the coarse-grain simulations and previous investigations [41].

3.3.2.3 Application 3: Characterizing Intermediate Filament Networks

Individual, isolated alpha-helices are rarely found in biology. Thus, the developed coarse-grain model can facilitate the investigation of hierarchical structures of proteins and protein filaments [42]. Indeed, AH-based protein networks constitute the

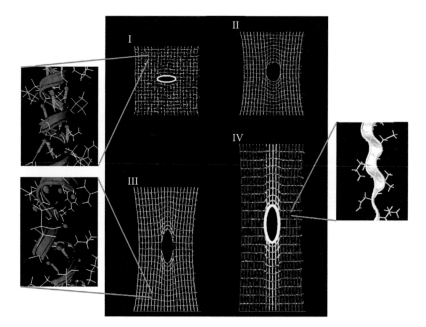

FIGURE 3.13 **(See color insert.)** Snapshots of protein network deformation, where coarse-grain representation of alpha-helix proteins was implemented. The deformation mechanism is characterized by the molecular unfolding of the alpha-helical protein domains, leading to the formation of large plastic yield regions, providing energy dissipation and preventing catastrophic failure. The blowups depict the atomistic structural arrangement of the alpha-helical protein domains based on the known correspondence with the coarse-grain mapping. (Adapted from Ackbarow, T., et al., *PLos ONE*, 4(6), e6015, 2009.)

intermediate filament structure in the cell's cytoskeleton and the nuclear membrane. Using a network of coarse-grain bead-spring structures, large meshes reflecting an assembly of intermediate filaments can be constructed and probed for mechanical properties and behavior (Figure 3.13).

It was found that the characteristic properties of alpha-helix-based protein networks are due to the particular nanomechanical properties of the protein constituents, enabling the formation of large dissipative yield regions around structural flaws, effectively protecting the protein network against catastrophic failure. The direct simulation of such large networks is only possible through the use of such multiscale coarse-grain models under discussion.

3.4   CASE STUDY III: MESOSCOPIC AGGREGATION OF FULLERENE-POLYMER CLUSTERS

Modification of nanoparticles by attachment of polymer chains can, in principle, allow manipulation of the geometry and interaction of particles on the nanoscale, allowing a tunable method of controlling their self-association [43,44]. A complete understanding of such interactions can result in unique binding properties or controlled self-assembly. Simulations implementing coarse-grain models have shown that nanoparticles tethered by polymer chains with various degrees of asymmetry, chain length, and polymer/particle interaction exhibit a rich spectrum of nanostuctures, including spherical, cylindrical, lamellar, sheetlike, and bicontinuous morphologies [45]. Experimental observations also show self-assembly of such tethered nanoparticles into spherical vesicles [46] and nanorods [47]. Here we focus on poly(ethylene oxide) (PEO)-grafted fullerenes (C_{60}-PEO nanoparticles). The thermodynamics of self-assembly of these nanoparticles, including the strong dependence of cluster distribution and average cluster size on concentration, indicate that the self-assembly process resembles the formation of wormy or spherical micelles, taking advantage of the water-soluble PEO and the hydrophobic fullerene.

Investigation of the formation and aggregation of such nanostructures is limited using full atomistic approaches, as there is a significantly large computational expense to compute the interactions of the nanoparticle (such as the carbon–carbon interactions of the fullerene) and the behavior of the polymer–nanoparticle interactions. Such computations are inconsequential, as it is known *a priori* that the mesoscopic structure is stable. Thus, coarse-graining methods are developed to investigate the subsequent hierarchical level of nanoparticle interactions and aggregation, circumventing the details of full atomistic behavior.

The coarse-graining approach to such systems has a threefold purpose:

1. To reduce the number of degrees of freedom to be simulated
2. To focus on the nanoparticle–nanoparticle interactions
3. To reproduce the molecular distributions and aggregations of the nanoparticles

Specifically, the goal of the coarse-grain model is to adequately reproduce the intramolecular and intermolecular structure of aqueous C_{60}-PEO systems, the distribution of PEO segments around the fullerenes, and the energetic landscape between nanoparticles.

It is noted that the intended mesoscopic investigation is not focused on mechanical behavior, and thus mechanical properties such as Young's modulus and bending stiffness, critical parameters in the previous potentials (Sections 3.2 and 3.3), are not part of the current coarse-grain model development. However, the C_{60}-PEO coarse-grain system provides an illustration of both the finer-trains-coarser multiscale paradigm, as well as the system-dependent approach to coarse-grain model development expressed throughout this chapter.

3.4.1 MODEL DEVELOPMENT

The coarse-grain model is parameterized based on fullerene–fullerene, fullerene–PEO, and PEO–PEO interactions. Of critical concern, as the nanoparticles aggregate in solution, is the integration of the influence of water on all interactions. Indeed, elimination of the explicit solvent results in the primary reduction of degrees of freedom. Fullerenes were condensed to a single-particle representation, while PEO chains were developed consisting of one particle per CH_2–O–CH_2 functional group (Figure 3.14). The coarse-graining approach for polymer systems is unique in the fact that polymer systems are typically governed by entropic effects (as opposed to the rigid structures of CNTs or tropocollagen). As such, the coarse-grain model focuses on interactions and molecular distributions rather than replication of specific structural or mechanical properties.

We define the energy of the system as

$$E_{\text{PEO-}C_{60}} = E_{\text{pair}},\tag{3.31}$$

such that

$$E_{\text{PEO-}C_{60}} = \sum_{\text{pair}} \phi_{\text{pair}}\tag{3.32}$$

(a)   (b)

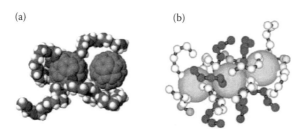

FIGURE 3.14 Representative snapshots of (a) two interacting C_{60}–PEO nanoparticles obtained from atomistic, explicit solvent simulations (water molecules and PEO chains on the right fullerene omitted for clarity) and (b) three coarse-grain C_{60}–PEO nanoparticle representations (PEO beads of the center nanoparticle shaded differently to differentiate adjacent molecules). (Reprinted with permission from Bedrov, D., Smith, G. D., and Li, L., *Langmuir*, 21, 5251–5255, 2005. Copyright 2005 American Chemical Society.)

Here, the coarse-grain potential is defined by the interactions alone. To account for bonding and angular geometry, the SHAKE algorithm was implemented [48] (which applies bond and angle constraints). As the intent of the coarse-grain model is not the mechanical behavior of the nanoparticles, less emphasis is placed on the accurate representation of molecular deformation, and application of the SHAKE algorithm is sufficient to maintain accurate geometry and polymer tether response during aggregation. Each interaction pair was investigated at the atomistic level separately, and can subsequently be considered part of the discussed atomistic test suite, with a focus on molecular interactions rather than molecular mechanical response. Atomistic investigations include

1. PEO–PEO and PEO–water intermolecular and PEO–PEO intramolecular interactions representing both hydrophilic and hydrophobic ether–water interactions [49]
2. Fullerene–fullerene interactions using full atomistic carbon–carbon potential to develop a coarse-grain C_{60}–C_{60} pair potential [50]
3. Water–carbon interactions [51] as the basis for water–fullerene interactions [52]

It is noted that each investigation focused on a particular molecular interaction, similar to previous case studies in which the atomistic test suites focused on individual molecular responses, such as stretching, bending, or adhesion. The interactions were then defined by

$$\phi_{\text{pair}} = \phi_{AB},\tag{3.33}$$

where ϕ_{pair} is derived from the atomistic test suite. The AB indices indicate the possible fullerene–fullerene, PEO–PEO, or PEO–fullerene interactions. As the atomistic studies did not determine all possible combinatorial parameters, standard mixing rules were implemented:

$$\varepsilon_{AB} = \sqrt{\varepsilon_A \varepsilon_B}\tag{3.34}$$

$$\sigma_{AB} = \frac{1}{2}\left(\sigma_A + \sigma_B\right)\tag{3.35}$$

Using the known atomistic interactions as well as the mixing rules, behavior was investigated between C_{60}–PEO nanoparticles in full atomistic and coarse-grain simulations. Nonbonded coarse-grain PEO–PEO interactions were parameterized to reproduce the PEO monomer–monomer interactions. Bonds and bends in the PEO chains were developed to match intramolecular correlations (end-to-end distance and radius of gyration). Parameterization of the coarse-grain potentials was then adjusted to reflect accurate radial distribution functions and coordination number of

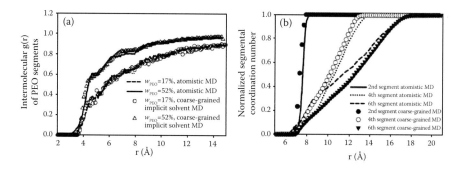

FIGURE 3.15 Full atomistic and coarse-grain results for model parameterization and validation for C_{60}–PEO nanoparticles. (a) PEO–PEO monomer intermolecular radial distribution function as obtained from atomistic, explicit solvent and coarse-grained, implicit solvent MD simulations of two PEO aqueous solutions. (b) Integrated coordination number of PEO segments in the C_{60}–PEO in an aqueous solution system as a function of their separation from the center of fullerene. (Reprinted with permission from Bedrov, D., Smith, G. D., and Li, L., *Langmuir*, 21, 5251–5255, 2005. Copyright 2005 American Chemical Society.)

the C_{60}–PEO system, i.e., the distribution of the polymer tethers about the fullerene (Figure 3.15).

Validation was carried out by calculation of the potential of mean force between two C_{60}–PEO nanoparticles via full atomistic (explicit solvent) and coarse-grain (implicit solvent) simulations. The coarse-grain model captures the most important features of the PMF, namely, the weak long-range attraction and strong short-range attraction. Again, we see how full atomistic investigations are applied to validate the behavior of a system component (as described in Section 2.3.5), after which the coarse-grain model can be extended to investigate larger systems.

3.4.2 MODEL APPLICATIONS

3.4.2.1 Application 1: Large Systems of Aqueous C_{60}–PEO Nanoparticles

The coarse-grain representation was implemented to compare the aggregation of 1000 bare fullerene or 1000 C_{60}–PEO nanoparticles in solution, with fullerene volume fractions ranging from 0.07 to 0.25 [43]. Full atomistic simulation of such systems would expend the majority of computational cost on the explicit solvent calculations, consisting of thousands of water molecules with negligible interactions with the nanoparticles, yet a large system is required to investigate the spatial and density effects of nanoparticle aggregation. It was shown that bare fullerenes form dense clusters with interactions between many fullerene neighbors, while the introduction of PEO tethers results in a polymer "hugging" phenomena (coverage of adjacent fullerenes by PEO chains). With an increase in surface coverage, the steric interactions between PEO chains and between PEO and fullerenes become dominant and shield the fullerene from interactions with other fullerenes, resulting in chain-like cluster configurations (Figure 3.16).

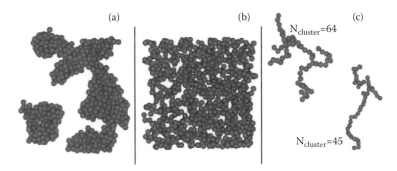

FIGURE 3.16   Representative coarse-grain simulation snapshots of (a) 1000 bare C_{60} fullerenes in aqueous solution (water not shown); (b) 1000 PEO-tethered C_{60} fullerenes, with equivalent volume fraction as (a) (water and PEO chains are not shown for clarity); and (c) two representative PEO–C_{60} clusters from the configuration shown in (b). (Reprinted with permission from Bedrov, D., Smith, G. D., and Li, L., *Langmuir*, 21, 5251–5255, 2005. Copyright 2005 American Chemical Society.)

3.4.2.2   Application 2: Systematic Variation of Polymer Architecture on C_{60}–PEO Nanoparticles

The effect of polymer architecture was investigated using the developed coarse-grain model, by manipulating the number and chain-length of the attached PEO tethers [44]. PEO-grafted fullerenes were comprised of a single tether of 60 repeat units (PEO functional groups), a three-arm star with 20 units per chain, or a six-arm star

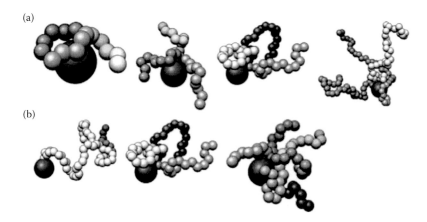

FIGURE 3.17   (See color insert.) Schematic overview of the types of coarse-grain nanoparticles employed in the clustering study. (a) Constant architecture motif, with each chain in a three-armed star with lengths of 5, 10, 20, and 40 PEO units, respectively. (b) Constant mass motif, where each nanoparticle has 60 total PEO units with arrangements in a linear, three-armed, or six-armed fashion, respectively. Note explicit solvent molecules are neither depicted nor required in the coarse-grain simulations. (Reproduced with permission of the PCCP Owner Societies. Copyright 2009. Hooper, J. B., Bedrov, D., and Smith, G. D., *Phys. Chem. Chem. Physics*, 11, 2034–2045, 2009.)

(a)   (b)   (c)

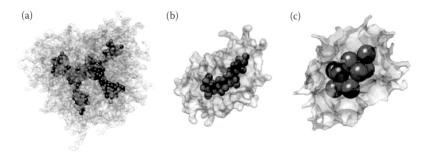

FIGURE 3.18 Representative snapshots for the constant mass series of clusters as depicted in Figure 3.17b: (a) Single, 60-unit PEO chains; (b) three-armed configuration, 20 units each; (c) six-armed configuration, 10 units each. The red spheres represent the fullerene core, while the surrounding polymer is depicted as a solvent-accessible isosurface. (Reproduced in part with permission of the PCCP Owner Societies. Copyright 2009. Hooper, J. B., Bedrov, D., and Smith, G. D., *Phys. Chem. Chem. Physics*, 11, 2034–2045, 2009.)

with 10 units per chain (Figure 3.17b). In addition, the influence of tether length of the three-chain configuration was investigated, from 5 to 40 units per chain (Figure 3.17a). Variation of number of PEO segments and segment length serve to control aggregate size and shape during self assembly.

It was concluded that higher molecular weight PEO (longer arms) and more compact PEO (more arms but constant total number of units) resulted in greater steric repulsion between fullerenes, engendering greater aggregate surface curvature and resulting in more spherically shaped aggregates (Figure 3.18).

It is emphasized that while the coarse-grained simulations were able to eliminate solvent degrees of freedom, the degrees of freedom of the tethered polymer chains cannot be excluded due to the nontrivial and nonisotropic contribution of the functional monomer groups into the interaction between C_{60}-PEO nanoparticles. Again, care must be taken to ensure such pertinent effects are included in any developed coarse-grain representation.

Such studies illustrate that chemical modifications of nanoparticles can result in unexpected interaction/self-assembly behavior due to interplay between nanoscale phenomena—phenomena that may be beyond the scope of atomistic investigations. The understanding of nanoparticle aggregations such as polymer-tethered fullerenes can facilitate the development of complex nanoscale structures with potential biomedical applications (i.e., drug delivery, binding to biological molecules) or advance polymer nanocomposites with superior mechanical properties.

3.5   SUMMARY AND CONCLUSIONS

Model formulation is dependent on the molecular system to be represented, as well as the intent of the coarse-grain simulations. In the model development for carbon nanotubes and tropocollagen molecules, a focus was on structure–property relations of secondary hierarchical structures such as CNT bundles or arrays and collagen fibrils. The intent of the investigation was to probe the mechanics of the larger hierarchical systems, and care was taken to accurately represent the mechanical response of the coarse-grain potential. Consequently, this lead to the integration of molecular softening and stiffening of carbon nanotubes and tropocollagen molecules within the coarse-grain response. The intention of the coarse-grain representation of alpha-helix protein domains, in contrast, was to extend well-known atomistic behavior to time and length scales inaccessible to full atomistic molecular dynamics. The integration of the unfolding behavior of an alpha-helix reduced the complex phenomena of hydrogen-bond rupture to a simple bond potential, via a unique "double well" potential formulation. In effect, the coarse-grain representation can accurately represent the response of two distinct configurations of the protein—both helical and unfolded—within the same model. For nanoparticle aggregation, the coarse-grain model development concentrated on polymer–nanoparticle interactions, with limited focus on mechanical properties. The intended application is not to produce a mechanical response, but rather to systematically study the effects of nanoparticle concentration and polymer architecture on self-assembly processes. As such, accurate mechanical properties and associated potentials were not required, further optimizing the computational efficiency of the simulations, required atomistic tests, and development of the coarse-grain model.

It is apparent that current coarse-graining methods are neither as accurate nor as predictive as all atom simulations. Future coarse-graining techniques can explore more rigorous parameterization methodologies for more accurate representations of the system. Indeed, full atomistic force fields and algorithms are constantly updated to improve the accuracy of results and versatility of applications. With an inevitable increase in computing power, larger and larger systems can be simulated via full atomistic representations, providing a counterargument for the development of any coarse-graining approaches. In addition, coarse-graining methods are inherently at a disadvantage due to their system-dependent nature—complex and diverse interactions must be described by a small number of parameters. The finer-trains-coarser multiscale paradigm implemented here attempts to illustrate the possible utility and advantages of coarse-graining methods through the delineation of component-level material properties and system-level mechanical behavior. The convergence of structure and mechanical properties is apparent in both natural and synthetic hierarchical systems, requiring new approaches to analyze and explicate the reciprocity of material properties, mechanical function, and structural arrangement. It is the contention of all multiscale methods that component-level behaviors can be adeptly represented by coarse-grain potentials, and thus serve to clarify structure-property relations without resorting to more complex models.

ACKNOWLEDGMENTS

This research was supported by the Army Research Office (W911NF-06-1-0291), by the National Science Foundation, (CAREER Grant CMMI-0642545 and MRSEC DMR-0819762), the Air Force Office of Scientific Research (FA9550-08-1-0321), the Office of Naval Research, (N00014-08-1-00844), and the Defense Advanced Research Projects Agency (DARPA) (HR0011-08-1-0067). M. J. B. acknowledges support through the Esther and Harold E. Edgerton Career Development Professorship.

REFERENCES

1. Buehler, M. J. 2006. Mesoscale modeling of mechanics of carbon nanotubes: Self-assembly, self-folding, and fracture. *Journal of Materials Research* 21(11):2855–2869.
2. Cranford, S. and M. J. Buehler. 2009. Mechanomutable carbon nanotube arrays. *International Journal of Materials and Structural Integrity* 3(2/3):161–178.
3. Buehler, M. J. 2006. Atomistic and continuum modeling of mechanical properties of collagen: Elasticity, fracture, and self-assembly. *Journal of Materials Research* 21(8): 1947–1961.
4. Buehler, M. J. 2008. Nanomechanics of collagen fibrils under varying cross-link densities: Atomistic and continuum studies. *Journal of the Mechanical Behavior of Biomedical Materials* 1(1):59–67.
5. Buehler, M. J. 2007. Molecular nanomechanics of nascent bone: Fibrillar toughening by mineralization. *Nanotechnology* 18:295102.
6. Baughman, R. H., A. A. Zakhidov, and W. A. de Heer. 2002. Carbon nanotubes—the route towards applications. *Science* 297:787–792.
7. Treacy, M. M. J., T. W. Ebbesen, and J. M. Gibson. 1996. Exceptionally high Young's modulus observed for individual carbon nanotubes. *Nature* 381:678–680.
8. Li, F. et al. 2000. Tensile strength of single-walled carbon nanotubes directly measured from their macroscopic ropes. *Applied Physics Letters* 77(20):3161.
9. Bhattacharjee, A. and M. Bansal. 2005. Collagen structure: The Madras triple helix and the current scenario. *IUBMB Life* 57(3):161–172.
10. Gautieri, A., M. J. Buehler, and A. Redaelli. 2009. Deformation rate controls elasticity and unfolding pathway of single tropocollagen molecules. *Journal of the Mechanical Behavior of Biomedical Materials* 2(2):130–137.
11. Sun, Y.-L. et al. 2004. Stretching type II collagen with optical tweezers. *Journal of Biomechanics* 37(11):1665–1669.
12. An, K.-N., Y.-L. Sun, and Z.-P. Luo. 2004. Flexibility of type I collagen and mechanical property of connective tissue. *Biorheology* 41:239–246.
13. Maskarinec, S. A. and D. A. Tirrell. 2005. Protein engineering approaches to biomaterials design. *Current Opinion in Structural Biology* 16:422–426.
14. Lorenzo, A. C. and E. R. Caffarena. 2005. Elastic properties, Young's modulus determination and structural stability of the tropocollagen molecule: A computational study by steered molecular dynamics. *Journal of Biomechanics* 38(7):1527–1533.
15. Buehler, M. J., F. F. Abraham, and H. Gao. Hyperelasticity governs dynamic fracture at critical length scales. *Nature* 426:141–146.
16. Buehler, M. J. and H. Gao. 2006. Dynamical fracture instabilities due to local hyperelasticity at crack tips. *Nature* 439:307–310.
17. Timoshenko, S. and G. H. Maccullough. 1940. *Elements of strength of materials*, 2nd ed. New York: D. Van Nostrand Company, Inc.
18. Buehler, M. J. et al. 2006. Self-folding and unfolding of carbon nanotubes. *Journal of Engineering Materials and Technology* 128:3–10.
19. Cohen, A. E. and L. Mahadevan. 2003. Kinks, rings, and rackets in filamentous structures. *Proceedings of the National Academy of Sciences* 100(21):12141–12146.
20. Qi, H. J. et al. 2003. Determinaton of mechanical properties or carbon nanotubes and vertically aligned carbon nanotube forests using nanoindentation. *Journal of Mechanics and Physics of Solids* 51:2213–2237.
21. Wilkie, I. C. 2005. Mutable collagenous tissue: Overview and biotechnological perspective. In *Echinodermata*, eds. W. E. G. Muller and V. Matranga, 221–250. Berlin–Heidelberg: Springer.
22. Rauch, F. and F. H. Glorieux. 2004. Osteogenesis imperfecta. *Lancet* 363:1377–1385.

23. Sillence, D. O., A. Senn, and D. M. Danks. 1979. Genetic heterogeneity in osteogenesis imperfecta. *Journal of Medical Genetics* 16:101–116.
24. Roughley, P. J., F. Rauch, and F. H. Glorieux. 2003. Osteogenesis imperfecta—clinical and molecular diversity. *European Cells and Materials* 5:41–47.
25. Beck, K. et al. 2000. Destabilization of osteogenesis imperfecta collagen-like model peptides correlates with the identity of the residue replacing glycine. *Proceedings of the National Academy of Sciences* 97(8):4273–4278.
26. Misof, K. et al. 1997. Collagen from the osteogenesis imperfecta mouse model (OIM) shows reduced resistance against tensile stress. *Journal of Clinical Investigation* 100: 40–45.
27. Gautieri, A. et al. 2009. Molecular and mesoscale mechanisms of osteogenesis imperfecta disease in collagen fibrils. *Biophysical Journal* 97:857–865.
28. Gautieri, A. et al. 2009. Single molecule effects of osteogenesis imperfecta mutations in tropocollagen protein domains. *Protein Science* 18:161–168.
29. Bertaud, J., Z. Qin, and M. J. Buehler. 2009. Atomistically informed mesoscale model of alpha-helical protein domains. *International Journal for Multiscale Computational Engineering* 7(3):237–250.
30. Bertaud, J. et al. 2009. Energy landscape, structure and rate effects on strength properties of alpha-helical proteins. *Journal of Physics: Condensed Matter* 22:035102.
31. Dietz, H. et al. 2006. Anisotropic deformation response of single protein molecules. *Proceedings of the National Academy of Sciences* 103(34):12724–12728.
32. Ackbarow, T. et al. 2007. Hierarchies, multiple energy barriers, and robustness govern the fracture mechanics of alpha-helical and beta-sheet protein domains. *Proceedings of the National Academy of Sciences* 104(42):16410–16415.
33. Karcher, H. et al. 2006. A coarse-grained model for force-induced protein deformation and kinetics. *Biophysical Journal* 90(8):2686–2697.
34. Ackbarow, T. and M. J. Buehler. 2007. Superelasticity, energy dissipation and strain hardening of vimentin coiled-coil intermediate filaments: Atomistic and continuum studies. *Journal of Materials Science* 42(21):8771–8787.
35. Bell, G. I. 1978. Models for the specific adhesion of cells to cells. *Science* 200(4342): 618–627.
36. Evans, E. 2001. Probing the relation between force, lifetime, and chemistry in single molecular bonds. *Annual Reviews in Biophysics and Biomolecular Structure* 30(1):105–128.
37. Evans, E. and K. Ritchie. 1997. Dynamic strength of molecular adhesion bonds. *Biophysical Journal* 72:1541–1555.
38. Walton, E. B., S. Lee, and K. J. Van Vliet. 2008. Extending Bell's model: How force transducer stiffness alters measures unbinding forces and kinetics of molecular complexes. *Biophysical Journal* 94:2621–2630.
39. Lantz, M. A. et al. 1999. Stretching the alpha-helix: A direct measure of the hydrogen-bond energy of a single-peptide molecule. *Chemical Physics Letters* 315(1):61–68.
40. Kageshima, M. et al. 2001. Insight into conformational changes of a single alpha-helix peptide molecule through stiffness measurements. *Chemical Physics Letters* 343(1):77–82.
41. Qi, H. J., C. Ortiz, and M. Boyce. 2006. Mechanics of biomacromolecular networks containing folded domains. *Journal of Engineering Materials and Technology* 128(4):509–518.
42. Ackbarow, T. et al. 2009. Alpha-helical protein networks are self-protective and flaw-tolerant. *PLos ONE* 4(6):e6015.
43. Bedrov, D., G. D. Smith, and L. Li. 2005. Molecular dynamics simulation study of the role of evenly spaced poly(ethylene oxide) tethers on the aggregation of C_{60} fullerenes in water. *Langmuir* 21:5251–5255.

44. Hooper, J. B., D. Bedrov, and G. D. Smith. 2009. The influence of polymer architecture on the assembly of poly(ethylene oxide) grafted C_{60} fullerene clusters in aqueous solution: A molecular dynamics simulation study. *Physical Chemistry Chemical Physics* 11:2034–2045.

45. Zhang, Z. et al. 2003. Tethered nano building blocks: Toward a conceptual framework for nanoparticle self-assembly. *Nano Letters* 3(10):1341–1346.

46. Zhou, S. et al. 2001. Spherical bilayer vesicles of fullerene-based surfactants in water: A laser light scattering study. *Science* 291:1944.

47. Sawamura, M. et al. 2002. Stacking of conical molecules with a fullerene apex into polar columns in crystals and liquid crystals. *Nature* 419:702.

48. Ryckaert, J. P., G. Ciccotti, and H. J. C. Berendsen. 1977. Numerical integration of the Cartesian equations of motion of a system with constraints: Molecular dynamics of n-alkanes. *Journal of Computational Physics* 23:327–341.

49. Smith, G. D., O. Borodin, and D. Bedrov. 2002. A revised quantum chemistry-based potential for poly(ethylene oxide) and its oligomers in aqueous solution. *Journal of Computational Chemistry* 23:1480.

50. Girifalco, L. A. 1992. Molecular properties of C_{60} in the gas and solid phases. *Journal of Physical Chemistry* 96(2):858–861.

51. Werder, T. et al. 2003. On the water-carbon interaction for use in molecular dynamics simulations of graphite and carbon nanotubes. *Journal of Physical Chemistry B* 107:1345–1352.

52. Li, L., D. Bedrov, and G. D. Smith. 2005. Repulsive solvent-induced interaction between C_{60} fullerenes in water. *Physical Review E* 71:011502.

53. Pilcher, H. R. 2003. Alzheimer's abnormal brain proteins glow. *Nature News* September 23, 2003.

54. Xu, Z., R. Paparcone, and M. J. Buehler. 2010. Alzheimer's Aβ(1–40) amyloid fibrils feature size dependent mechanical properties. *Biophysical Journal* 98(10):2053–2062.

55. Monticelli, L. et al. 2008. The MARTINI coarse-grained force field: Extension to proteins. *Journal of Chemical Theory and Computation* 4:819–834.

56. Marrink, S. J., A. H. de Vries, and A. E. Mark. 2004. Coarse grained model for semiquantitative lipid simulations. *Journal of Physical Chemistry B* 108:750–760.

57. McClain, D. et al. 2007. Electrostatic shielding in patterned carbon nanotube field emission arrays. *Journal of Physical Chemistry C* 111(20):7514–7520.

58. Yang, J. and L. Dai. 2003. Multicomponent interposed carbon nanotube micropatterns by region-specific contact transfer and self-assembling. *Journal of Physical Chemistry B* 107:12387–12390.

59. Buehler, M. J., S. Keten, and T. Ackbarow. 2008. Theoretical and computational hierarchical nanomechanics of protein materials: Deformation and fracture. *Progress in Materials Science* 53:1101–1241.

60. Qin, Z., L. Kreplak, and M. J. Buehler. 2009. Hierarchical structure controls nanomechanical properties of vimentin intermediate filaments. *PLos ONE* 4(10):e7294.

61. Tang, Y., R. Ballarini, M. J. Buehler, and S. J. Eppell. 2010. Micromechanisms of deformation of collagen fibrils under uniaxial tension. *Journal of the Royal Society, Interface* 7(46), 839.

4 Coarse Molecular-Dynamics Analysis of Structural Transitions in Solid Materials

Dimitrios Maroudas, Miguel A. Amat, and Ioannis G. Kevrekidis

CONTENTS

4.1 INTRODUCTION

Structural transformations of crystalline solids and other condensed-matter systems underlie a broad class of technologically significant problems. A fundamental atomic-scale understanding is required to determine the onsets of such structural transitions, i.e., the conditions under which materials undergo transitions as a result of thermal, mechanical, or chemical driving forces. Accurate theoretical prediction of such structural transition onsets is a major challenge that lies at the heart of condensed-matter physics, materials science, and the entire spectrum of engineering sciences. Toward this end, atomistic simulation methods provide ideal means for mechanistic analysis and quantitative prediction.

Direct atomistic dynamical simulation methods, such as classical molecular dynamics (MD), have been used to probe the dynamical response of materials over extremely fine time scales (picoseconds to nanoseconds) and to analyze materials' structural evolution at full atomic resolution. The level of detail offered by such simulations is often beyond that achievable by state-of-the-art experimental techniques.

Furthermore, the fine time-scale information generated from these atomistic simulations can be used to identify and elucidate the mechanisms that govern the underlying physicochemical processes, which determine the materials response to thermal, mechanical, or chemical impulses.

Currently, the main problem associated with the implementation of direct atomistic dynamical simulation methods, such as MD, is their severe limitations to simulate large molecules or condensed-matter systems over observable time scales while preserving full atomic resolution; on the other hand, computational supercells containing up to several billion atoms (i.e., spanning substantial length scales at atomic resolution) have been used for simulating short-time dynamics. Monitoring the long-term dynamics of complex molecules or condensed-matter systems is of fundamental importance for the kinetic analysis of rate processes characterized by time scales that are slower by orders of magnitude than the nanosecond scales that can be accessed routinely by MD. Such slow time scales govern the kinetics of rate processes ranging from protein folding to the structural response of heterogeneous materials with fine microstructure during their synthesis, processing, or function. The kinetics of rare events also limits the study by direct atomistic simulation of the long-term behavior of such complex systems due to the limited sampling of order-parameter space that it allows.

Proper sampling of the order-parameter space of complex molecules and condensed-matter systems provides the means to construct their underlying free-energy landscapes, i.e., their free-energy surfaces over order-parameter space. Access to this fundamental thermodynamic signature of a material system allows important thermodynamic and kinetic information about the system to be obtained. For example, from the free-energy landscape, stable states can be distinguished from unstable ones and the location of the corresponding transition points can be determined. Knowledge of these stability regions (corresponding to local minima or free-energy wells) can be used to assess the relative stability of different thermodynamic states and knowledge of the paths (in order-parameter space) that connect them allows for the calculation of the free-energy barriers between stable phases; these barriers can be used to determine the corresponding escape rates from free-energy wells. Nevertheless, the time-scale limitations of direct atomistic simulation methods make the construction of free-energy landscapes particularly challenging.

The coarse molecular-dynamics (CMD) approach provides an excellent means to circumvent such challenges. The purpose of this chapter is to demonstrate the capabilities of the CMD approach to predict accurately and in a computationally efficient manner the onset of thermally and mechanically induced structural transitions in condensed matter. The rest of this chapter is structured as follows. The theoretical foundation of the CMD approach is outlined in Section 4.2. Applications of the CMD approach in the analysis of thermally and mechanically induced structural transitions in crystalline materials or ordered layers are presented in Sections 4.3 through 4.5. The melting transition of a model crystalline silicon slab is analyzed in Section 4.3 and the melting point is predicted. A pressure-induced polymorphic transition in a model nickel crystal is analyzed in Section 4.4 and the critical pressure for the transition is calculated. A thermally induced order-to-disorder transition in a krypton monolayer physisorbed onto a graphite substrate is studied in Section 4.5

and the transition temperature is determined. Finally, our main conclusions are summarized in Section 4.6.

4.2   COARSE MOLECULAR DYNAMICS

The CMD method was originally developed by Hummer and Kevrekidis [1] as an attempt to circumvent shortcomings in obtaining and analyzing the evolution of slow coarse-grained variables of complex dynamical systems. Toward this end, the projection operation formalism of Zwanzig [2] allows, in principle, the relation of microscopic dynamics to such slow evolution; yet, the exact formulas for the corresponding noise and memory terms are practically inaccessible. CMD circumvents the evaluation of such terms by estimating *on the fly* the thermodynamic driving forces for the slow evolution, as well as their local dynamics. The method relies on the existence of one or more coarse variables that contain all the necessary information about the long-term dynamics of the system's state. Here, in the cases of structural transitions that we are focusing on, it will be shown that a single such variable, ψ, suffices. In addition, the CMD method requires that the $\psi(t)$ evolution, where t is time, must be slow and attracting. This means that there exists a one-dimensional attracting manifold, parameterized by ψ, such that the statistics of the remaining variables of the system become quickly "slaved" to ψ and then evolve slaved to ψ. In the cases presented here, it should be emphasized that the coarse variable is chosen based on knowledge of the system at hand; however, powerful techniques that rely on modern data mining tools are being developed, such as those based on diffusion maps [3], to extract systematically good coarse observables without depending on any *a priori* knowledge of the system or making any assumptions.

CMD is an equation-free technique that is based on two transformations: *lifting* and *restricting* [1,4–7]. The former mapping (lifting) involves consistently initializing the system with a specific value of the coarse variable, $\psi(t = 0) = \psi_0$, whereas the latter mapping (restricting) involves monitoring the evolution of the coarse-grained observable, $\psi(t)$. Detailed descriptions of CMD computational implementations are given in Sections 4.3 through 4.5 on a case-by-case basis. An appealing feature of CMD that makes it very convenient to use is that it can be wrapped around existing molecular simulation codes; most of the computational effort involved is spent on executing a large number of short bursts of conventional molecular simulations.

CMD is based on the description of the evolution of the probability density, $P(\psi,t)$ to find the system at state ψ in time t, where $\psi(t)$ is an appropriate coarse-grained observable that describes the state of the system [1,7]. When the corresponding stochastic process is Markovian and invariant with respect to a shift in time, t, the evolution of $P(\psi,t)$ can be described by the Fokker-Planck (FP) equation

$$\frac{\partial}{\partial_t} P(\psi,t) = -\frac{\partial}{\partial \psi}\left\{\left[v(\psi) - \frac{\partial}{\partial \psi}D(\psi)\right]P(\psi,t)\right\}, \qquad (4.1)$$

where $v(\psi) = \langle[\psi(t + \tau) - \psi(t)]\rangle/\tau$ and $D(\psi) = \langle[\sigma^2(t + \tau) - \sigma^2(t)]\rangle/(2\tau)$, evaluated in the limit of $\tau \rightarrow 0$, are the drift velocity and the diffusion coefficient, respectively. The brackets $\langle \rangle$ denote ensemble averaging, and $\sigma^2(t)$ denotes the variance or spread of the coarse variable, ψ, at time t; this is equivalent to $D(\psi) = \langle [\psi(t + \tau) - \psi(t)]^2 \rangle / (2\tau)$ in the $\tau \to 0$ limit. Integrating the equilibrium version of the FP equation and using the equilibrium ansatz $P_{eq}(\psi) = \exp[-G(\psi)/k_B T]$, where k_B is Boltzmann's constant, T is temperature, and $G(\psi)$ is the effective free energy of the system at state ψ, yields

$$\frac{G(\psi)}{k_B T} = -\int \frac{\upsilon(\psi')}{D(\psi')} d\psi' + \ln(D(\psi)) + C, \qquad (4.2)$$

with C being an integration constant; the equilibrium version of the FP equation is derived by setting to zero the probability density flux, the divergence (in ψ-space) of which is taken in the right-hand side of Equation 4.1.

A number of CMD-based analyses have been conducted to study nonequilibrium phenomena in complex systems and their relation to their thermodynamic state. Most of these studies have been limited to either small atomic and molecular systems or to the use of stochastic simulations, such as kinetic Monte Carlo (KMC) and equilibrium Monte Carlo (MC) simulations with the propagation through MC steps playing the role of time stepping. Such examples include the CMD study by Hummer and Kevrekidis of the dynamics of alanine [1] and the CMD study by Sriraman et al. of water molecules filling or emptying carbon nanotubes [6]. In these studies, on-the-fly coarse-grained information estimated from many short and properly initialized independent replica MC/KMC simulations was used to identify the existing basins of attraction or stable phases, as well as to find the transition points in the physical behavior of the complex systems under consideration. Other examples include the study through KMC simulation of line-defect motion in impure crystalline solids [8] and of micelle formation [4,5] based on both equilibrium MC and KMC simulations. In the following sections, we discuss three condensed-matter applications of CMD to demonstrate the viability of the CMD method in predicting structural transition onsets, as well as the corresponding stable and unstable states.

4.3   MELTING OF A SILICON SLAB

The thermodynamic melting point marks the onset of the solid-to-liquid transition; this is defined as the temperature at given pressure for which the Gibbs free energies of the solid and liquid phases of a material are equal. Traditionally, computational methods for calculating thermodynamic melting points have followed two approaches: equilibrium (phase coexistence) calculations and nonequilibrium techniques. There are several phase-coexistence techniques available to determine the thermodynamic melting temperature, T_m. Specifically, in the method developed by Landman et al. [9], solid and liquid phases in coexistence are created artificially, whereas the methods by Broughton and Li [10] and Lutsko et al. [11] are based on computing the Gibbs free energy of both solid and liquid phases as a function of temperature. On the other hand, the development of nonequilibrium methods based on MD simulations was motivated largely by experimental efforts [12,13], which provided direct measurements of the velocity of the liquid interface of molten silicon

(Si) produced during pulsed laser annealing experiments. The methods developed by Kluge and Ray [14], Phillpot et al. [15], and Lutsko et al. [16] are excellent examples of such nonequilibrium techniques. These techniques use a slab supercell (or a bicrystal model) containing an equilibrated solid material at a temperature below melting. The supercell is then suddenly perturbed to a temperature well above melting. This perturbation creates a melting front that is nucleated at the slab's surfaces (or the grain boundary of the bicrystal) and propagates toward the solid core at a temperature-dependent velocity. As the perturbed temperature approaches T_m, the melting-front propagation velocity tends to zero.

In this section, the above mentioned surface-initiated (i.e., heterogeneously nucleated) melting is used as a representative nonequilibrium structural transition of a condensed-matter phase to demonstrate how to (1) extract the underlying effective free-energy landscape in the thermodynamic limit and (2) obtain the melting temperature, T_m, corresponding to the onset of the structural transition under consideration. This task is carried out by selecting an appropriate coarse variable, a structural order parameter that describes the state of the system, running multiple short MD simulations, and processing their results as described within the CMD framework.

The model consists of a slab supercell with 34 planes parallel to the surface plane and containing 50 atoms each; the free-surface planes of the slab are taken to be normal to the [001] crystallographic direction. The interatomic interactions are described by the many-body Tersoff potential (T3) for Si [17]. The equations of motion are integrated using a fifth-order Gear predictor–corrector algorithm with a time step of 0.5 fs and the temperature is kept constant by velocity rescaling at each time step. The order-to-disorder transition that each plane undergoes as melting proceeds is monitored by a planar order parameter, ξ, based on the planar structure factor [15,16]: $0 \leq \xi \leq 1$; $\xi = 1$ corresponds to a perfect crystalline solid plane; $\xi = 0$ corresponds to a molten plane; and $\xi = 1/2$ corresponds to the interface between the liquid and the solid. As melting proceeds, this information can be translated into the number of melted planes as a function of time, which, when normalized with the total number of planes in the system, constitutes our choice of coarse variable, ψ, which describes properly the state of our slab system. By definition, $\psi = 0$ corresponds to a perfect crystalline solid, whereas $\psi = 1$ corresponds to a melt. For comparison purposes, the T_m for this model was determined [18] using the method introduced by Phillpot et al. [15] and Lutsko et al. [16], where a melt/crystal propagation front is created and the temperature-dependent front propagation velocity, v_p, is monitored. Extrapolation of v_p to zero yields $T_m = 2562 \pm 10$ K [18]. It should be emphasized that as $T \rightarrow T_m$, the very slow interfacial propagation speed in conjunction with the increased amplitude of the fluctuations of the dynamical variables make the accurate determination of v_p very demanding computationally, through analysis of extremely long MD trajectories.

The CMD approach was implemented by setting the system to the temperature of interest and using a lattice parameter corresponding to the zero-pressure isobar; it is important to keep the system at zero pressure to avoid development of thermomechanical stresses. The dependence of the lattice parameter on T can be obtained by carrying out an isothermal–isobaric MD simulation [19]. At $t = 0$, the system is initialized/forced to satisfy a prescribed value of ψ. For example, choosing a value of

$\psi = 4/34$ corresponds to setting an initial configuration with four melted planes out of the total of 34 planes. The initialization of the system satisfying this coarse description ("lifting" transformation) is nonunique [7]; in this work, it is carried out using a plane-by-plane Metropolis MC-type scheme with an added bias potential. This bias potential has the form of a harmonic potential, $U_{\mathrm{har}} = K(\xi - \xi_{\mathrm{obi}})^2$, with a stiff spring constant ($K = 5 \times 10^4$) and it allows for fast sampling toward the objective values at each plane; ξ denotes the running value of the planar order parameter at the plane of interest and ξ_{obj} denotes its objective value.

Upon successful initialization consistent with the desired ψ (i.e., melting the desired number of planes and setting the new interface locations), one proceeds by time stepping through MD for a 200-ps horizon, while recording points in the coarse variable trajectory, $\psi(t)$, every 0.5 ps. This process is repeated by choosing values of ψ between 4/34 and 32/34 at increments of 2/34. At each initial coarse-variable value of choice, 25 independent replicas are generated and the trajectories are monitored along with their respective variance as they proceed to relax the system from its imposed initialization. By computing the slopes of the averaged coarse-variable evolution and its variance, the local effective drift velocity and diffusion coefficient of Equation 4.1 are estimated. Figure 4.1 shows a schematic outline of the implemented CMD procedure.

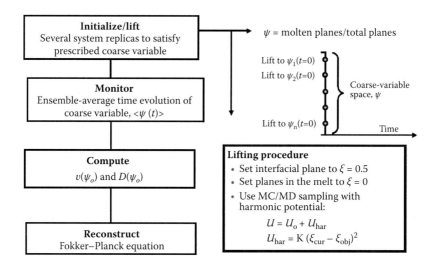

FIGURE 4.1 Schematic outline of the CMD procedure with emphasis on the lifting scheme for studying the thermodynamic melting of the model crystalline silicon system. The coarse-variable space, $0 \le \psi \le 1$, is mapped by implementing an MC-based sampling scheme with an added quadratic penalty function (harmonic potential) that takes into account the current and target values of the planar order parameter, ξ, required to satisfy chosen values of the coarse variable, ψ_0. This is done in a plane-by-plane fashion and involves setting the target values of ξ to the different planes: $\xi_{\mathrm{obj}} = 0$ (melted plane) and $\xi_{\mathrm{obj}} = 0.5$ (interfacial plane). Following this initialization, the constraint is released and the ensemble averaged coarse-variable evolution, $\langle \psi(t) \rangle$, is monitored to obtain the drift velocity, $v(\psi_0)$, and the diffusion coefficient, $D(\psi_0)$, of the underlying Fokker–Planck equation.

The results shown here span a T range from 2540 to 2580 K at increments of 5 K. At each T, the system is initialized at chosen values of ψ over the interval $0 \leq \psi \leq 1$. Figure 4.2 shows the evolution of the coarse variable, $\psi(t)$, for various initial conditions $\psi_0 = \psi(t = 0)$ at $T = 2540$ and 2580 K. In Figure 4.2a, $T = 2540$ K, it is shown that in almost all of the cases, $\psi(t)$ drifts toward a solid state characterized by $\psi < 0.5$; melting occurs only when the system is initialized at values very close to $\psi_0 = 1$. In Figure 4.2b, $T = 2580$ K, it is shown that $\psi(t)$ always drifts toward a molten state, a clear indication that $T > T_m$. It is important to note that the time interval of 200 ps chosen for the MD simulations is only a small fraction of the time required for the evolution of the slowest coarse variables to attain steady state.

At each T, analyzing the differently initialized coarse-variable trajectories yields the diffusion coefficient and the drift velocity as functions of the initial value of the coarse variable, ψ_0. With $v(\psi)$ and $D(\psi)$ available, Equation 4.1 can be reconstructed and integrated to give the result of Equation 4.2, which can be used to generate the effective free-energy landscapes shown in Figure 4.3. Each such landscape exhibits two (thermodynamic potential) wells: one corresponding to the solid state and another corresponding to the molten state.

Drawing from thermodynamic coexistence criteria, the free-energy difference between the bottoms of the two wells, ΔG_{wells} (as indicated by the two horizontal dashed lines at $T = 2540$ K in Figure 4.3) is related to the departure from the equilibrium phase coexistence (melting) temperature, T_m. Two regions are identified corresponding to temperatures above and below the thermodynamic melting temperature ($T > T_m$ and $T < T_m$, respectively), along with their corresponding activation barriers for the melting transition. A plot of $\Delta G_{wells}/kT$ as a function of T is shown in Figure 4.4. The temperature for which this free-energy difference goes to zero

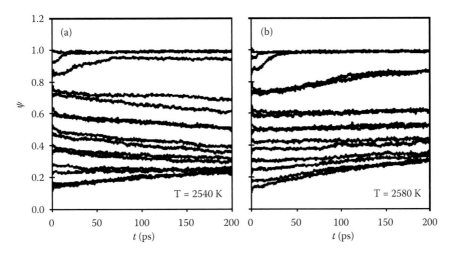

FIGURE 4.2   Evolution profiles of the coarse variable, $\psi(t)$, initialized at different values, ψ_0, for the Si slab model at temperatures of (a) 2540 and (b) 2580 K, i.e., below and above T_m, respectively. The evolution exhibits drift toward the "two potential wells," the solid and the melt. (Adapted from Amat, M. A., Kevrekidis, I. G., and Maroudas, D., *Phys. Rev. B*, 74, 132201, 2006.)

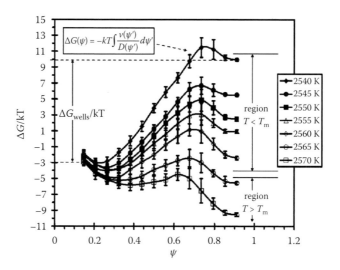

FIGURE 4.3 Effective free-energy surface as a function of the coarse variable, ψ, for the Si slab model at various temperatures around the melting transition. Thermodynamic potential wells centered around low and high (*half-well*) coarse-variable values, ψ, represent the crystalline and molten state of the system, respectively. (Adapted from Amat, M. A., Kevrekidis, I. G., and Maroudas, D., *Phys. Rev. B*, 74, 132201, 2006.)

corresponds to T_m. A linear fit to the CMD results in conjunction with the phase coexistence criterion ($\Delta G_{wells} \rightarrow 0$) yields $T_m = 2564$ K, in excellent agreement with our T_m prediction based on the melting front propagation velocity calculations [18].

In summary, the implementation of the CMD approach provides an accurate and computationally efficient method for constructing the effective free-energy

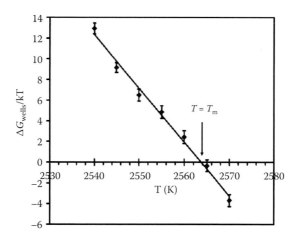

FIGURE 4.4 Temperature dependence of the scaled free-energy difference between the two potential-well minima, yielding the melting temperature, T_m, for the Si slab model. (Adapted from Amat, M. A., Kevrekidis, I. G., and Maroudas, D., *Phys. Rev. B*, 74, 132201, 2006.)

landscapes that govern the melting transition in condensed-matter systems. More specifically, the predictions of this approach for the melting transition of the model silicon system used in this study are in excellent agreement with those of "traditional" methods; these include results from analysis of long MD simulations, as well as values reported in the literature for the thermodynamic melting of the Si model described by this Tersoff potential (T3) based on solid–liquid coexistence analysis: these values are 2547 ± 22 K [20] and 2567 K [21]. More importantly, this agreement demonstrates the power of the CMD approach in establishing the connection between the nonequilibrium evolution of the coarse variable and its underlying effective free-energy gradients, as ultimately captured in the effective free-energy landscapes, such as those shown in Figure 4.3. Such thermodynamic information allows for identification of important features in the landscape, which relate to the inherent stability of the system. As it has been shown elsewhere [1,4,5], having access to the effective free energy allows for the extraction of kinetic information relevant to the rates of exchange between stable basins. Finally, it is important to note the connection between the atomistic and coarse-grained descriptions of the system, which is established through the definition of the (system-size dependent) coarse variable. Although 34 planes in the slab supercell model have been sufficient to correctly capture the transition onset for the interatomic potential employed, it is emphasized that the larger the system size (more planes), the better the coarse graining of ψ and the finer the increments used when sampling its space.

4.4 POLYMORPHIC TRANSITIONS IN METALLIC CRYSTALS

The ability to predict stable crystalline phases under applied mechanical loading is crucial in understanding polymorphic transitions in crystalline solids. Although the MD method developed by Parrinello and Rahman [22,23] (PR-MD) has been a major contribution in the analysis of solid–solid transformations under stress, it requires (if used on its own) tedious analysis of long transient trajectories near critical points for predictions of polymorphic transition onsets. In addition, the PR-MD predictions tend to "overshoot" with respect to the actual thermodynamic transition point [24], which has been attributed to the absence of sites (lattice defects) in the MD supercell required to initiate the heterogeneous nucleation of the phase into which the crystal transforms [24]. Due to this effect, PR-MD may miss known intermediate phases [25,26] rendering the method applicable only to transitions characterized by very low barriers.

To address such challenges, recent developments, namely metadynamics [27–30] and CMD [31] have provided systematic alternatives to mere analysis of conventional MD trajectories. Metadynamics is based on time stepping of a coarse-grained variable in response to a driving force according to a biased thermodynamic potential, through a superposition of Gaussians, that evolves as sampling of coarse-variable space proceeds. This procedure eventually fills the wells of the free-energy surface and drives the system out of the corresponding local minima. On the other hand, as discussed in Sections 4.2 and 4.3, CMD runs the "untouched" MD code (i.e., without modifying/biasing the thermodynamic potential during the run), with only initializing "at will" a proper coarse-grained variable. This initialization ensures thorough sampling of coarse-variable space and bypasses problems associated with crossing between states that are separated by high-energy barriers.

Here, we demonstrate the capabilities of CMD as an efficient computational approach to (1) locate stable crystalline phases and (2) accurately predict the transformation onset in polymorphic transitions in crystals. Specifically, the focus is on determining the loading condition, expressed by a critical pressure $P = P_c$, at which the onset of a transition from a body-centered cubic (bcc) to a hexagonal close-packed (hcp) phase occurs in a crystal under hydrostatic loading. To validate the CMD approach, the work of Zhao et al. [32] has been chosen as a reference; in that study, PR-MD simulations and lattice-statics calculations were used to determine the stability limits of different Morse-model crystalline phases under hydrostatic loading at low temperature. In particular, the case of the Morse-Ni crystal is portrayed, where a bcc phase is stabilized under compression [32]. Results also are reported in terms of stretch factors, $\lambda = (\rho_0/\rho)^{1/3}$, where ρ_0 and ρ are the densities of the cubic crystal at zero pressure and at the pressure of interest, respectively.

The MD simulation model consists of a cubic supercell with 1458 atoms arranged in a bcc lattice. The equations of motion are obtained through the Parrinello–Rahman ansatz, where the supercell geometry is described by a matrix, $\mathbf{h}$, whose columns correspond to the three vectors, $\mathbf{a}$, $\mathbf{b}$, and $\mathbf{c}$, that define the edges of the supercell [22,23]. Cell rotations are avoided by setting the subdiagonal elements of $\mathbf{h}$ to zero and the temperature is kept constant using a Berendsen thermostat [33] with a time constant equal to the integration time step. The results presented here are for a temperature $T = 1$ K; for the pressure-induced transitions of interest, thermal activation effects that determine the kinetic rates of these structural transitions are significant even at a very low (but finite, $T > 0$ K) temperature close to $T = 0$ K [32].

Implementation of the CMD approach offers certain major advantages. Specifically, the method provides the means to compute the drift velocity and diffusion coefficient in the FP equation and to construct, through Equation 4.2, an effective free energy, G_{eff}, that can be used to assess the relative stability between phases. In the case of pressure-induced structural transitions, these G_{eff} landscapes can be used to determine the transformation onset, expressed by a critical pressure (P_c).

The change in the supercell geometry as the crystal transforms from, e.g., bcc to hcp, is reflected in changes among the three angles, α, β, and γ, which are formed by vectors $\mathbf{b}$ and $\mathbf{c}$, $\mathbf{c}$ and $\mathbf{a}$, and $\mathbf{a}$ and $\mathbf{b}$, respectively. The coarse variable chosen for the study of the bcc-to-hcp transformation is $\psi \equiv (\alpha - 90°) + (\beta - 90°) + (\gamma - 90°)$; a value of $\psi = 0°$ corresponds to a perfect bcc lattice. Clearly, the choice of coarse variable is not unique. Other scalar variables (combinations of matrix $\mathbf{h}$ elements) also constitute good coarse-variable choices, and the results of this CMD study do not change with the different coarse-variable choices made [31].

In this particular application, lifting is tailored to explore coarse-variable space through deformations covering a range $|\psi_0| \leq 5°$. The only angle contributing to the coarse variable, ψ_0, is α, while β and γ are kept at 90°. This is achieved by forcing all of the $\mathbf{h}$ elements to zero except for h_{11}, h_{22}, h_{33}, and h_{13}, while imposing an additional constraint that brings the system to the prescribed value of ψ_0. The lifting scheme is outlined in Figure 4.5 and is illustrated geometrically in Figure 4.6. As shown in Figure 4.6, the deformation is carried out in the [010] direction and the only angle

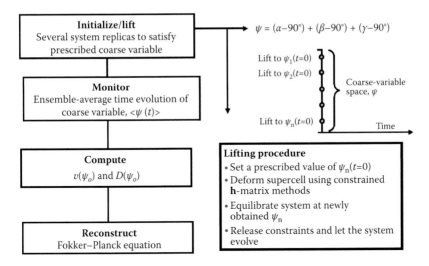

$\psi = (\alpha - 90°) + (\beta - 90°) + (\gamma - 90°)$

Initialize/lift
Several system replicas to satisfy prescribed coarse variable

Monitor
Ensemble-average time evolution of coarse variable, $\langle \psi(t) \rangle$

Compute
$v(\psi_o)$ and $D(\psi_o)$

Reconstruct
Fokker–Planck equation

Lift to $\psi_1(t=0)$
Lift to $\psi_2(t=0)$

Lift to $\psi_n(t=0)$

Coarse-variable space, ψ

Time

Lifting procedure
• Set a prescribed value of $\psi_n(t=0)$
• Deform supercell using constrained **h**-matrix methods
• Equilibrate system at newly obtained ψ_n
• Release constraints and let the system evolve

FIGURE 4.5 Schematic outline of the CMD procedure implemented, with emphasis on the lifting scheme, for studying the pressure-induced bcc-to-hcp polymorphic transformation of the Ni crystal model system. The coarse-variable space is mapped by deforming the supercell at different levels (e.g., $0° \leq \psi_0 \leq 5°$) and proper initialization is achieved by applying constraints to the **h**-matrix during the lifting stage. Following this initialization, the constraints are released and the ensemble averaged coarse-variable evolution, $\langle \psi(t) \rangle$, is monitored to obtain the drift velocity, $v(\psi_0)$, and the diffusion coefficient, $D(\psi_0)$, of the underlying Fokker–Planck equation.

contributing to the coarse variable is α, while $\beta = \gamma = 90°$. The constraints are satisfied at each time step during lifting and are implemented by simply modifying the components of the vectors **a**, **b**, and **c** accordingly. This scheme is nonunique and could be implemented through other methods, such as employing Lagrange multipliers. The lifted states, satisfying $\psi = \psi_0$, are equilibrated for a period of about 9 ps. At the end of this period, all angular constraints are released and the (unconstrained)

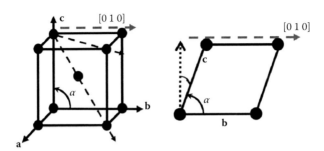

FIGURE 4.6 Illustration of the lifting scheme implemented for the pressure-induced polymorphic transition problem, involving a cell deformation through a change in the angle α formed by vectors **b** and **c**. The dashed lines labeled [010] indicate the direction of the lifting or the crystallographic direction of the cell deformation.

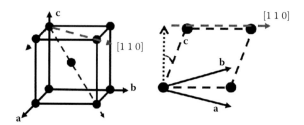

FIGURE 4.7   Illustration of an alternative lifting scheme for the pressure-induced polymorphic transition problem. Here, the dashed lines labeled [110] represent the direction in which the unit cell is to be deformed.

system is allowed to evolve. During its evolution, $\psi(t)$ is monitored over a time-horizon of 3 ps and used to obtain the values of $\upsilon(\psi_0)$ and $D(\psi_0)$ needed to construct the G_{eff} landscapes at each applied pressure.

An alternative choice of coarse variable is illustrated in Figure 4.7. Here, a search along other crystallographic directions is implemented and can identify other existing stable phases, including unknown ones. In general, this search can be extended to any three-angle combination as well. One can implement a lifting scheme based on such a coarse variable with initially noncubic supercells and repeat such angle-based searches in coarse-variable space for stable and unstable phases. Upon lifting the system over representative points in parameter space and reconstruction of the Fokker–Planck equation, the G_{eff} landscape can be obtained and used to determine the transformation onset.

Figure 4.8 shows representative results for the low-temperature evolution of the coarse variable, $\psi(t)$, for various initial conditions ψ_0. Figures 4.8a and 4.8b correspond to reduced pressures of $P = 1360$ and $P = 1340$, respectively; $P \equiv P'r_0^3/D$, where P' is dimensional pressure and r_0 and D are Morse-potential parameters [32]. At both pressures, initialization at coarse-variable values over the range $2° \leq \psi_0 \leq 5°$ result in trajectories that are invariably "attracted" to the same final state, $\psi(t) = 3.3°$, which corresponds to an hcp phase. On the other hand, initializing the system at coarse variables near $\psi_0 = 0°$ ($\psi_0 \leq 0.25°$) yields trajectories that (1) fall back to the original bcc structure (at higher P), or (2) drift away from bcc (at lower P). Analyses of the entire ensemble of trajectories (25 copies at each choice of ψ_0) reveal the existence of three different regions: one attracting region at $2° \leq \psi_0 \leq 5°$ (stable hcp), a second region at values of ψ_0 near zero (stable/unstable bcc) that can be attracting depending on the applied pressure, and a third region at values of $\psi < 2°$ corresponding to intermediate states.

Figure 4.9a shows the G_{eff} landscape corresponding to the conditions of Figure 4.8a. Inset (1) reveals a small thermodynamic-potential well at ψ values near $\psi = 0°$, which corresponds to the stable bcc phase. The G_{eff} landscapes were computed at several applied pressure levels within the range $1320 \leq P \leq 1400$. Based on the relative depth of the wells, it was found that the hcp phase is energetically much more favored than the bcc one. In addition, it was found that the well corresponding to

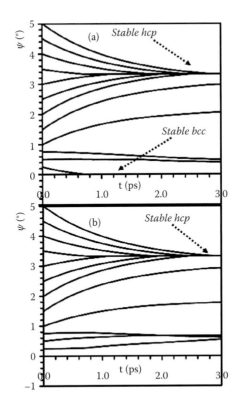

FIGURE 4.8 Evolution of the ensemble-averaged value of the coarse variable, $\psi(t)$, initialized at different values, ψ_0, at reduced pressures of (a) 1360 (above P_c) and (b) 1340 (below P_c) at $T = 1$ K. (Reprinted with permission from Amat, M. A., Kevrekidis, G., and Maroudas, D., *Appl. Phys. Lett.*, 90, 171910. Copyright 2007, American Institute of Physics.)

the bcc phase disappears at reduced pressures $P \leq 1345$. The slope $\Delta(\Delta G_{eff}/kT)/\Delta\psi$ also was computed at values near $\psi = 0°$ and reveals positive/negative values in the presence/absence of the bcc well; see Inset (2) of Figure 4.9a. This change in the sign of the computed slope is related to changes in the stability of the bcc phase: a zero-slope value marks the transformation onset. In Figure 4.9b, the dotted line crossing the ordinate at zero defines two regions: The bcc phase is stable above the dotted line and unstable below. Interpolation using the points immediately above and below this dotted line yields a transformation onset of $P_c = 1348 \pm 3$, which corresponds to a stretch factor of $\lambda_c = 0.7492$. This result is in excellent agreement with the values reported in Zhao, Maroudas, and Milstein [32] and validates our CMD approach; the work of Zhao, Maroudas, and Milstein [32] employed conventional MD simulations involving very long MD trajectories (ns-scale runs at each P). Finally, it should be pointed out that, as shown in the inset to Figure 4.9a, the CMD approach also identifies the transition state, *TS*, of the polymorphic transformation.

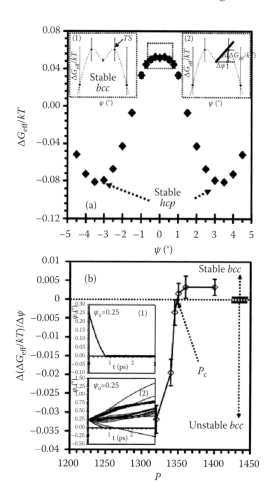

FIGURE 4.9 (a) Effective free-energy landscape as a function of the coarse variable ψ at $P = 1360$. Wells centered at $\psi = 0°$ and $\psi = 3.3°$ correspond to bcc and hcp phases, respectively. Insets (1) and (2) correspond to magnifications of the boxed region of the plot with (2) illustrating the slope computation. (b) Slope of $\Delta G_{eff}/k_B T$ with respect to ψ in the vicinity of $\psi = 0°$ as a function of P. Insets (1) and (2) show representative coarse trajectories of the 25-copy ensemble initialized at $\psi_0 = 0.25°$ for $P = 1360$ and 1340, respectively. Inset (1) shows a typical attractive evolution toward a stable bcc phase (at $P > P_c$), whereas inset (2) shows a nonattractive evolution at $P < P_c$. (Reprinted with permission from Amat, M. A., Kevrekidis, G., and Maroudas, D., *Appl. Phys. Lett.*, 90, 171910. Copyright 2007, American Institute of Physics.)

4.5 ORDER-TO-DISORDER TRANSITIONS IN PHYSISORBED LAYERS OF NOBLE-GAS ADSORBATES ON GRAPHITE

The final application of the CMD method that we discuss addresses thermally induced order-to-disorder and disorder-to-order transformations exhibited by inert gases adsorbed onto graphite substrates. The CMD approach developed for the analysis of such transitions was presented in Amat et al. [34]. The CMD method is designed to function as a wrapper around an existing MD model based on a holding-potential form for krypton (Kr) atoms physisorbed on graphite, which has been described in the literature [35–37]. The specific system chosen to illustrate the development and implementation of this CMD method consists of a monolayer of Kr atoms physisorbed onto a graphite substrate surface, which is known to undergo a thermally induced order-to-disorder transition around 130 K [38,39]. The CMD-based prediction of this transition onset is validated by comparisons with results based on conventional MD techniques for the same system, which have been reported by Amat et al. [34].

The order parameter ξ is defined as

$$\xi = \left(\frac{1}{N^2}\right)\left|\sum_{j=1}^{N}\exp\left(i\mathbf{k}\cdot\mathbf{q}_j\right)\right|^2,$$ (4.3)

and is chosen as the relevant coarse variable, $\psi \equiv \xi$, for describing the state of the system (i.e., commensurate/ordered versus noncommensurate/disordered). In Equation 4.3, $\mathbf{q}_j$ is a 2D vector that corresponds to the position vector of the jth Kr atom projected onto the surface plane, $\mathbf{k}$ is a reciprocal lattice vector of the reciprocal lattice associated with the Bravais lattice spanned by the commensurate $\sqrt{3} \times \sqrt{3}$ phase of the physisorbed layer [34], and N is the number of Kr atoms in the computational supercell.

This coarse variable is then initialized at will to map its entire space $\psi \in [0,1]$ at increments of $\Delta\psi = 0.05$. This initialization is achieved by using a Metropolis MC-based scheme with a quadratic penalty function (a harmonic potential) that takes into account the difference between the running and target values of the coarse variable, ψ and ψ_0, respectively, using a stiff and variable spring constant, K, initialized at $K = 500$ (in reduced units) as illustrated in Figure 4.10. The variation in the value of K is introduced by dividing its constant value by $|\Delta\psi_n - \Delta\psi_c| + \delta$, where $\Delta\psi_n = (\psi_n - \psi_{obj})^2$, $\Delta\psi_c = (\psi_c - \psi_{obj})^2$, and δ is a small value ($\sim 10^{-5}$). The subscripts n, c, and obj refer to the newly MC-generated trial value of the coarse variable, its previous value before the trial move, and its objective value, respectively. After the target coarse-variable value, ψ_0, is satisfied, the system is equilibrated at this new state. Each sampled point in coarse-variable space is comprised of an ensemble of replicas and is represented by $\langle\psi_0\rangle = \langle\psi(t = 0)\rangle$. Here, the brackets $\langle\rangle$ indicate an ensemble average over 25 independent replicas of the system.

Following this initialization (lifting) procedure, all constraints are released and the system is allowed to evolve using conventional MD, i.e., through a short burst of the atomistic time stepper. The averaged coarse-variable evolution, $\langle\psi(t)\rangle$, is then monitored over a time horizon of 200 ps, from which the drift velocity, $v(\psi_0)$, and the diffusion coefficient, $D(\psi_0)$, are obtained and used to reconstruct the underlying Fokker–Planck equation. At temperatures near the transition onset, simulation runs of 200 ps are considered to be quite short with respect to the long-time coarse-variable evolution. This time-scale assessment is supported by the results of Amat et

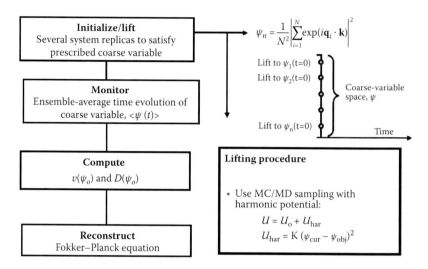

FIGURE 4.10 Schematic outline of the CMD procedure implemented with emphasis on the lifting scheme for studying order-to-disorder transitions in the Kr-on-graphite system. The coarse-variable space, $0 \leq \psi \leq 1$, is mapped by implementing a MC-based sampling scheme with a quadratic penalty function (harmonic potential) that takes into account the current and target values of the coarse variable. Following this initialization, the constraint is released and the ensemble averaged coarse-variable evolution, $\langle \psi(t) \rangle$, is monitored to obtain the drift velocity, $v(\psi_0)$, and the diffusion coefficient, $D(\psi_0)$, of the underlying Fokker–Planck equation. (Reprinted with permission from Amat, M. A., Arienti, M., Fonoberov, V. A., Kevrekidis, I. G., and Maroudas, D., *J. Chem. Phys.*, 129, 184106. Copyright 2008, American Institute of Physics.)

al. [34]. Specifically, the MD-calculated values of the Kr self-diffusion coefficient at the temperatures of 127 K and 128 K, which are very close to the transition temperature (i.e., $T \sim T_t$), were practically equal; this reveals that the MD simulation period of 4 ns that was used by Amat et al. [34] was not sufficiently long to distinguish between these two diffusion-coefficient values due to the critical slowdown near the transition onset. The choice of the planar order parameter as the coarse variable at temperatures equal to or higher than 90 K, where a small fraction of Kr atoms is promoted over the first physisorbed layer, could affect the overall quality of the coarse variable; it was found, however, that this coarse variable choice remains satisfactory for describing the state of the system [34].

Representative coarse-variable evolution plots after releasing the lifting constraints are shown in Figure 4.11. Figure 4.11a shows coarse trajectories, $\langle \psi(t) \rangle$, initiated at different ψ_0 values at a temperature of 100 K, which is well below the transition temperature, T_t. It is evident that the system evolves toward the ordered state, which is represented by high values of the coarse variable, ($\psi \cong 1$). Figure 4.11b shows representative coarse trajectories at a temperature that is well above T_t, $T = 130$ K. In this case, the tendency of the coarse-variable evolution is toward the disordered state, which is represented by low values of the coarse variable, ($\psi \cong 0$). This qualitative analysis of the coarse variable serves as an efficient tool to bracket the location of the transition onset to within a narrower temperature range. Generation

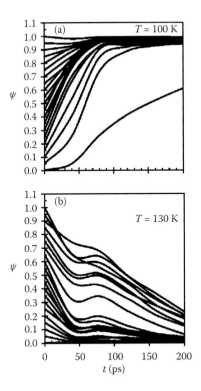

FIGURE 4.11   Ensemble averaged coarse-variable evolution, $\langle \psi(t) \rangle$, after lifting at a temperature (a) well below ($T = 100$ K) and (b) well above ($T = 130$ K) the transition temperature, T_t. The coarse trajectories in (a) show that the system evolves to its ordered state ($\psi \rightarrow 1$), whereas those in (b) show that the coarse evolution drifts toward the disordered state ($\psi \rightarrow 0$). The observed crossing between some of the curves in the coarse-variable evolution may imply that a second coarse variable becomes important. This issue could be due to the small fraction of atoms promoted away from the original monolayer. (Reprinted with permission from Amat, M. A., Arienti, M., Fonoberov, V. A., Kevrekidis, I. G., and Maroudas, D., *J. Chem. Phys.*, 129, 184106. Copyright 2008, American Institute of Physics.)

of at-will-initiated coarse trajectories at temperatures between 115 and 130 K along with the corresponding monitoring and analysis required to obtain the drift velocity, $v(\psi_0)$, and the diffusion coefficient, $D(\psi_0)$, and use of Equation 4.2 allow for the generation of the corresponding underlying effective free-energy landscapes. These landscapes, $\Delta G/k_B T$ as a function of ψ, are shown in Figure 4.12 over the temperature range of interest.

The coarse-variable definition stems from the assumption that the state of the system at temperatures well below and above the transition temperature, T_t, is represented by a one-dimensional free-energy landscape, $\Delta G(\psi)/k_B T$, consisting of two (half) wells that are centered at coarse-variable values of one and zero and correspond to the ordered (commensurate) and the disordered (noncommensurate) states, respectively. This trend is illustrated in Figure 4.12, where the lowest and highest temperatures examined are 115 K and 130 K, respectively. The broken solid lines

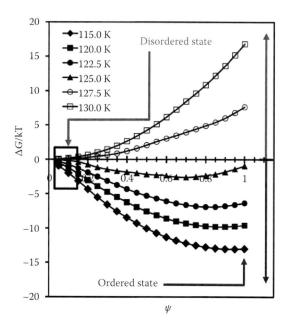

FIGURE 4.12 Effective free-energy landscapes for temperatures over the range 115 K ≤ T ≤ 130 K. The broken solid lines point to the bottom of the thermodynamic potential wells corresponding to the ordered and disordered states, respectively. The solid vertical lines are used to designate the regions corresponding to the ordered and disordered states, respectively. The square box shown encloses the region used to calculate the slopes that have been employed in the determination of the transition temperature, T_t. (Reprinted with permission from Amat, M. A., Arienti, M., Fonoberov, V. A., Kevrekidis, I. G., and Maroudas, D., *J. Chem. Phys.*, 129, 184106. Copyright 2008, American Institute of Physics.)

depict the bottoms of the effective free-energy wells corresponding to the ordered and disordered states, respectively. As shown in Figure 4.12, the location of the effective free-energy well corresponding to the ordered state (0.5 < ψ ≤ 1) becomes shallower and shifts toward lower values of ψ as the temperature is increased.

At temperatures T ≥ 127.5 K, a drastic change is observed resulting in effective free-energy wells that are now centered at $\psi = 0$. Guided by this observation, the various effective free-energy landscapes can be assigned into one of two regions: a region at temperatures between 115 K and 125 K corresponding to ordered states and designated by the downward pointing vertical line, and a region at temperatures T ≥ 127.5 K corresponding to the disordered state and designated by the upward pointing vertical line. It is important to note that the shape of the effective free-energy landscapes corresponds to either the "solid" or the "melt" phase as expected from a second-order transition.

Analysis of the effective free-energy landscapes at coarse-variable values near zero allows for calculation of the slope of $\Delta G/k_B T$ with respect to ψ at each temperature studied. This slope calculation is illustrated in the inset to Figure 4.13, where negative and positive slope values are related to the presence and absence of ordered states, respectively. These slope values, $\Delta(\Delta G/k_B T)/\Delta\psi$, are then plotted as a function

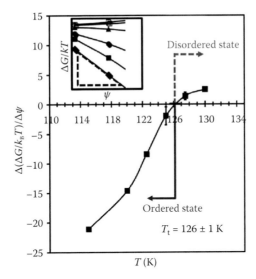

FIGURE 4.13 Temperature dependence of the slope of the effective free energy, $\Delta G/k_B T$, with respect to the coarse variable, ψ, used for the bracketing and determination of the transition onset, T_t. The inset corresponds to the box shown in Figure 4.12 and is used to highlight the slope computation. (Reprinted with permission from Amat, M. A., Arienti, M., Fonoberov, V. A., Kevrekidis, I. G., and Maroudas, D., *J. Chem. Phys.*, 129, 184106. Copyright 2008, American Institute of Physics.)

of temperature and are shown in Figure 4.13. The two regions above and below the transition onset are characterized by positive and negative slope values, respectively, and are designated by broken dashed and solid lines, respectively. Such ordered–disordered to negative–positive correspondence between the state of the system and the slope of the effective free-energy surface with respect to the coarse variable ψ allows for determining the transition onset in the limiting case where the slope becomes equal to zero. This is obtained by interpolation and yields a temperature of 126 K for the commensurate-to-fluid (order-to-disorder) transition for the Kr-on-graphite system. This criterion is based on thermodynamic arguments and yields results that agree well with the results of the phase-transition analysis based on conventional MD simulations [34]. Alternatively, the transition onset could be located by implementing a Newton–Raphson iterative procedure in "variable and parameter" space, as has been shown elsewhere [6].

4.6 SUMMARY AND CONCLUSIONS

The CMD method that was presented in this chapter gives a direct connection between macroscopic or observable materials properties and the underlying atomistic processes. This connection was demonstrated with the study of three drastically different classes of structural transitions in condensed matter: thermally induced melting of crystalline solids, stress-induced polymorphic solid–solid transformations of metallic crystals, and thermally induced order-to-disorder and disorder-to-order transitions of rare-gas monolayers physisorbed on graphite. The procedure implemented toward predicting the different transition onsets [18,31,34] consisted of (1) the identification of a coarse variable that describes properly the thermodynamic state of the system, (2) the generation of short bursts of judiciously initialized fine-scale (MD) simulations from which the relevant coarse-scale information was obtained on the fly, and (3) the construction of an effective free-energy landscape describing the relevant features about the inherent stability of the different states. Specifically, the three problems that were addressed in this chapter establish the capabilities of CMD to successfully relate short-time nonequilibrium processes of a condensed-matter system to the underlying thermodynamic information. In particular, it was demonstrated that having access to the effective free-energy landscape allows for locating stable basins, unstable phases, and transition points, which are obtained through analyses of the independent replica fine-scale simulations of the system of interest.

Also, it was demonstrated that CMD makes the passage of information between fine and coarse descriptions possible through the lifting and restricting transformations. This is equivalent to implementing two mappings: one from a high-dimensional space, i.e., a fine-scale description that corresponded to atomic-scale descriptions for the problems addressed in this chapter, to a relevant low-dimensional space, i.e., a coarse-scale description (restricting) and one in the opposite direction, such as from a coarse-scale to a fine-scale description (lifting). In particular, one of the major advantages of CMD compared to other multiscale techniques is contained within the lifting scheme. Lifting enables the generation of nonunique distributions of atomistic/fine-scale information, which are consistent with specific values of the coarse variable. This one-to-one correspondence between the fine-scale (e.g., atomistic) and coarse-scale (macroscopic) descriptions of a system's state allows for initialization of the system at will to explore the entire coarse-variable space, irrespective of the energy barriers that may separate different states of the system. Mapping-on-demand makes accessible regions of coarse-variable space, which either require exhaustive and computationally expensive sampling based on conventional fine-scale techniques in order to be reached, or are simply beyond the reach of current fine-scale simulation capabilities. As demonstrated here, ideal candidates for study through such CMD approaches include crystalline materials and other condensed-matter systems that undergo structural transformations upon heating or under stress.

Although the applications chosen in this chapter focused on condensed-matter problems that exhibit relatively short relaxation times, this choice is not a requirement. CMD also can be applied to study slower dynamical phenomena, as well as soft condensed-matter systems such as polymers, complex fluids, and biomolecules. Furthermore, the CMD methodological framework is not restricted to classical MD; it can be implemented with atomic-scale methods (time steppers) that are coarse-grained in time, such as kinetic MC simulations, and with descriptions of interatomic interactions that are based on higher levels of theory than classical MD, such as Car–Parrinello MD [40] and other *ab initio* MD methods.

In closing, regarding computational implementation, it should be mentioned that, unlike other techniques, such as transition path sampling [41,42] or thermodynamic integration, CMD does not require an *a priori* knowledge of the final state reached in the transition path, nor does it require modification of the underlying potential-energy surface for efficient sampling, as is done in metadynamics or hyperdynamics [43,44]. It was also found that CMD implementation provided important computational gains over conventional MD techniques. Specifically, in the cases of structural transitions described herein, CMD resulted in boost factors of ≥ 5 in computational performance [18,31,34]; this performance assessment was made with respect to accurate conventional MD results based on sampling through long (several nanoseconds) MD trajectories. Moreover, the CMD approach is inherently well suited for massively parallel algorithm implementation where additional computational benefits can be achieved.

ACKNOWLEDGMENTS

This work was supported by the National Science Foundation through grant numbers CTS-0205584, ECS-0317345, CTS-0417770, and CBET-0613501, by the U.S. Department of Energy through CMPD, and by DARPA.

REFERENCES

1. G. Hummer and I. G. Kevrekidis. 2003. *Journal of Chemical Physics* 118:10762.
2. R. Zwanzig. 2001. *Nonequilibrium statistical mechanics,* New York: Oxford University Press. 2001.
3. See, e.g., B. Nadler, S. Lafon, R. C. Coifman, and I. G. Kevrekidis. 2006. *Applied and Computational Harmonic Analysis* 21:113.
4. D. I. Kopelevich, A. Z. Panagiotopoulos, and I. G. Kevrekidis. 2005. *Journal of Chemical Physics* 122:044907.
5. D. I. Kopelevich, A. Z. Panagiotopoulos, and I. G. Kevrekidis. 2005. *Journal of Chemical Physics* 122:044908.
6. S. Sriraman, I. G. Kevrekidis, and G. Hummer. 2005. *Physical Review Letters* 95:130603.
7. I. G. Kevrekidis, C. W. Gear, J. M. Hyman, P. G. Kevrekidis, O. Runborg, and K. Theodoropoulos. 2003. *Communications in Mathematical Sciences* 1:715.
8. M. Haataja, D. J. Srolovitz, and I. G. Kevrekidis. 2004. *Physical Review Letters* 92:160603.
9. U. Landman, W. D. Luedtke, R. N. Barnett, C. L. Cleveland, M. W. Ribarsky, E. Arnold, S. Ramesh, H. Baumgart, A. Martinez, and B. Khan. 1986. *Physical Review Letters* 56:155.
10. J. Q. Broughton and X. P. Li. 1987. *Physical Review B* 35:9120.
11. J. F. Lutsko, D. Wolf, and S. Yip. 1988. *Journal of Chemical Physics* 88:6525.
12. G. J. Galvin, M. O. Thompson, J. W. Mayer, R. B. Hammond, N. Paulter, and P. S. Peercy. 1982. *Physical Review Letters* 48:33.
13. See, e.g., J. Y. Tsao and P. S. Peercy. 1987. *Physical Review Letters* 58:2782.
14. M. D. Kluge and J. R. Ray. 1989. *Physical Review B* 39:1738.
15. S. R. Phillpot, J. F. Lutsko, D. Wolf, and S. Yip. 1989. *Physical Review B* 40:2831.
16. J. F. Lutsko, D. Wolf, S. R. Phillpot, and S. Yip. 1989. *Physical Review B* 40:2841.
17. J. Tersoff. 1989. *Physical Review B* 39:5566.
18. M. A. Amat, I. G. Kevrekidis, and D. Maroudas. 2006. *Physical Review B* 74:132201.
19. M. P. Allen and D. J. Tildesley. 1987. *Computer simulations of liquids.* Oxford: Clarendon.
20. S. J. Cook and P. Clancy. 1993. *Physical Review B* 47:7686.
21. S. Yoo, X. C. Zeng, and J. R. Morris. 2004. *Journal of Chemical Physics* 120:1654.
22. M. Parrinello and A. Rahman. 1980. *Physical Review Letters* 45:1196.
23. M. Parrinello and A. Rahman. 1981. *Journal of Applied Physics* 52:7182.
24. K. Mizushima, S. Yip, and E. Kaxiras. 1994. *Physical Review B* 50:14952.

25. P. Focher, G. L. Chiarotti, M. Bernasconi, E. Tosatti, and M. Parrinello. 1994. *Europhysics Letters* 26:345.

26. I. Souza and J. L. Martins. 1997. *Physical Review B* 55:8733.

27. R. Martonak, A. Laio, and M. Parrinello. 2003. *Physical Review Letters* 90:075503.

28. A. Laio and M. Parrinello. 2002. *Proceedings of the National Academy of Sciences of the USA* 99:12562.

29. R. Martonak, D. Donadio, A. R. Oganov, and M. Parrinello. 2006. *Nature Materials* 5:623.

30. R. Martonak, A. Laio, M. Bernasconi, C. Ceriani, P. Raiteri, and M. Parrinello. 2005. *Zeitschrift für Kristallographie* 220:489.

31. M. A. Amat, I. G. Kevrekidis, and D. Maroudas. 2007. *Applied Physics Letters* 90:171910.

32. J. Zhao, D. Maroudas, and F. Milstein. 2000. *Physical Review B* 62:13799.

33. H. J. C. Berendsen, J. P. M. Postma, W. F. van Gunsteren, A. DiNola, and J. R. Haak. 1984. *Journal of Chemical Physics* 81:3684.

34. M. A. Amat, M. Arienti, V. A. Fonoberov, I. G. Kevrekidis, and D. Maroudas. 2008. *Journal of Chemical Physics* 129:184106.

35. W. A. Steele. 1973. *Surface Science* 36:317.

36. W. E. Carlos and M. W. Cole. 1980. *Surface Science* 91:339.

37. N. D. Shrimpton and W. A. Steele. 1991. *Physical Review B* 44:3297.

38. D. M. Butler, J. A. Litzinger, and G. A. Stewart. 1980. *Physical Review Letters* 44:466.

39. E. D. Specht, A. Mak, C. Peters, M. Sutton, R. J. Birgeneau, K. L. D'Amico, D. E. Moncton, S. E. Nagler, and P. M. Horn. 1987. *Zeitschrift für Physik B Condensed Matter* 69:347.

40. R. Car and M. Parrinello. 1985. *Physical Review Letters* 55:2471.

41. C. Dellago, P. G. Bolhuis, and D. Chandler. 1998. *Journal of Chemical Physics* 108:9236.

42. C. Dellago, P. G. Bolhuis, and D. Chandler. 1999. *Journal of Chemical Physics* 110:6617.

43. A. F. Voter. 1997. *Physical Review Letters* 78:3908.

44. A. F. Voter. 1997. *Journal of Chemical Physics* 106:4665.

5 Multiscale Modeling Approach for Studying MDH-Catalyzed Methanol Oxidation

Nirmal Kumar Reddy Dandala, A. P. J. Jansen, and Daniela Silvia Mainardi

CONTENTS

5.1 MULTISCALE MODELING OF BIOLOGICAL SYSTEMS

Biological processes can be considered as complex networked systems that involve events that occur on a variety of time and length scales ranging from nano- to meso- and even macro-scales, thus making these processes fundamental challenges to model. Multiscale modeling is an effective tool to deal with these complex biochemical processes by integrating various physical techniques ranging from atomistic to macroscopic simulations (Figure 5.1) [1]. Particular modeling techniques can be

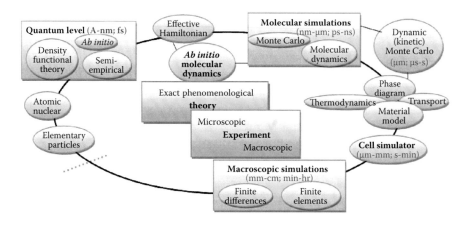

FIGURE 5.1 The big computational spectrum considers various space and time scales ranging from the modeling of nuclear systems through quantum level to molecular level simulations, transport and cell simulations, and macroscopic simulations.

used at each scale; however, a combination of various computational tools is usually required in order to effectively model an entire biological system, which can be treated as a cascade of individual subsystems.

Common multiscale modeling techniques fall into two categories, "series" and "parallel" [2]. Multiscale modeling in "series" refers to the methodology where individual models are developed to represent subsystems at various time and length scales, exchanging the information between them. Simulations are run independently at each scale in such a technique. On the other hand, multiscale modeling in "parallel" refers to the case where various models for various subsystems are integrated in the same simulation, i.e., treating the different scales of the system simultaneously [2]. This chapter will focus on serial multiscaling and, thus, a description of methods relevant at each scale will be provided.

In the microscopic regime (picometers to angstrom length scale), atomistic simulations use quantum mechanical (QM) methods such as *ab initio*, semiempirical, and density functional theory (DFT) to provide accurate structural and energetic information in small (containing less than one hundred atoms) molecular systems. The main role of these atomic models in a serial multiscale scheme is to provide parameters for larger time and length scales. Structural information and some electronic properties can be calculated fairly accurately using the Hartree–Fock *ab initio* methods or the semiempirical methods [3] and a greater accuracy can be achieved by using DFT [4], which provides a better description of electronic and geometrical properties through the consideration of a correlation functional that is completely neglected by Hartree–Fock. DFT reduces the many-body electron problem treated by the wave function to a single electron problem treated by the electron density, and therefore it is computationally cheaper than correlated *ab initio* methods such as configuration interaction and Møller–Plesset perturbation theory. Thus, DFT methods have become powerful tools in calculating not only structural and energetic molecular information but also other quantities, such as activation energies of chemical and biological reactions. QM methods in general are useful in modeling smaller site-specific systems, like reactions at a small region within an enzyme active site (limited to a few hundreds of atoms). Larger systems (containing more than one hundred atoms) for which atomic resolution is still important are usually modeled using molecular mechanics (MM), if only relaxed geometries are needed, or molecular dynamics (MD) if the dynamics of the process is relevant. Both techniques, MM and MD, rely on the parameterization of bonding and nonbonding interactions via force fields [6]. These methods are operative over wide length- and time-scale ranges, while still dealing with individual atoms.

In the macroscopic spatial regime (millimeters to meters length scale), modeling has been developed around continuum equations solved by finite element analysis (FEA), finite difference methods and Monte Carlo methods (Figure 5.1) [7]. Particularly, the dynamic (kinetic) Monte Carlo (kMC) methods are designed to deal with a large range of scales and require parameters that can be obtained from experiments or simulations from smaller scales. In a typical kMC simulation, the short time dynamics and the interactions between species are represented by discrete hops from one configuration to another based on random probabilities. These simulations are usually conducted on a grid that acts as a platform for species to interact and these interactions are dependent on a set of predefined rules and the initial and final atomic configurations [8].

The main focus of this chapter is on the description of the integration of DFT and MM information on a kMC method, leading to a proposed multiscale approach to study enzyme-assisted methanol oxidation. The actual processes in the example selected for this chapter are obtained using kMC, a technique described in Section 5.2. DFT and MM are used to determine the necessary parameters and the details on how they were used are described after the case study example is introduced in Section 5.4.

5.2 MONTE CARLO MODELING

A multiscale approach is proposed to investigate the methanol oxidation mechanism by methanol dehydrogenase (MDH) enzyme, implemented in a coarse-grained dynamic (kinetic) Monte Carlo program that permits the investigation of time scales up to the range of seconds and length scales up to meters. The kMC program developed by Johan Lukkien and A. P. J. Jansen, Carlos 4.1, is described here [8,9]. Carlos is a general purpose program used to model surface reactions in two dimensions and it offers the user the flexibility to specify almost any kind of reactions in the input file. To find out which reactions are possible, Carlos uses pattern recognition where each reaction can be specified as a change in species sites on a reactive surface (grid). A detailed description of how the reactions are specified and the interpretation of the reaction code by Carlos are discussed in later sections.

The approach presented here starts by describing the system of catalytic reactions by a stochastic model where an MDH catalytic surface (reactive region of the enzyme active site) is created for methanol oxidation to take place, and the most commonly used lattice-gas models are considered to define the enzyme catalytic surface. In this model, the surface is defined by a discrete, regular two or three dimensional array of lattice points called *sites* [8–10] (Figure 5.2). A label is assigned to each grid point that defines the occupation state of the corresponding site, which are updated according to certain rules. These sites with distinctive characteristic values and the lattice together are termed a *configuration* [8–10]. Thus, a reaction can be seen as a change from one configuration to another. The species are added through a reservoir onto the lattice and they can diffuse, react, associate, and dissociate according to a set of probabilities and rules. The system undergoes dynamic evolution as the simulation proceeds until the time step ends or a quasi-equilibrium steady state is reached.

Steady-state Monte Carlo simulations (where the system does not vary with time) [8–10] follow a simple pattern (Figure 5.2). First, a lattice site is chosen randomly and probabilities are calculated to determine the possibility of a reaction at that site. The reactions may or may not occur with the species in the neighboring sites and the occupation numbers are changed according to the probabilities, which denote the rate constants. Once the reaction probabilities are calculated, the next site is chosen and the same aforementioned procedure is repeated. After each grid site has been visited at least once, on average, the simulation completes one Monte Carlo time step. On the other hand, when dealing with dynamically evolving systems (which vary with time), there is a certain uncertainty in evaluating the reaction probabilities uniquely with respect to choosing the reaction steps. In the particular implementation described here, the Monte Carlo methods by Fichthorn and Weinberg are used [11] to simulate the dynamical interpretation by obtaining statistical averages. Time is incorporated as a variable during the Monte Carlo simulation to get the real time evolution of the system. Microscopic reaction rates are needed as input and these can be obtained either from experiments or *ab initio* calculations and, as a result, kMC methods yield time evolution of individual reaction rates (difficult to determine

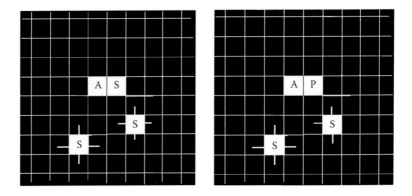

FIGURE 5.2 Lattice (black grid) showing the sites occupied by substrate molecules (S denotes the molecule that will be converted to a product). After following a random walk, substrate molecules reach the most reactive region of the enzyme active site (A), and they are converted into product molecules (P).

experimentally), which compete with each other as functions of external variables such as pressure and temperature.

Since Monte Carlo is a probabilistic approach, the rates are specified as probabilities and the time evolution of surface configuration is given by a master equation (Equation 5.1), which is derived from first principles [9].

$$\frac{d P(c,t)}{d t} = \sum_{c' \neq c} [P(c',t)k_{c',c} - P(c,t)k_{cc'}], \tag{5.1}$$

where $P(c,t)$ denotes the probability of finding the system in configuration c at time t, and $k_{cc'}$ is the transition probability of the reaction to transfer from configurations c to c'.

Several methods have been developed for the numerical implementation of the master equation. The first reaction method (FRM) [9], which is appropriate for cases where reaction constants vary with time, is used here. According to this method, when the system is in a given configuration, c, the set of all possible reactions is determined and a time of occurrence, $t_{c'c}$, is generated for each reaction, i, compatible with configuration c, according to Equation 5.2.

$$\exp\left[-\int_{t}^{t_{c'c}} k_{c'c}^{i}(t') \cdot dt \right] = r \tag{5.2}$$

In Equation 5.2, $k_{c'c}^{i}$ is the time-dependent rate of reaction i, and r is a random number selected uniformly in the interval (0,1). Then, the reaction with the smallest time is selected, the configuration is changed accordingly, and the time is incremented to t. Finally, the set of possible reactions is generated according to the new configuration, c', and the procedure is repeated. The microscopic rate, k_i, is related to macroscopic parameters as stated by Equation 5.3:

$$k_i = v_i \exp\left[-\frac{E_{ai}}{RT} \right], \tag{5.3}$$

where v_i is the prefactor, and E_{ai} is the activation energy of a reaction, i.

5.3   MODELING OF AN ENZYME-ASSISTED REACTION

The general problem of bridging the gap between time and length scales is one of the challenges when dealing with the modeling of large biological systems. Modeling enzyme-assisted reactions has always been a difficult task due to the complexities involved [12–14]. Biological systems are highly disordered and their modeling is generally treated using diffusion-limited aggregate clusters (DLA) [15–17] and percolation thresholds [18,19]. Most of the enzymatic reactions follow the so-called Michaelis–Menten mechanism, which is based on mass action laws that originate from the Smoluchowski's approach of diffusion-limited reactions [20]. Monte Carlo modeling has been an ideal choice to deal with these biological processes as the random sampling techniques help in modeling the complex processes effectively by spending most of the simulation time in configurations that are more probable, and minimizing or neglecting those configurations in the configuration space that are highly improbable. Reaction mechanisms of substrate molecules immobilized on a two-dimensional lattice representing the enzyme's surface have deserved some attention in the literature [20–22]. This is an important assumption since considering the enzyme to be immobilized on a two-dimensional surface, as opposed to a three-dimensional surface, greatly decreases the complexity of the system.

Monte Carlo simulations have been used to model three-dimensional systems as well. The flexibility and robustness of the Monte Carlo technique provide the advantage of monitoring the tendencies of stochastic processes containing diffusional and kinetic parameters [23]. One of the noted contributions in this field is due to Hughes Berry [20], who conducted Monte Carlo simulations of an isolated Michaelis–Menten enzyme reaction in two dimensions following a fractal kinetics and spatial segregation approach to model biological membranes. This author used Monte Carlo simulations to model the evolution of an isolated enzymatic system that consists of an enzyme (E), substrate (S), and product (P) molecules, and enzyme-substrate complexes (ES). Further immobile "obstacles" (O) were placed randomly on the lattice to represent the macromolecular crowding (enzyme amino acids not directly involved in the reactions) that occurs in a biological system. Other works talk about macromolecular crowding and its affect on the enzyme kinetics, such as the work by Agarwal et al. [21] that discusses the effect of macromolecular crowding on the rate of diffusion-limited reactions. These authors modeled the aqueous phase of cell cytoplasm, which is crowded with macromolecules, and reported the effect of macromolecular crowding (obstacle density) on the enzyme kinetics. Nag et al. [22] followed a similar approach to model the kinetics of enzymatic reactions with patterned substrates immobilized on a surface to investigate the rate of turnover of substrate molecules during enzyme catalysis.

Another interesting work on modeling enzymatic reactions is by Houquiang Li et al., who modeled the reaction mechanism between substrate molecules and enzyme active sites using fractal kinetics [24]. They have considered the enzyme surface to be rough (which can be represented as a fractal model) and assumed it to be a DLA cluster. The substrate molecules followed a random walk on this DLA cluster toward the active site by the shortest available distance. The significant problem these authors have encountered is the distribution of active sites on the surfaces of the enzyme. The active site distribution is given by Equation 5.4,

$$D_{AS}(\alpha, M) \approx \exp\left\{\frac{-A}{\ln M}[\alpha - \alpha_0(M)]^2\right\}, \tag{5.4}$$

which is actually derived from the analytic solution proposed by Lee and Stanley [25] of the growth-site probability distributions for a family of hierarchical models for DLA clusters. In Equation 5.4, M is the cluster mass of enzymes and α and A are constants. This is followed by another set of equations. Equation 5.5 is the fractal dimension of the enzyme active site distribution,

$$D_g = \frac{\ln\left[\dfrac{(b-\theta)(b-\theta+1)}{2-2\theta}\right]}{\ln b}, \tag{5.5}$$

where b is the linear size or the radius of gyration that is given by Equation 5.6,

$$b = \sum \frac{[(x-x_{cm})^2 + (y-y_{cm})^2]}{M}, \tag{5.6}$$

where x_{cm} and y_{cm} are the coordinates of the center of mass and θ is the number of empty layers of enzyme clusters.

In all the aforementioned studies reported in the literature, the kinetics of enzymatic reactions is analyzed either by considering two dimensional membranes or by immobilizing the enzyme on two dimensional lattices; this latest approach will be illustrated next.

5.4 METHANOL DEHYDROGENASE ENZYME

As a case study, the modeling of methanol oxidation by methanol dehydrogenase (MDH) is described. MDH enzyme (Figure 5.3) is a pyrroloquinoline quinine (PQQ)-containing protein found in the intracytoplasmic membrane of gram-negative methanotrophic bacteria [26,27], which are associated with the rhizosphere of paddy plants [28]. The crystal structure of MDH from *Methylobacterium extorquens* [29,30] and from *Methylophilus methylotrophus W3A1* [31,32] has been characterized as having

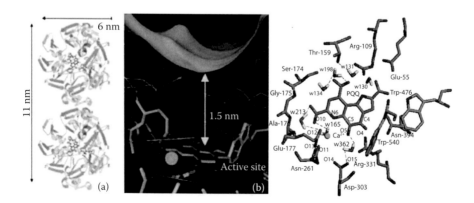

FIGURE 5.3 (See color insert.) (a) MDH enzyme from the entry 1W6S (*Methylobacterium Extorquens W3A1*) of the Protein Data Bank. (b) View of the active site of MDH (left) and distribution of amino acids in the active site (right).

an $\alpha_2\beta_2$ tetrameric structure (Figure 5.3a) with molecular masses of 62 and 8 kDa for α and β, respectively. The α subunit assumes the shape of a superbarrel with eight β-sheets arranged with radial symmetry (the "propeller fold") and is held together by novel tryptophan docking motifs. Its two active sites (one in each identical unit) contain a Ca^{2+} ion, a PQQ molecule—which serves as its redox cofactor, and several other amino acids [29,33] (Figure 5.3b). The PQQ molecule is found buried in the interior of the super barrel within a chamber and forms a means of communication with the exterior through a shallow funnel-shaped depression in the surface that is hydrophobic in nature.

The oxygen atoms of the PQQ molecule are hydrogen-bonded to the residues GLU55, ARG109, THR159, SER174, ARG331, and ASN394, while the calcium is bonded to the O5, N6, and O10 atoms of PQQ, the O11 of ASN261, and the O12 and O13 of GLU177 (Figure 5.3b). The calcium cation plays a major role in the catalytic activity not only by contributing to the formation of the enzyme-substrate complex but also by decreasing the pK_a of methanol substrate and polarizing the oxygen at C5 (Figure 5.3b) [26,27]. The water molecules are also found to play a part in the catalytic activity as the X-ray crystallographic studies suggest that water molecules W362, W165, and W213 form hydrogen bonds with the calcium cation, PQQ, and GLU177 in the active site, while the left upper portion of PQQ is formed by the water molecules W130, W131, W134, and W198.

5.4.1 METHANOL OXIDATION MECHANISM BY METHANOL DEHYDROGENASE ENZYMES

MDH has the ability to oxidize methanol and other alcohols to their corresponding aldehydes. It particularly oxidizes methanol to formaldehyde through concomitant reduction of PQQ to $PQQH_2$ [26,27]. The electron released through the reduction of PQQ to $PQQH_2$ is transported through the shallow hydrophobic funnel-shaped depression in the surface to cytochrome c_L [34], the MDH natural electron mediator.

Four possible mechanisms for methanol oxidation by MDH have been proposed in the literature, the addition-elimination (A-E) [26,27,35,36], the hydride transfer (H-T) [26,27,37–39], the addition-elimination-protonation [40], and the modified hydride transfer [41] mechanisms (Figure 5.4). The preferred methanol electrooxidation mechanism by MDH has been a devoted topic of research and several experimental as well as theoretical studies have been carried out to elucidate its nature.

The A-E is a three-step mechanism that involves a proton transfer from methanol to ASP 303, which is an active site base (Figure 5.4a). This catalytic base at the MDH active site subtracts a proton (H16) from methanol and initiates the oxidation reactions. This addition of the proton leads to the formation of a covalent hemiketal intermediate resulting in the linkage of the oxyanion (O16$^-$) in the methanol molecule to the C5 of PQQ. The second step involves the elimination of proton (H16) from ASP303 and transfer to O5 of PQQ. The final step contains a second proton (H17) transfer from the methanol molecule to the O4 of PQQ, resulting in the formation of formaldehyde (HCHO) [26,27]. Frank et al. conducted some experimental studies and suggested the high affinity of C5 carbonyl of the isolated PQQ molecule toward

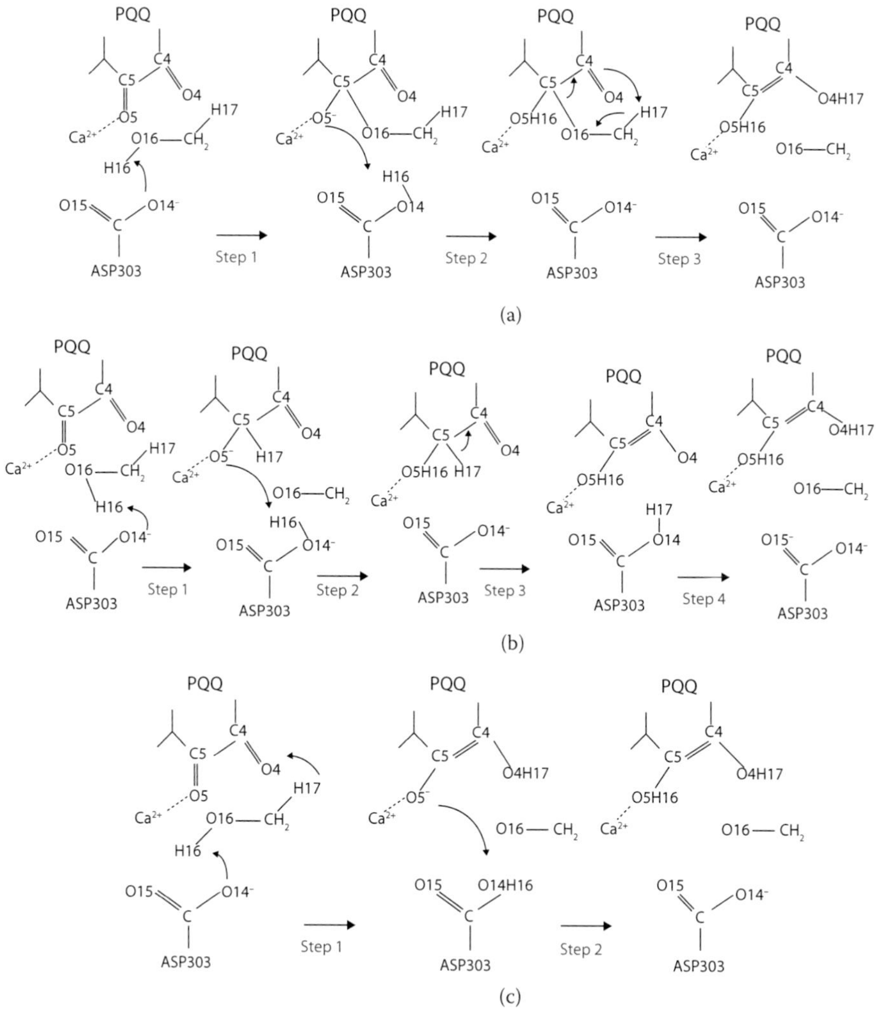

FIGURE 5.4 Three of the four proposed mechanisms (*from top A-E, H-T, AND MODIFIED H-T*) that may occur in the oxidation of methanol by MDH[27].

the nucleophilic reagents [42]. This supported the formation of a covalent PQQ-substrate complex (hemiketal intermediate) in favor of the A-E mechanism. The formation of this hemiketal intermediate has been confirmed through absorption and fluorescence spectroscopic studies, all in favor of the A-E over the H-T mechanism [43,44]. Theoretical and spectroscopic studies have also been carried out by Itoh et al. who showed that PQQ systems in organic solutions can oxidize certain alcohols to their corresponding aldehydes following an A-E mechanism [45,46]. They reported the formation of hemiketal and hemiacetal intermediates through spectral changes, thus supporting the A-E mechanism.

The hydride transfer (H-T) is a four-step mechanism (Figure 5.4b) that involves two proton transfers in the first step, where a H16 and H17 are transferred from the methanol molecule to ASP303 and C5 of PQQ respectively, resulting in the formation of formaldehyde. In the second step, there is a proton (H16) transfer from ASP303 to the C5 of PQQ. The third step involves the proton (H17) transfer from the C5 of PQQ to ASP303, and the final step consists of the transfer of H17 from ASP303 to the O4 of PQQ [26,27]. Xia et al. conducted crystallographic studies on the three-dimensional structure of a new crystal form of MDH obtained from *Methylophilus W3A1* in the presence of the substrate using data recorded at a synchrotron and reported results in favor of the H-T mechanism over the A-E mechanism [32]. The proximity of the substrate binding site to PQQ and to the side chain of ASP297 (Figure 5.4c) has been analyzed and further analysis suggests that the methanol hydroxyl group is closer to the PQQ C5 atom (3.1 Å) than the methyl group (3.9 Å), in favor of the A-E mechanism [32]. The H-T mechanism involves the nucleophilic addition of the putative methoxide to the C5 carbonyl of PQQ followed by an intramolecular ret-roene reaction. Zheng et al. examined that pathway by locating the transition state for this retroene reaction using the HF/3-21 G (d) level of theory [47]. These authors have also carried out energy calculations using a hybrid DFT at the B3LYP/3-21G(d) theory level and compared the results with X-ray crystallographic data to observe that the methanol oxidation by MDH follows the H-T mechanism. Recently, Kay and coworkers conducted EPR studies on substrate binding to PQQ-Ca^{+2} in ethanol dehydrogenase and found a strong coordination of the substrate to the Ca^{+2} cation, which is unlikely in the A-E mechanism [48].

In 2007, Leopoldini et al. carried out DFT calculations to calculate the intrinsic reaction coordinate (IRC) pathways for the A-E and H-T mechanisms employing the hybrid B3LYP correlation functionals with a 6-31+G* basis set for the C, H, N, and O atoms, except for Ca (LANL2DZ was used) [40]. In addition to the PQQ cofac-tor, the Ca^{+2} ion and the amino acids in coordination with it (GLU171, ASN261, and ASP303) and five other amino acids in the proximity of PQQ have also been consid-ered (ARG331, GLU55, ARG109, THR153, and SER174) (Figure 5.3b). Their calcu-lations yielded energy barriers for the rate-determining step (cleavage of C_{met}-H17) to be higher than the general requirements of an enzymatic catalytic process (15–20 kcal/mol). Thus, these authors proposed an alternative mechanism, the "addition-elimination-protonation" (Figure 5.4c), in which the cleavage of the C–H bond in the substrate occurs before the protonation of the PQQ cofactor by the ASP303. They found this more reliable as the activation energy in the rate-determining step came down to 16.0 kcal/mol and 11.1 kcal/mol in the gas phase and protein environments, respectively. Other alternate mechanisms for the methanol oxidation by MDH were reported by Idupulapati et al., who carried out IRC pathway calculations using a combined DFT-molecular dynamics approach level (BLYP/DNP) on various MDH models. They proposed two alternate mechanisms, a modified addition-elimination mechanism and a modified hydride transfer mechanism. According to the proposed third step of the original A-E, the cleavage of the C_{met}–H17 bond and the transfer of H17 to PQQ with concomitant breakage of the C5–O16 bond occur in a concerted fashion. But the modified addition-elimination mechanism considers this to occur in two steps, where in the first step the cleavage of the C_{met}–H17 bond occurs, and in the second step the H17 is transferred to the PQQ, thereby reducing it. The modified hydride transfer mechanism proposes that the hydride (H17) transfers directly to O4 instead of C5 of PQQ, unlike the original H-T mechanism, which involves the transfer of H17 from C5 to O14 of ASP303 and from ASP303 to O4 of PQQ. This eliminates the third and fourth steps of the original H-T mechanism, thereby making it a two-step process. The modified H-T mechanism is the one considered for the case study presented here (Figure 5.4c).

5.5 COMPONENTS OF THE MULTISCALE MODEL

5.5.1 DENSITY FUNCTIONAL THEORY

As shown by Equation 5.3, activation energies are required in order to calculate the rate constants for the processes to be modeled. For the MDH-assisted methanol oxidation case study presented here, the activation energies calculated by Idupulapati et al. [49], who carried out IRC pathway calculations using a combined DFT/MD approach on various MDH models [50], are considered. DFT calculations by Idupulapati et al. [49] were conducted using the Becke–Lee–Yang–Par (BLYP) exchange correlation functional in combination with the double numerical with polarization (DNP) basis set, as employed in the DMOL³ module of Materials Studio® software [51,52]. The activation energies for the various steps in the modified H-T mechanism (Figure 5.4c) were thus calculated by these authors [49].

5.5.2 TRANSITION STATE THEORY

The calculation of rate constants from the activation energies from Idupulapati et al. [49] is conducted using transition state theory (TST) for the prediction of pre-exponential factors and Equation 5.3 [53]. TST is a powerful and effective way to connect the kinetic with the thermodynamic properties of a reacting system. It states that the reactants need to overcome an energy barrier in order to form products and it postulates an equilibrium (Boltzmann) energy distribution that relates the reaction rate of a reaction with its Gibbs free energy by the following expression:

$$k = \frac{k_B T}{h} \exp\left(-\frac{\Delta G}{RT}\right), \qquad (5.7)$$

where k is the rate constant (s^{-1}), K_B is the Boltzmann's constant (3.29×10^{-24} cal/K), h is the Planck's constant (6.626×10^{-34} J s), T is the absolute temperature (298.15 K at room temperature), R is the universal gas constant (8.314 J K^{-1} mol^{-1}), and ΔG is the Gibbs free energy of activation [53].

5.5.3 MOLECULAR MECHANICS

Molecular mechanics (MM) is a computer simulation technique that, instead of solving the electronic Schröedinger equation for obtaining accurate structural and energetic information of a molecular system like QM methods, it fits the real potential energy surface of the system using a force field [54]. The force field is a mathematical representation that provides the information about the bond-stretching energy, the angle-bending energy, the torsion, and the interactions between bonded and non-bonded atoms, among other more complex interactions such as out-of-plane bending, cross terms, polarization, and salvation [54]. Thus, approximate minima points corresponding to energy-minimized conformations of a molecular system can be obtained by minimizing the force field. In this work, MM simulations are used to obtain optimum configurations of MDH immobilized over graphite surfaces. The DISCOVER module with the COMPASS force field is used as available in Materials Studio software [51].

In the case study described here, a multiscale modeling in series approach is followed, where the activation energies (from which the rates of reactions are calculated) and configuration of MDH enzyme are obtained at the microscopic level and transferred to a kinetic Monte Carlo simulation that is operative on longer scales.

5.6 IMPLEMENTATION OF THE MULTISCALE APPROACH TO STUDY MDH-ASSISTED METHANOL OXIDATION

5.6.1 MOLECULAR MECHANICS FOR LATTICE PREPARATION

The models discussed in this chapter deal with diffusion of substrate molecules in 2-D media and, therefore, the case of an enzyme immobilized on a 2-D surface is considered. The orientation of MDH on the immobilized surface is pivotal as it is required to obtain the projections of the various amino acids present in MDH on the 2-D lattice. Thus, to obtain the optimum configuration, MM simulations were performed by considering various orientations of MDH on a graphite surface (Figure 5.5 shows the three best conformations found), and minimization calculations were run in DISCOVER module with a COMPASS force field as applied in Materials Studio software [51].

The MM minimization calculations give information on the final potential energies (Table 5.1) and the lowest energy configuration (configuration 3) was then selected as the reference model from which the positions of amino acids were considered. The energy variation with the number of iterations during the simulation (Figure 5.6) gives an estimate of the oscillation in overall energy and the convergence criteria.

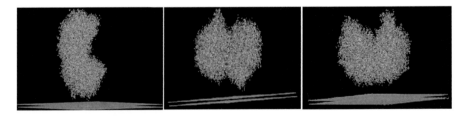

FIGURE 5.5 Three of the various configurations of MDH immobilized on a graphite surface minimized in DISCOVER using a COMPASS force field: (from left to right) Config 1, Config 2, and Config 3.

TABLE 5.1

Potential Energies Obtained by Running Molecular Mechanics Simulations on Three of the Various Configurations of MDH Immobilized on a 2-D Graphite Surface

Configuration	Relative Potential Energy (kcal/mol)
Config 1	440.1
Config 2	1187.6
Config 3	0

Note: Configurations taken from Figure 5.5. Relative energies are computed with respect to the most stable conformation.

5.6.2 ADSORPTION MODELS FOR DOCKING SITE INFORMATION

The docking site information of methanol on MDH is also important, as it is required to know the entrance and exit points of methanol molecules while running the Monte Carlo simulations on 2-D lattices. For this purpose, an adsorption locator tool is used in which the methanol molecules are placed randomly at various positions on the MDH molecule and adsorption energies are calculated based on a Metropolis Monte Carlo algorithm.

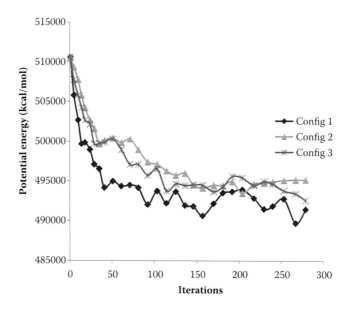

FIGURE 5.6 Variation of potential energies of the three configurations (from Figure 5.5) with the number of iterations gives an estimation of the convergence criteria while running the MM simulations.

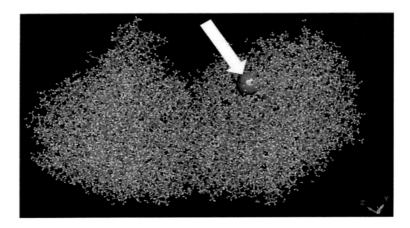

FIGURE 5.7 **(See color insert.)** Methanol molecule (pointed arrow) docked at the binding pocket of MDH.

The minimum energy configuration of MDH obtained from MM simulations (discussed in Section 5.6.1) is considered to run the adsorption calculations. Methanol molecules are placed around the MDH (Figure 5.8) at a minimum adsorption distance of 3 Å. The temperature range used for the adsorption calculations is 293 K to 318K (20 °C to 45 °C), which is appropriate for MDH.

Initially, one methanol molecule is used (Figure 5.7) and then, gradually, the number of methanol molecules is increased to ten. When the number of methanol molecules is increased, the MDH is attacked from almost all directions (Figure 5.8), which implies that the docking of methanol molecules to MDH is random and that the methanol can dock at various sites on MDH.

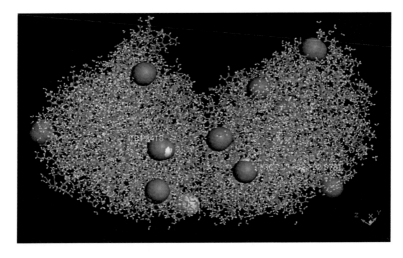

FIGURE 5.8 **(See color insert.)** Ten methanol molecules attacking the MDH at various regions.

5.6.3   KINETIC MONTE CARLO MODELING OF MDH-ASSISTED METHANOL
OXIDATION CONTAINING MULTISCALE COMPONENTS

After obtaining an optimum configuration of MDH immobilized on a 2-D surface, the next step is exploring the kinetics and construction of the lattice representing the enzyme active site, where methanol oxidation will take place according to the mechanism depicted in Figure 5.4c. It is important first to determine the assumptions and boundary conditions of the system, since the reaction mechanics are relative to these aspects. The following assumptions were made to describe the model and to reduce the complexity involved:

1. The enzyme is assumed to be immobilized on a 2-D lattice.
2. The motion of substrate molecules follows a random walk on the surface of the lattice.
3. The motion of the substrate molecules is restricted to the 2-D plane onto which the positions of the obstacles are projected.
4. There are no interactions between the individual substrate molecules.
5. There are no interactions between the individual substrate molecules and the obstacles.
6. Once the substrate molecules start a random walk, they move toward the active site by the shortest available distance.
7. The excluded volume condition is maintained, i.e., at any instant of time one lattice site cannot be occupied by more than one more molecule of the same or different species.

The well known Michaelis–Menten equilibrium enzyme kinetics equation [55] is considered:

$$E + S \rightarrow ES \rightarrow E + P, \tag{5.8}$$

where E denotes the enzyme, S denotes the substrate, ES is the intermediate complex, and P is the product. Since the reaction is assumed to be diffusion-limited, the formation of an intermediate AS is not taken into account and the substrate is almost instantaneously converted to product at the active site (A). Therefore, Equation 5.8 reduces to a simpler reaction between the active site (A) and the substrate (S) molecule:

$$A + S \xrightarrow{k_{(t)}} A + P \tag{5.9}$$

Though the species are assumed to be noninteracting, macromolecular crowding is an aspect that needs to be taken into account since it has a direct affect on the enzyme kinetics. Thus, obstacles (O) are distributed on the lattice to represent the macromolecular crowding that occurs in biological systems. To carry out the preliminary simulation tests, we used a 64×64 lattice with cyclic boundary conditions. An important aspect to consider while modeling enzymatic reactions is the distribution of the substrate and obstacles. Previous works reported on enzyme kinetics modeling have used random sampling to distribute the species on the lattice and their reports showed that the enzyme kinetics were influenced by aspects like obstacle density. In this work, a biased approach is used in sampling the lattice and investigating its effect on the kinetics of methanol oxidation by MDH.

Works by Hughes Berry [20] and Agarwal et al. [21] have distributed the obstacles randomly on the lattice, but in this work some bias is introduced by having a particular distribution for the obstacles on the lattice. The center of masses for all the amino acids in MDH were calculated and plotted with the active site amino acids as the center, which is, in a way, similar to the coarse-graining method where the molecules are represented as beads. The representation of the MDH active site as beads is shown in Figure 5.9.

The rate constants of the modified H-T process (Figure 5.4c) are given as input at the start of each simulation and these are calculated from the activation energies obtained through the DFT/MD calculations of Idupulapati et al. [49]. The obstacle and substrate concentrations are specified at the start of each simulation and the simulation time is counted in Monte Carlo steps. For each simulation, the obstacle concentration is fixed while the substrate concentration decreases as the simulation proceeds due to the conversion of substrate molecules to product. The simulation starts off with the substrate molecules (*red*) diffusing (*making a random walk*) on the lattice with the obstacles distributed (*black*) on it. Active site amino acids (*green*) are distributed at the center of the lattice and when the substrate molecules come in contact with any one of the active site amino acids, a product (*blue*) is formed that also diffuses on the lattice (Figure 5.10). The substrate molecule (*red*) entering the active site regime to react can be seen in Figure 5.10. Through this model, it is not only possible to visualize the enzymatic reactions that take place inside the enzyme

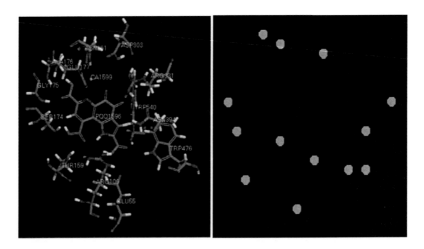

FIGURE 5.9 **(See color insert.)** Active site cluster in the MDH enzyme (left) showing the various amino acids and the active site model showing the amino acids plotted as dots on the lattice according to their centers of mass (right).

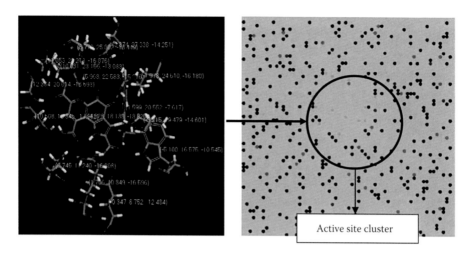

FIGURE 5.10   (*See color insert.*) The active site cluster (*green dots*) plotted on the 2-D lattice with substrate (*red dots*) and product (*blue dots*) molecules making a random walk. The black dots represent the obstacles.

but also to get useful information, such as the concentration profiles of substrate and product with respect to Monte Carlo time, obstacle density, reaction rate constants, and the positions of obstacles on the lattice.

The diffusion properties were studied by calculating the root mean square distance (rms) covered by the diffusing substrate and product molecule, which is related to the diffusivity by the equation:

$$\langle R^2 \rangle = 4Dt^{\alpha}, \tag{5.10}$$

where $\langle R^2 \rangle$ denotes the rms, D the diffusivity, and t the time.

The exponent α can be obtained as the slope of the log–log graphs plotted between $\langle R^2 \rangle$ and t. α defines the diffusive behavior of the system and does depend on the type of lattice (rough/smooth) and other factors like the obstacle concentrations. The diffusive behavior of the substrate or product can be analyzed by calculating the average diffusion length for time t as

$$\langle R^2 \rangle = \frac{1}{N_p t} \sum_{i=1}^{N_p(t)} \left\{ [x - x_i(t)]^2 + [y - y_i(t)]^2 \right\}, \tag{5.11}$$

where (x_0, y_0) is the position of the active site at the center of the lattice, $(x_i(t), y_i(t))$ is the position of the substrate or product at time t, and $N_p(t)$ is the number of products at time t. Thus, by calculating factors like α, we can actually know the type of diffusion (normal or anomalous) that methanol molecules undergo to reach the active site.

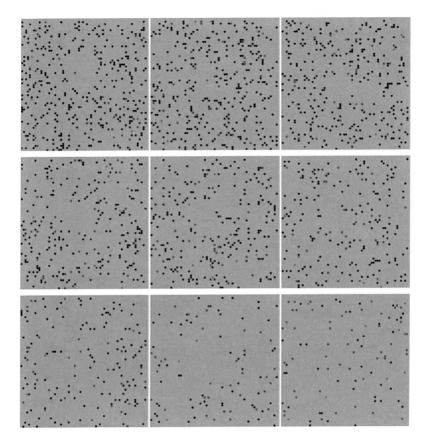

FIGURE 5.11 **(See color insert.)** Various lattice configurations showing the substrate (*red*) and product (*blue*) molecules with decreasing obstacle concentration (*black dots*) taken at $t = 1500$ MC steps.

The diffusive behavior of the random walkers depends on the obstacle density, so simulations are carried out for various obstacle concentrations. Through these models, the influence of obstacle density (the number of amino acids in MDH) on the diffusion of methanol toward the active site is investigated. This would give useful information like the variations in the rate of product formation or substrate consumption as the number of obstacles is changed. The variation of substrate and product concentrations with reducing obstacle density can be seen in Figure 5.11.

5.7   CONCLUSION

The models discussed in this chapter are means to investigate the diffusive properties of methanol molecules inside the MDH, from the docking site to the active site center. However, if the active site cluster is carefully observed (Figure 5.3b), the active site amino acids are indistinguishable from one another, which is what is actually happening inside the enzyme.

From the reaction mechanisms that were proposed (Figure 5.4), it is evident that only ASP303, the PQQ cofactor, and the Ca^{+2} atoms are involved in the reaction, while the other amino acids make up the environment of the active site. In order to make this a more accurate model, a new 2-D lattice needs to be developed that would represent each amino acid of the active site cluster separately. In this way, the amino acids can be distinguished from one another. Moreover, the reaction specification also needs to be changed. The oxidation of methanol should take place only when the methanol molecules come in contact with the three specific molecules and not with the other amino acids in the active site. The idea is to impart specificity, which is usually observed in enzymatic reactions, to the reaction. However, this requires a deeper understanding of the reaction mechanisms happening at the active site through which greater accuracy could be imparted to the system.

Further studies on this include a much deeper investigation into the reaction mechanisms of methanol oxidation by MDH. The approach discussed so far is more qualitative rather than quantitative and the simulations are based on various assumptions (see Section 5.6.3). The enzyme-assisted reactions were modeled in a 2-D grid, but it should be realized that enzymatic reactions occur in 3-D spaces in living systems. So an interesting approach would be to extend these simulations to a 3-D space by constructing a 3-D lattice that would accommodate all the species of MDH in their exact positions as they would exist in nature. Another approach would be to represent the lattice in fractals so that it would represent the MDH surface and then evaluate the fractal kinetics of methanol oxidation by MDH.

Other than the diffusion properties, the reaction mechanism at the active site can also be modeled using Monte Carlo simulations. Models can be developed such that the active site is more specific with respect to the amino acids and the reaction of methanol with the amino acids can also be monitored. Our goal would be to obtain an optimum lattice model that would give us the obstacle and product concentrations that may help us in further modeling an artificial enzyme like MDH, which would not only increase the power output when used in a bio fuel cell but also give us the advantage of specificity and the ability to control the oxidation reaction of methanol.

REFERENCES

1. V. Tozzini. 2010. Multiscale modeling of proteins. *Accounts of Chemical Research* 43:220–230.
2. G. S. Ayton, W. G. Noid, and G. A. Voth. 2007. Multiscale modeling of biomolecular systems: In serial and in parallel. *Current Opinion in Structural Biology* 17:192–198.
3. R. Car. 2002. Introduction to density-functional theory and ab-initio molecular dynamics. *Quantitative Structure-Activity Relationships* 21:97–104.
4. M. G. Marques and E. K. U. Gross. 2004. Time-dependent density functional theory. *Annual Review of Physical Chemistry* 55:427–455.
5. F. Jensen. 2007. *Introduction to computational chemistry*. Chichester, UK: Wiley.
6. J. W. Ponder and D. A. Case. 2003. Force fields for protein simulations. *Advances in Protein Chemistry* 66:27–85.
7. R. M. Nieminen. 2002. From atomistic simulation towards multiscale modelling of materials. *Journal of Physics Condensed Matter* 14:2859–2876.

8. J. J. Lukkien, J. P. L. Segers, P. A. J. Hilbers, R. J. Gelten, and A. P. J. Jansen. 1998. Efficient Monte Carlo methods for the simulation of catalytic surface reactions. *Physical Review E — Statistical Physics, Plasmas, Fluids, and Related Interdisciplinary Topics* 58:2598–2610.

9. R. M. Nieminen and A. P. J. Jansen. 1997. Monte Carlo simulations of surface reactions. *Applied Catalysis A: General* 160:99–123.

10. A. P. J. Jansen. 1995. Monte Carlo simulations of chemical reactions on a surface with time-dependent reaction-rate constants. *Computer Physics Communications* 86:1–12.

11. K. A. Fichthorn and W. H. Weinberg. 1991. Theoretical foundations of dynamical Monte Carlo simulations. *The Journal of Chemical Physics* 95:1090–1096.

12. B. A. Scalettar, J. R. Abney, and C. R. Hackenbrock. 1991. Dynamics, structure, and function are coupled in the mitochondrial matrix. *Proceedings of the National Academy of Sciences of the United States of America* 88:8057–8061.

13. A. P. Minton. 2001. The influence of macromolecular crowding and macromolecular confinement on biochemical reactions in physiological media. *Journal of Biological Chemistry* 276:10577–10580.

14. A. S. Verkman. 2002. Solute and macromolecule diffusion in cellular aqueous compartments. *Trends in Biochemical Sciences* 27:27–33.

15. T. A. Witten and L. M. Sander. Diffusion-limited aggregation, a kinetic critical phenomenon. *Physical Review Letters* 47:1400–1403.

16. T. Rage, V. Frette, G. Wagner, T. Walmann, K. Christensen, and T. Sun. 1996. Construction of a DLA cluster model. *European Journal of Physics* 17:110–115.

17. S. Schwarzer, J. Lee, S. Havlin, H. E. Stanley, and P. Meakin. 1991. Distribution of growth probabilities for off-lattice diffusion-limited aggregation. *Physical Review A* 43:1134–1137.

18. J. W. Essam. 1980. Percolation theory. *Reports on Progress in Physics* 43:833–912.

19. S. S. Manna and J. Naeem. 1991. Growth probability distribution in percolation. *Journal of Physics A: General Physics* 24:1593–1601.

20. H. Berry. 2002. Monte Carlo simulations of enzyme reactions in two dimensions: Fractal kinetics and spatial segregation. *Biophysical Journal* 83:1891–1901.

21. M. Agrawal, S. B. Santra, R. Anand, and R. Swaminathan. 2008. Effect of macromolecular crowding on the rate of diffusion-limited enzymatic reaction. *Pramana - Journal of Physics* 71:359–368.

22. A. Nag, T. Zhao, and A. R. Dinner. 2007. Effects of nonproductive binding on the kinetics of enzymatic reactions with patterned substrates. *Journal of Chemical Physics* 126:035103.

23. A. Gil, J. Segura, J. A. G. Pertusa, and B. Soria. 2000. Monte Carlo simulation of 3-D buffered Ca^{2+} diffusion in neuroendocrine cells. *Biophysical Journal* 78:13–33.

24. H. Li and F. Wang. 1996. Protein conformation and enzymatic kinetics. In *Fractal geometry in biological systems*, eds. P. M. Iannaccone and M. Khokha, 57–123. Boca Raton, FL: CRC Press.

25. J. Lee, S. Havlin, and H. E. Stanley. 1992. Analytic solution of the growth-site probability distribution for structural models of diffusion-limited aggregation. *Physical Review A* 45:1035.

26. C. Anthony. 2000. Methanol dehydrogenase, a PQQ-containing quinoprotein dehydrogenase. *Subcellular Biochemistry* 35:73–117.

27. C. Anthony and P. Williams. 2003. The structure and mechanism of methanol dehydrogenase. *Biochimica et Biophysica Acta - Proteins and Proteomics* 1647:18–23.

28. S. K. Dubey, P. Padmanabhan, H. J. Purohit, and S. N. Upadhyay. 2003. Tracking of methanotrophs and their diversity in paddy soil: A molecular approach. *Current Science* 85:92–95.

29. M. Ghosh, C. Anthony, K. Harlos, M. G. Goodwin, and C. Blake. 1995. The refined structure of the quinoprotein methanol dehydrogenase from Methylobacterium extorquens. *Structure* 3:177–187.

30. P. R. Afolabi, F. Mohammed, K. Amaratunga, O. Majekodunmi, L. Dales, R. Gill, D. Thompson, B. Cooper, P. Wood, M. Goodwin, and C. Anthony. 2001. Site-directed mutagenesis and X-ray crystallography of the PQQ-containing quinoprotein methanol dehydrogenase and its electron acceptor, Cytochrome c(L). *Biochemistry* 40:9799–9809.

31. S. White, G. Boyd, F. S. Mathews, Z. X. Xia, W. W. Dai, Y. F. Zhang, and V. L. Davidson. 1993. The active site structure of the calcium-containing quinoprotein methanol dehydrogenase. *Biochemistry* 32:12955–12958.

32. Z. X. Xia, Y. N. He, W. W. Dai, S. A. White, G. D. Boyd, and F. S. Mathews. 1999. Detailed active site configuration of a new crystal form of methanol dehydrogenase from Methylophilus W3A1 at 1.9 A resolution. *Biochemistry* 38:1214–1220.

33. P. A. Williams, L. Coates, F. Mohammed, R. Gill, P. T. Erskine, A. Coker, S. P. Wood, C. Anthony, and J. B. Cooper. 2005. The atomic resolution structure of methanol dehydrogenase from Methylobacterium extorquens. *Acta Crystallographica Section D: Biological Crystallography* 61:75–79.

34. S. L. Dales and C. Anthony. The interaction of methanol dehydrogenase and its cytochrome electron acceptor. *Biochemical Journal* 312:261–265.

35. N. B. Idupulapati and D. S. Mainardi. 2008. A DMol3 study of the methanol addition-elimination oxidation mechanism by methanol dehydrogenase enzyme. *Molecular Simulation* 34:1057–1064.

36. M. C. Pirrung. 1997. Modeling of the chemistry of quinoprotein methanol dehydrogenase. Oxidation of methanol by calcium complex of coenzyme PQQ via addition-elimination mechanism. *Chemtracts* 10:828–830.

37. X. Zhang, S. Y. Reddy, and T. C. Bruice. 2007. Mechanism of methanol oxidation by quinoprotein methanol dehydrogenase. *Proceedings of the National Academy of Sciences of the United States of America* 104:745–749.

38. S. Y. Reddy, F. S. Mathews, Y. J. Zheng, and T. C. Bruice. 2003. Quinoprotein methanol dehydrogenase: A molecular dynamics study and comparison with crystal structure. *Journal of Molecular Structure* 655:269–277.

39. C. Anthony. 2001. Pyrroloquinoline quinone (PQQ) and quinoprotein enzymes. *Antioxidants and Redox Signaling* 3:757–774.

40. M. Leopoldini, N. Russo, and M. Toscano. 2007. The preferred reaction path for the oxidation of methanol by PQQ-containing methanol dehydrogenase: Addition-elimination versus hydride-transfer mechanism. *Chemistry — A European Journal* 13: 2109–2117.

41. A. Heller. 2004. Miniature biofuel cells. *Physical Chemistry Chemical Physics* 6:209–216.

42. J. Frank, Jr., S. H. Van Krimpen, P. E. J. Verwiel, J. A. Jongejan, A. C. Mulder, and J. A. Duine. 1989. On the mechanism of inhibition of methanol dehydrogenase by cyclopropane-derived inhibitors. *European Journal of Biochemistry* 184:187–195.

43. A. J. J. Olsthoorn and J. A. Duine. 1998. On the mechanism and specificity of soluble, quinoprotein glucose dehydrogenase in the oxidation of aldose sugars. *Biochemistry* 37:13854–13861.

44. J. Frank Jzn, M. Dijkstra, J. A. Duine, and C. Balny. 1988. Kinetic and spectral studies on the redox forms of methanol dehydrogenase from Hyphomicrobium X. *European Journal of Biochemistry* 174:331–338.

45. S. Itoh, H. Kawakami, and S. Fukuzumi. 1998. Model studies on calcium-containing quinoprotein alcohol dehydrogenases. Catalytic role of Ca^{2+} for the oxidation of alcohols by coenzyme PQQ (4,5-dihydro-4,5-dioxo-1H-pyrrolo2,3- f.quinoline-2,7,9-tricarboxylic acid). *Biochemistry* 37:6562–6571.

46. S. Itoh, H. Kawakami, and S. Fukuzumi. 2000. Development of the active site model for calcium-containing quinoprotein alcohol dehydrogenases. *Journal of Molecular Catalysis B: Enzymatic* 8:85–94.

47. Y. J. Zheng and T. C. Bruice. 1997. Conformation of coenzyme pyrroloquinoline quinone and role of Ca^{2+} in the catalytic mechanism of quinoprotein methanol dehydrogenase. *Proceedings of the National Academy of Sciences of the United States of America* 94:11881–11886.

48. C. W. M. Kay, B. Mennenga, H. Görisch, and R. Bittl. 2004. Characterisation of the PQQ cofactor radical in quinoprotein ethanol dehydrogenase of Pseudomonas aeruginosa by electron paramagnetic resonance spectroscopy. *FEBS Letters* 564:69–72.

49. D. S. M. N. B. Idupulapati. 2009. Methanol electro-oxidation by methanol dehydrogenase enzymatic catalyst: A computational study. In *Modern Aspects of Electrochemistry*, eds. P. B. Balbuena and V. Subramanian. New York: Springer Science.

50. N. B. Idupulapati. 2010. Quantum chemical modeling of methanol oxidation mechanisms by methanol dehydrogenase enzyme: Effect of substitution of calcium by barium in the active site. *Journal of Physical Chemistry A* 114(4):1887–1896.

51. Accelrys, Inc. 2006. *Materials Studio*. San Diego: Accelrys, Inc.

52. Accelrys, Inc. 2003. *DMOL3 User Guide*. San Diego: Accelrys, Inc.

53. J. B. Foresman. 1996. *Exploring chemistry with electronic structure methods*. Pittsburgh, PA: Gaussian, Inc.

54. A. R. Leach. 2001. *Molecular modeling: Principles and applications*. London: Prentice Hall.

55. A. Cornish-Bowden. 2004. *Fundamentals of enzyme kinetics*. London: Portland Press, Ltd.

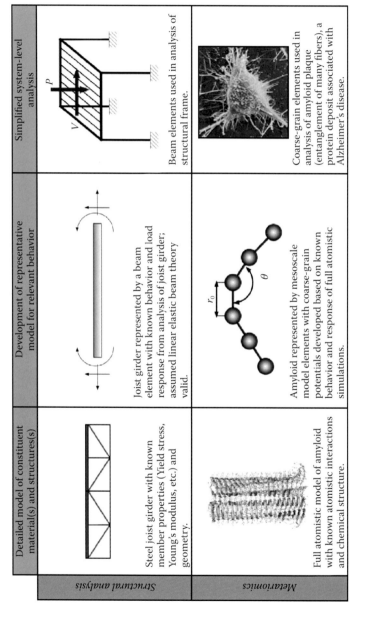

	Detailed model of constituent material(s) and structures(s)	Development of representative model for relevant behavior	Simplified system-level analysis
Structural analysis	Steel joist girder with known member properties (Yield stress, Young's modulus, etc.) and geometry.	Joist girder represented by a beam element with known behavior and load response from analysis of joist girder; assumed linear elastic beam theory valid.	Beam elements used in analysis of structural frame.
Metariomics	Full atomistic model of amyloid with known atomistic interactions and chemical structure.	Amyloid represented by mesoscale model elements with coarse-grain potentials developed based on known behavior and response of full atomistic simulations.	Coarse-grain elements used in analysis of amyloid plaque (entanglement of many fibers), a protein deposit associated with Alzheimer's disease.

FIGURE 2.2 Analogous comparison of system simplification between structural analysis and materiomics. Detailed model constituents and their structural arrangement are analyzed to parameterize a coarse-grain representative model, maintaining relevant behaviors and implemented for simplified analysis. Here, we see the transition of steel joist girders to beam elements to structural frame analysis, paralleled by a full atomistic model of an amyloid, a coarse-grain mesoscopic representation, and the system-level amyloid plaque (aggregation of thousands of amyloids). (Image of amyloid plaque reprinted with permission from Macmillan Publishers Ltd., Pilcher, H.R., *Nature News*, 2003. Copyright 2003.)

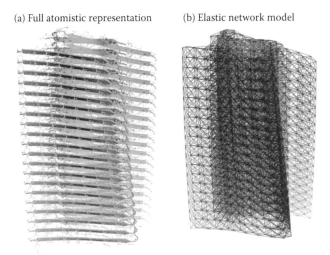

(a) Full atomistic representation   (b) Elastic network model

FIGURE 2.3   (a) Full atomistic representation and (b) elastic spring network representation of an amyloid fibril. Here, springs are connected to all neighbor atoms within a cutoff of 10 Å, and spring stiffnesses are assigned according to an exponentially decaying function (Equation 2.4). Model was implemented to determine normal vibration modes and structure stiffness. (Model images courtesy of Dr. Z. Xu, Massachusetts Institute of Technology.)

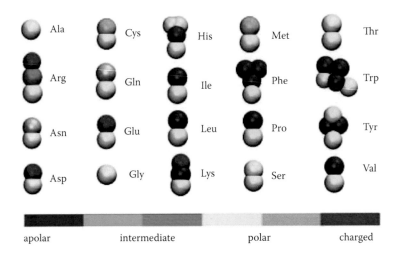

FIGURE 2.4   Coarse-grained representations of all amino acid types for MARTINI force field formulation for proteins. Different colors represent different particle types consisting of four main types of interaction sites: polar, nonpolar, apolar, and charged. (Reprinted with permission from Monticelli, L., et al., *J. Chem. Theory Comput.*, 4, 819–834, 2008. Copyright 2008 American Chemical Society.)

FIGURE 2.5 Simulation snapshot of DPPC/cholesterol bilayer structure with MARTINI coarse-grain model representation. Cholesterol molecules are displayed in green, with red hydroxyl groups. The DPPC lipid tails are shown in silver. Lipid head groups are displayed in purple and blue. Such a complex system can only be simulated via a coarse-graining approach. (Reprinted with permission from Marrink, S.J., et al., *J. Phys. Chem. B*, 108, 750–760, 2004. Copyright 2007 American Chemical Society.)

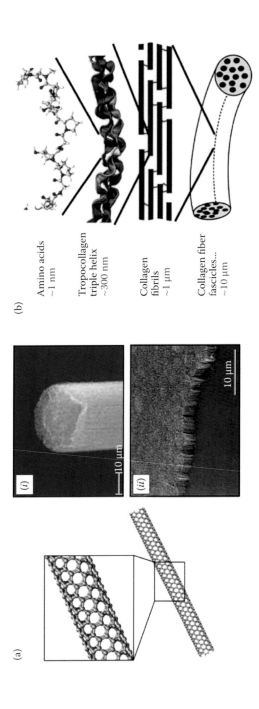

FIGURE 3.1 Overview of carbon nanotube and tropocollagen systems. (a) Depiction of full atomistic representation of (5,5) carbon nanotube. Subplots show higher order hierarchical arrangements, including (i) SEM of single bundle of carbon nanotubes (Reprinted with permission from McClain, D. et al., *J. Phys. Chem. C* 111(20), 7514–7520, 2007. Copyright 2007 American Chemical Society.) and (ii) SEM micrograph of vertically aligned carbon nanotube array (Reprinted with permission from Yang, J., and Dai, L., *J. Phys. Chem. B*, 107, 12387–12390, 2003. Copyright 2003 American Chemical Society.). (b) Schematic view of some of the hierarchical features of collagen, ranging from the amino acid sequence level at nanoscale up to the scale of collagen fibers with lengths on the order of 10 μm. (From Buehler, M.J., Keten, S., and Ackbarow, T., *Prog. Mater. Sci.*, 53, 1101–1241, 2008.) The coarse-grain model development discussed here is focused on the behavior of tropocollagen molecules (component level) and their role in the mechanical behavior and properties of collagen fibrils (system level).

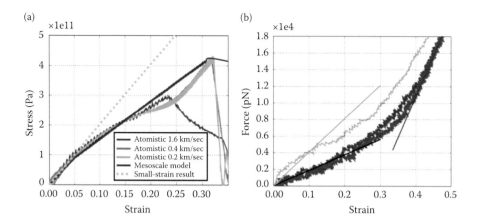

FIGURE 3.3 Full atomistic simulation results for the axial stretching of (a) a single-walled carbon nanotube (From Buehler, M.J., *J. Mater. Res.* 21(11), 2855–2869, 2006. With permission.) and (b) a tropocollagen molecule (From Buehler, M.J., *J. Mater. Res.*, 21(8), 1947–1961, 2006. With permission.). The nanotube results depict a softening behavior as the carbon bonds yield at high strain, while the tropocollagen results depict a stiffening behavior, as the molecule undergoes extension of the helical structure before direct straining of the protein backbone.

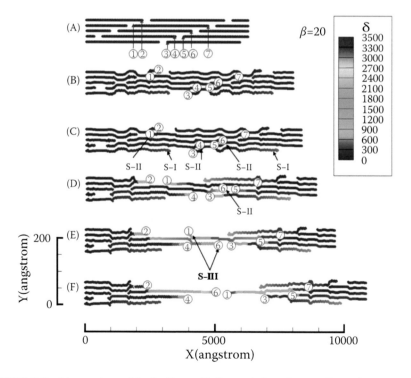

FIGURE 3.8 Mesoscale model of collagen fibril, consisting of a two-dimensional array of ultralong coarse-grain tropocollagen molecules. The snapshots show the molecular structure as the fibril undergoes tensile deformation, where the color is defined by the magnitude of the slip vector [61]. A detailed analysis of the molecular deformation mechanisms suggests that intermolecular slip plays a major role in mediating large tensile strains in collagen fibrils leading up to failure, following a significant elastic regime. (Adapted from Tang, Y., Ballarini, R., Buehler, M. J., and Eppell, S. J., *J. Roy. Soc. Interface*, 7(46), 839, 2010.)

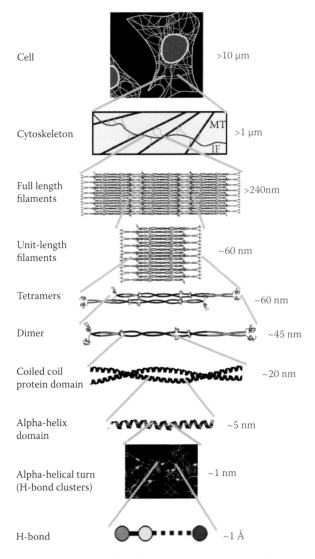

Cell		>10 μm
Cytoskeleton	MT / IF	>1 μm
Full length filaments		>240nm
Unit-length filaments		~60 nm
Tetramers		~60 nm
Dimer		~45 nm
Coiled coil protein domain		~20 nm
Alpha-helix domain		~5 nm
Alpha-helical turn (H-bond clusters)		~1 nm
H-bond		~1 Å

FIGURE 3.9 Schematic depicting hierarchical structure of alpha-helix protein-based intermediate filaments (IFs), which provide structural tensegrity to the cytoskeleton of cellular membranes. Over seven levels of hierarchy are transcended, from hydrogen bonds to alpha-helical turns, alpha-helical proteins (which are the focus of coarse-graining discussed here), dimers (coiled-coiled protein domain), tetramers, unit-length filaments, and full-length filaments to the cellular level. (Adapted from Qin, Z., Kreplak, L., and Buehler, M.J., *PLos ONE*, 4(10), e7294, 2009.)

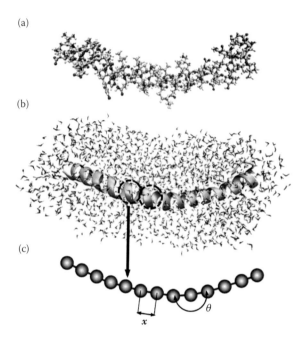

(a)

(b)

(c)

θ

x

FIGURE 3.10 Schematic of coarse-graining procedure, in which full atomistic representation is replaced by a mesoscopic bead-spring model. A pair of beads represents one turn in the alpha-helix (also called a convolution), and thus 3.6 residues with the corresponding mass. (a) Full atomistic representation depicting all atoms and bonds; folded states of the turns are stabilized by the presence of hydrogen bonds between residues (not shown); water molecules not shown for clarity. (b) Ribbon representation of protein, illustrating alpha-helical folded conformation of backbone chain and individual convolutions; explicit solvent (water molecules) shown. (c) Developed coarse-grain representation, with a single bead per convolution; need for explicit solvent eliminated in coarse-grain model, as effects are integrated into coarse-grain potentials.

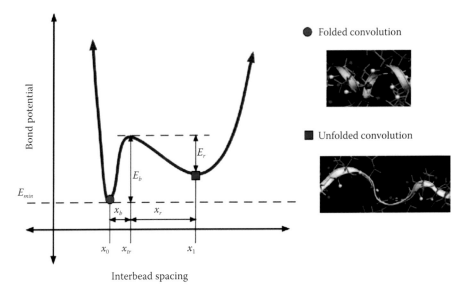

FIGURE 3.11 Double-well profile of the bond-stretching potential of the coarse-grain model, representing the energy landscape associated with the unfolding of one convolution (see Equation 3.41). The values of the equilibrium states, x_0 and x_1, energy barriers, E_b and E_r, and the transition state, x_{tr}, are obtained from geometric analysis of the alpha-helix geometry, as well as the full atomistic simulations. The transition state (local energy peak) corresponds to the breaking of hydrogen bonds between convolutions of the alpha-helix. After failure of these weak bonds, the convolution unfolds to a second equilibrium state with a large inter-particle distance. Under further loading, the covalent bonds begin to stretch, which leads to a second increase of the potential at large deformation.

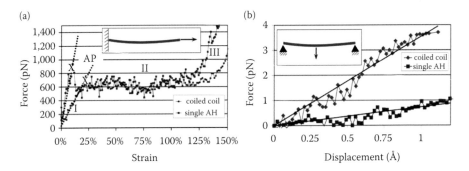

FIGURE 3.12 Full atomistic tests of alpha-helical protein (single-helix and coiled-coiled conformations) used to parameterize the coarse-grain model. (a) Force-strain results of direct tension simulations. The first regime (I) consists of a linear increase in force, until a strain of approximately 13% for the single alpha-helix, noted as the angular point (AP), which corresponds to the rupture and unfolding of an alpha-helix convolution. The second regime (II) represents the unfolding of the helix under approximately constant force. The third regime represents a nonlinear increase in strain due to backbone stretching of the protein. (b) Force-displacement results of three-point bending simulations. The slope of the curve is proportional to the bending stiffness. Only the single alpha-helix values were used in the development of the coarse-grain representation. (Adapted from Ackbarow, T., and Buehler, M.J., J. Mater. Sci., 42(21), 8771–8787, 2007.)

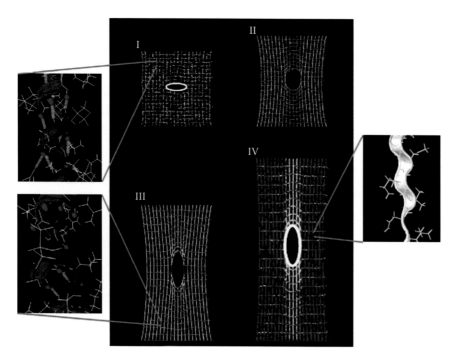

FIGURE 3.13 Snapshots of protein network deformation, where coarse-grain representation of alpha-helix proteins was implemented. The deformation mechanism is characterized by the molecular unfolding of the alpha-helical protein domains, leading to the formation of large plastic yield regions, providing energy dissipation and preventing catastrophic failure. The blowups depict the atomistic structural arrangement of the alpha-helical protein domains based on the known correspondence with the coarse-grain mapping. (Adapted from Ackbarow, T., et al., *PLos ONE*, 4(6), e6015, 2009.)

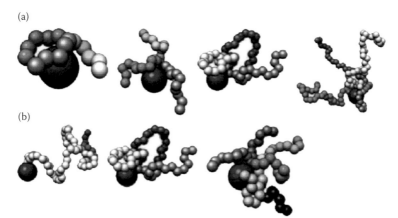

FIGURE 3.17 Schematic overview of the types of coarse-grain nanoparticles employed in the clustering study. (a) Constant architecture motif, with each chain in a three-armed star with lengths of 5, 10, 20, and 40 PEO units, respectively. (b) Constant mass motif, where each nanoparticle has 60 total PEO units with arrangements in a linear, three-armed, or six-armed fashion, respectively. Note explicit solvent molecules are neither depicted nor required in the coarse-grain simulations. (Reproduced with permission of the PCCP Owner Societies, copyright 2009. Hooper, J.B., Bedrov, D., and Smith, G.D., *Phys. Chem. Chem. Physics*, 11, 2034–2045, 2009.)

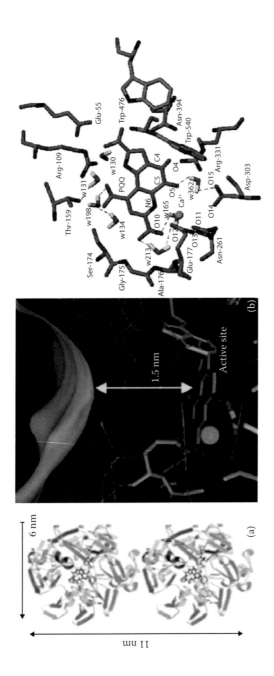

FIGURE 5.3 (a) MDH enzyme from the entry 1W6S (*Methylobacterium Extorquens W3A1*) of the Protein Data Bank. (b) View of the enzyme's active site and the X-ray crystal structure of amino acids in the active site.

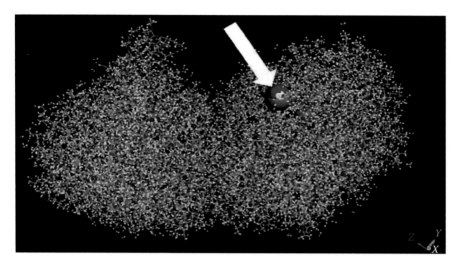

FIGURE 5.7 Methanol molecule (pointed arrow) docked at the binding pocket of MDH.

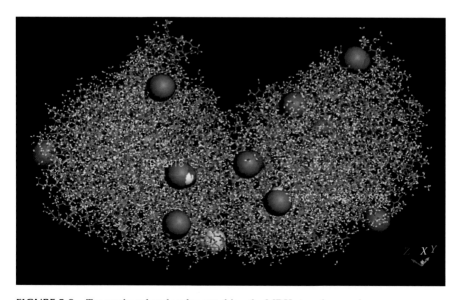

FIGURE 5.8 Ten methanol molecules attacking the MDH at various regions.

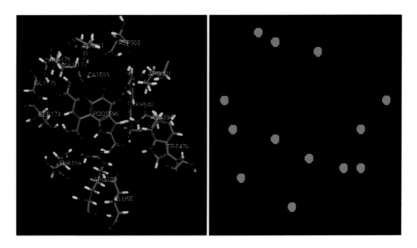

FIGURE 5.9 Active site cluster in the MDH enzyme (left) showing the various amino acids and the active site model showing the amino acids plotted as dots on the lattice according to their centers of mass (right).

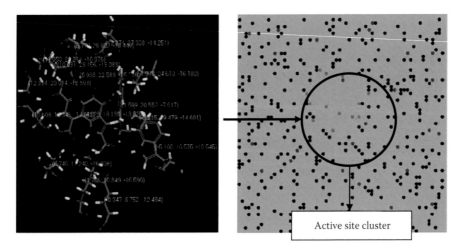

Active site cluster

FIGURE 5.10 The active site cluster (*green dots*) plotted on the 2-D lattice with substrate (*red dots*) and product (*blue dots*) molecules making a random walk. The black dots represent the obstacles.

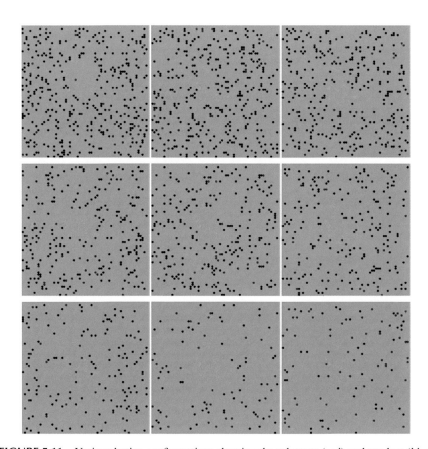

FIGURE 5.11 Various lattice configurations showing the substrate (*red*) and product (*blue*) molecules with decreasing obstacle concentration (*black dots*) taken at $t = 1500$ MC steps.

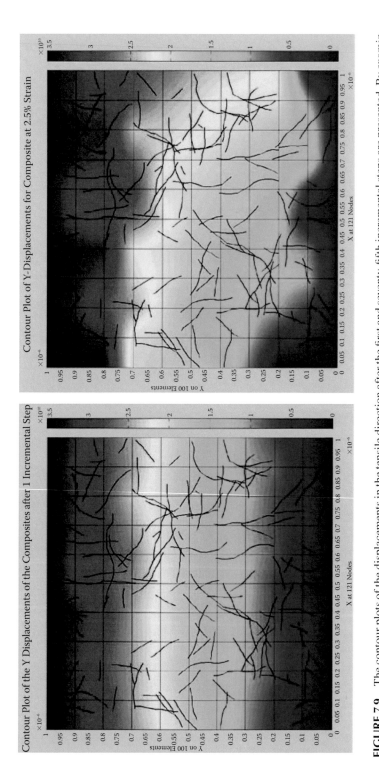

FIGURE 7.9 The contour plots of the displacements in the tensile direction after the first and seventy-fifth incremental steps are presented. By examining the deformation before and after the onset of nonlinear polymer behavior, the role of the nanotubes in mechanical reinforcement is evident.

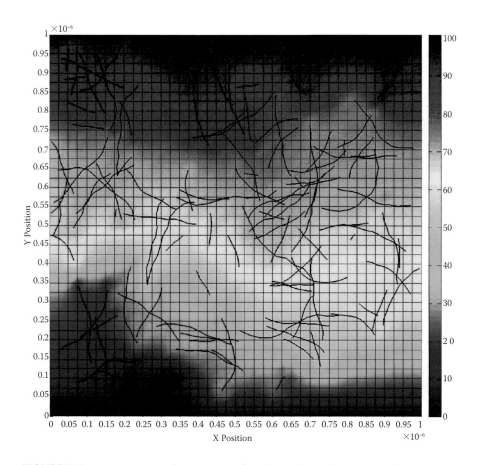

FIGURE 7.10 A contour plot of temperatures from the nonlinear thermal model is presented.

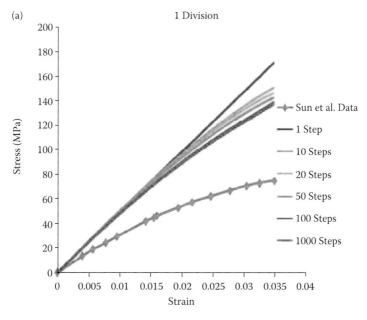

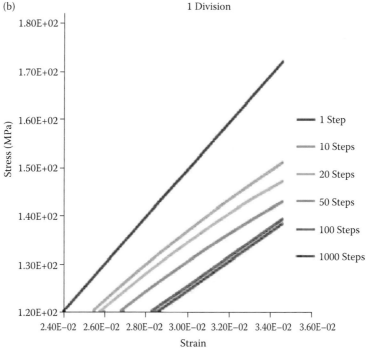

FIGURE 7.11 (a) The mesh convergence analysis for the model with 100 incremental steps is shown for 1, 10, 20, 40, 60, and 80 divisions of the RVE. (b) A closer view of the model results at high strains identifies 60 divisions as sufficient mesh refinement.

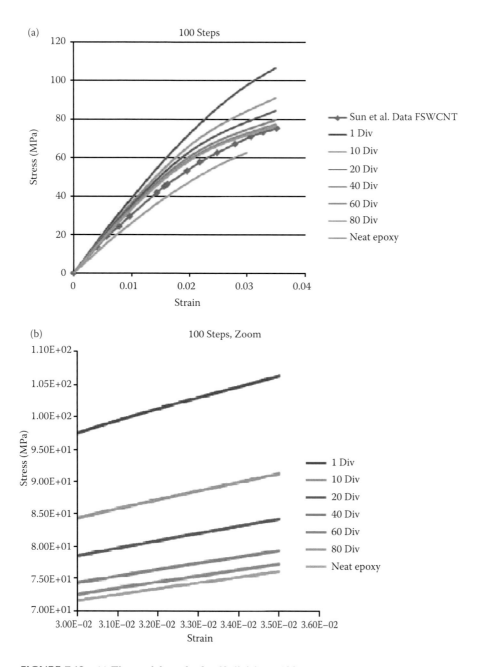

FIGURE 7.12 (a) The model results for 60 divisions, 100 steps are compared to the results of Sun et al. (b) A close-up view of the stress–strain curve high strain values is shown. (Reproduced from Sun, L. et al., *Carbon* vol. 46, Feb. 2008, pp. 320–328.)

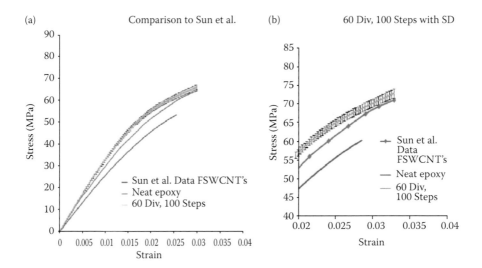

FIGURE 7.13 The maximum and minimum RVEs out of the 500 RVEs sampled in the MCMCA are compared to the experimental results of Sun et al. to highlight the span of possible outcomes for random nanotube geometries. (Reproduced from Yu, M., Lourie, O., Dyer, M.J., Moloni, K., Kelly, T.F., and Ruoff, R.S. *Science* vol. 287, Jan. 2000, pp. 637–640.)

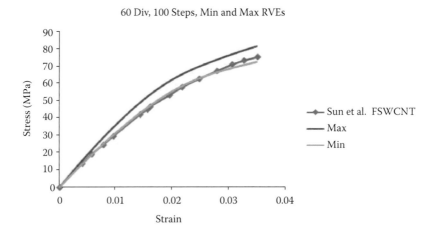

FIGURE 7.14 The model results for 60 divisions, 100 steps are compared to the results of Zhu et al. The center blue F-SWCNT line represents the experimental data. (Reproduced from Zhu, J., Kim, J., Peng, H., Margrave, J., Khabashesku, V., and Barrera, E. *Nano Letters* vol. 3, Aug. 2003, pp. 1107–1113.)

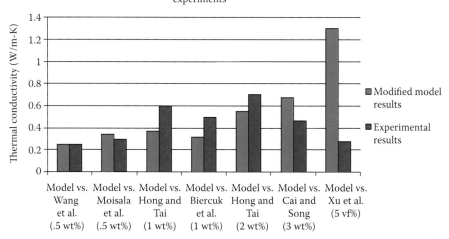

FIGURE 7.17 A comparison of the modified model's thermal conductivity results to various experiments is presented. Substantially better agreement with experimental results is clearly evident.

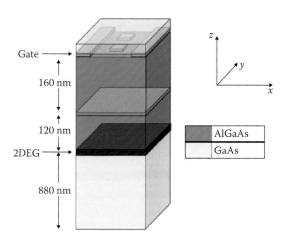

FIGURE 9.4 The heterojunction composed of GaAs/AlGaAs, where the edges are defined by gates (gray planes on surface). Silicon doping is depicted by the thin layer and electron gas that lies 280 nm below the surface. The dielectric constants of the materials are assumed to be the same, $\kappa = 12{:}4$, and the crystal is surrounded by air in five directions, except the substrate direction.

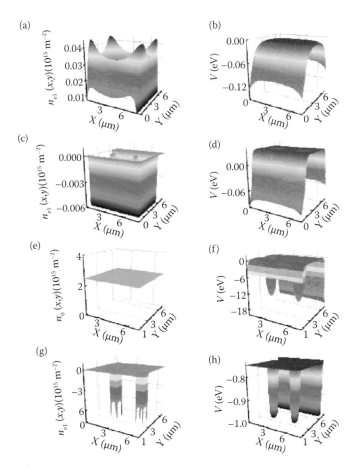

FIGURE 9.5 3D density (left) and potential (right) profiles calculated at zero temperature considering the structure shown in Figure 1.4. The inner (square) contacts deplete electrons at the center. Interestingly, one also observes charge fluctuations at the substrate.

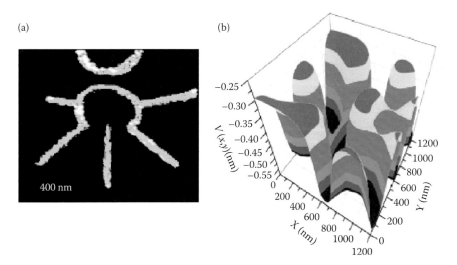

FIGURE 9.8 (a) Scanning electron microscopy picture of the oxidation-defined quantum dot (R. Haug's group, Hannover University). (b) The self-consistently calculated potential profile.

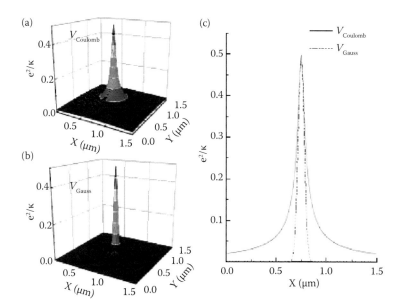

FIGURE 9.18 (a) A single Coulomb and (b) a Gaussian impurity located at the center of a 1.5-μm × 1.5-μm unit cell, approximately 30 nm above the electron gas ($z = z0 = 0$). The short-range behaviors are similar, while the long-range parts are strongly different. Potential profiles projected through the center (x, $y = 0.75$ μm), for the Coulomb (solid black line) and Gaussian (broken red line) impurities.

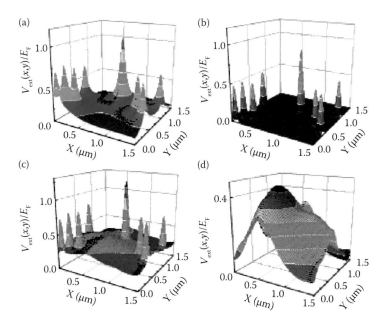

FIGURE 9.19 External potential generated 30 nm below a plane containing 10 (a) Coulomb and (b) Gaussian donors. The range of the Gaussian potential is determined by the spacer thickness. (c) The long-range part of the Coulomb potential profile, where only the lowest two Fourier components are back-transformed to configuration space. (d) The Gaussian potential profile plus the long-range part of the Coulomb potential in order to compare the different potential landscapes.

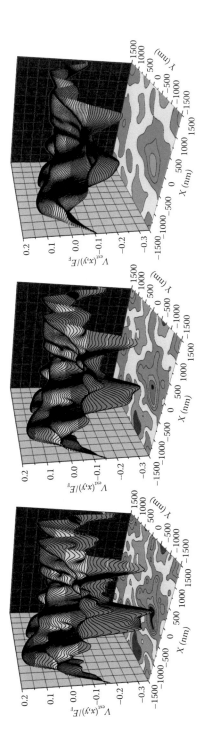

FIGURE 9.20 Electrostatic potential profile generated by the random distribution of 120 ions at $z = 90$ nm (upper), $z = 120$ nm (middle), and $z = 200$ nm (lower). Similar to Figure 9.19, damping is due solely to the dielectric material in between 2DES and the donor layer.

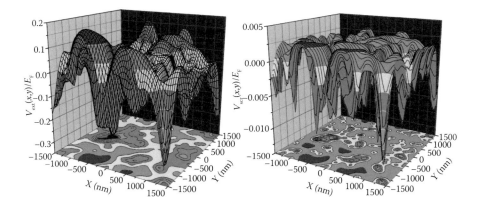

FIGURE 9.23 The external potential generated by random distribution of 90 ions, residing 120 nm above the 2DES (left). Damping is due solely to the dielectric spacer. Screened potential by the 2DES at $T = 0$ and $B = 0$, for the given external potential, within the linear screening regime (right).

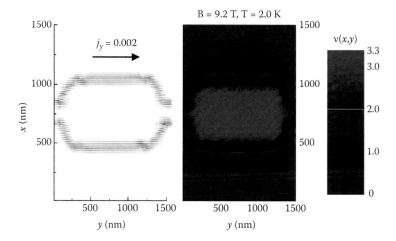

FIGURE 9.33 Spatial distribution of the current density (left) and filling factor (right). One-to-one correspondence between the current and the incompressible strips (black contour) can be clearly seen. The higher current densities in the left panel point to the formation of hot-spots. System parameters are given at the figure labels, whereas current density is given in units of cyclotron energy.

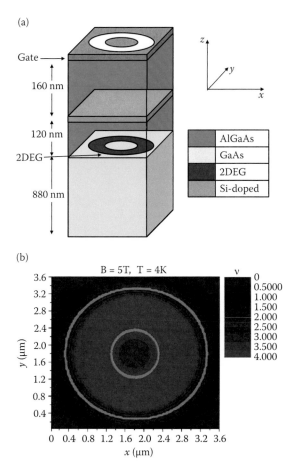

(a)

Gate

160 nm

120 nm

2DEG

880 nm

	AlGaAs
	GaAs
	2DEG
	Si-doped

(b)

B = 5T, T = 4K

ν
0
0.5000
1.000
1.500
2.000
2.500
3.000
3.500
4.000

y (μm)

x (μm)

FIGURE 9.35 Corbino disc is defined by inner and outer spherical gates or etching 280 nm below the surface. The spatial distribution of the local filling factors, where incompressible strips are highlighted by gray color. Note that strips carry equilibrium edge currents; however, there is no excess current imposed as in the Hall bar geometry.

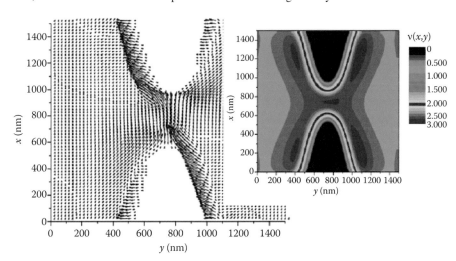

x (nm)

y (nm)

ν(x,y)
0
0.500
1.000
1.500
2.000
2.500
3.000

x (nm)

y (nm)

FIGURE 9.36 The spatial distribution of the classical current in the close vicinity of a gate-defined QPC (right) and the local filling factor distribution, where two incompressible strips come close to each other. The conductance is not quantized; however, interference may occur due to partitioning between edge states.

6 First-Principles Alloy Thermodynamics

Axel van de Walle

CONTENTS

6.1 INTRODUCTION

Over the past few decades, density functional theory (DFT) [1–4] methods have demonstrated high accuracy in calculations of structural and energetic properties in a wide variety of materials. While calculations at absolute zero have traditionally been the focus of this *ab initio* (or first-principles) literature, there is considerable and growing interest in obtaining finite-temperature properties, such as equilibrium phase diagrams and, more generally, thermodynamic properties. The main difficulty associated with this endeavor is that highly accurate quantum-mechanical electronic structure calculations are currently limited in terms of size of the simulation cell and in the number of distinct atomic configurations that can be feasibly sampled within a reasonable computing time frame. On the other hand, calculations of thermodynamic properties require, by definition, large system sizes and extensive sampling of phase space. Fortunately, this impasse can be resolved by coupling DFT energy methods with statistical mechanical models [5–14]. This chapter provides an overview of the methodologies underlying first-principles calculations of alloy phase stability and provides examples intended to illustrate the capabilities and accuracy of these methods.

6.2 OVERVIEW

At a given temperature (T) and pressure (P), thermodynamic stability is governed by the magnitude of the Gibbs free energy (G):

$$G = E - TS + PV, \tag{6.1}$$

where E, S, and V denote energy, entropy, and volume, respectively. The formal statistical mechanical procedure for calculating G from first-principles is well-defined. Quantum-mechanical calculations can be performed to compute the energy $E(s)$ of different microscopic states (s) of a system, which then must be summed up in the form of a partition function (Z):

$$Z = \sum_s \exp[-E(s)/k_B T] \tag{6.2}$$

The Gibbs free energy is then derived as $G = F + PV$, where F is the Helmholtz free energy, given by $F = ETS = k_B T \ln Z$, where k_B is Boltzmann's constant.

Figure 6.1a illustrates, for the case of a disordered crystalline binary alloy, the nature of the disorder characterizing a representative finite-temperature atomic structure. This disorder can be characterized in terms of the *configurational* arrangement of the elemental species over the sites of the underlying parent lattice, coupled with the *displacements* characterizing positional disorder in each arrangement. In principle, the sum in Equation 6.2 extends over all configurational and displacive states accessible to the system, a phase space that is astronomically large for a realistic system size. In practice, the methodologies of atomic-scale molecular dynamics (MD) and Monte Carlo (MC) simulations, coupled with thermodynamic integration techniques (e.g., Frenkel and Smit [15]) reduce the complexity of a free energy calculation to a more tractable problem of sampling on the order of several to tens of thousands representative states.

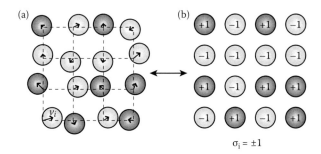

FIGURE 6.1 (a) Disordered crystalline alloy. The state of the alloy is characterized both by the atomic displacements v_i and the occupation of each lattice site. (b) Mapping of the real alloy onto a lattice model characterized by occupation variables σ_i describing the identity of atoms on each of the lattice sites.

Electronic DFT provides an accurate quantum-mechanical framework for calculating the relative energetics of competing atomic structures in solids, liquids, and molecules for a wide range of material classes. Due to the rapid increase in computational cost with system size, however, DFT calculations are typically limited to structures containing several hundred atoms, whereas *ab initio* MD simulations are practically limited to time scales of less than a nanosecond. For liquids or compositionally-ordered solids, where the time scales for structural rearrangements (displacive in the latter case, configurational and displacive in the former) are sufficiently fast, and the size of periodic cells required to accurately model the atomic structure are relatively small, DFT-based MD methods have found direct applications in the calculation of finite-temperature thermodynamic properties [16,17]. For crystalline solids containing both positional and concentrated compositional disorder, however, direct applications of DFT to the calculation of free energies remains intractable; the time scales for configurational rearrangements are set by solid-state diffusion, ruling out direct application of MD, and the necessary system sizes required to accurately model configurational disorder are too large to permit direct application of DFT as the basis for MC simulations. Effective strategies have nonetheless been developed for bridging the size and time-scale limitations imposed by DFT in the first-principles computation of thermodynamic properties for disordered solids. These approaches involve exploitation of DFT methods as a framework for parameterizing coarse-grained statistical models that serve as efficient "effective Hamiltonians" in direct simulation-based calculations of thermodynamic properties.

6.3 THERMODYNAMICS OF ORDERED ALLOYS

In an ordered solid, thermal fluctuations take the form of electronic excitations and lattice vibrations and, accordingly, the free energy can be written as $F = E_0 + F_{elec} + F_{vib}$, where E_0 is the absolute-zero total energy and F_{elec} and F_{vib} denote electronic and vibrational free energy contributions, respectively. This section describes the calculation of the electronic and vibrational contributions most commonly considered in phase-diagram calculations under the assumption that electron–phonon interactions are negligible (i.e., F_{elec} and F_{vib} are simply additive).

To account for electronic excitations, electronic DFT can be extended to nonzero temperatures by allowing for partial occupations of the electronic states [18]. Within this framework, assuming that both the electronic charge density and the electronic density of states can be considered temperature-independent, the electronic contribution to the free energy $F_{elec}(T)$ at temperature T can be decomposed as

$$F_{elec}(T) = E_{elec}(T) - E_{elec}(0) - TS_{elec}(T), \qquad (6.3)$$

where the electronic band energy $E_{elec}(T)$ and the electronic entropy $S_{elec}(T)$ are respectively given by

$$E_{elec}(T) = \int f_{\mu,T}(\varepsilon)\varepsilon g(\varepsilon)\, d\varepsilon, \qquad (6.4)$$

$$S_{elec}(T) = -k_B \int s(f_{\mu,T}(\varepsilon))g(\varepsilon)\,d\varepsilon,$$

(6.5)

where $g(\varepsilon)$ is the electronic density of states obtained from a density functional calculation,

$$s(x) = x \ln x + (1-x)\ln(1-x),$$

(6.6)

and $f_{\mu,T}(\varepsilon)$ is the Fermi distribution when the electronic chemical potential is equal to μ,

$$f_{\mu,T}(\varepsilon) = \left(1 + \exp\left(\frac{(\varepsilon-\mu)}{k_B T}\right)\right)^{-1}$$

(6.7)

The chemical potential μ is the solution to $\int f_{\mu,T}(\varepsilon)g(\varepsilon)\,d\varepsilon = n_e$, where n_e is the total number of electrons. Under the assumption that the electronic density of states near the Fermi level is slowly varying relative to $f_{\mu,T}(\varepsilon)$, the equations for the electronic free energy reduce to the well-known Sommerfeld model, an expansion in powers of T whose lowest-order term is

$$F_{elec}(T) = -\frac{\pi^2}{6} k_B^2 T^2 g(\varepsilon_F),$$

(6.8)

where $g(\varepsilon_F)$ is the zero-temperature value of the electronic density of states at the Fermi level (ε_F). An example of application of this methodology can be found in the study of Wolverton and Zunger [19].

The above formalism would, in principle, be exact if an exact DFT could be implemented. However, further approximations, such as the local density approximation (LDA), are typically used in practical implementations of DFT [20] and limit the accuracy of $F_{elec}(T)$. This issue is exacerbated in magnetic systems, where it becomes necessary to explicitly sample the various spin configurations to obtain a reliable description of the system's thermodynamics. In this approach, DFT is used to obtain ground-state energies subject to local magnetic moment constraints that impose various specific spin configurations. Energy differences between different spin configurations can then be used to infer effective spin interactions to be included in an effective coarse-grained Heisenberg model, whose thermodynamic analysis is considerably more tractable, although still not trivial [21].

The quantum treatment of lattice vibrations in the harmonic approximation provides a reliable description of thermal vibrations in many solids for low to moderately high temperatures [22,23]. To describe this theory, consider an infinite periodic system with n atoms per unit cell and let $u\binom{l}{i}$ for $i = 1,\ldots,n$ denote the displacement of atom i in cell l away from its equilibrium position and let M_i be the mass of atom i.

Within the harmonic approximation, the potential energy U of this system is entirely determined by (i) the potential energy (per unit cell) of the system at its equilibrium position E_0 and (ii) the *force constants tensors* $\Phi\begin{pmatrix} ll' \\ ij \end{pmatrix}$ whose components are given, for $\alpha, \beta = 1,2,3$, by

$$\Phi_{\alpha,\beta}\begin{pmatrix} ll' \\ ij \end{pmatrix} = \frac{\partial^2 U}{\partial u_\alpha \begin{pmatrix} l \\ i \end{pmatrix} \partial u_\beta \begin{pmatrix} l' \\ j \end{pmatrix}}, \tag{6.9}$$

evaluated at $u\begin{pmatrix} l \\ i \end{pmatrix} = 0$ for all l,i. Such a harmonic approximation to the Hamiltonian of a solid is often referred to as a Born–von Kármán model.

The thermodynamic properties of a harmonic system are entirely determined by the frequencies of its normal modes of oscillation, which can be obtained by finding the eigenvalues of the so-called $3n \times 3n$ dynamical matrix of the system:

$$D(\mathbf{k}) = \sum_l e^{i2\pi(k \cdot l)} \begin{pmatrix} \dfrac{\Phi\begin{pmatrix} 0 & l \\ 1 & 1 \end{pmatrix}}{\sqrt{M_1 M_1}} & \cdots & \dfrac{\Phi\begin{pmatrix} 0 & l \\ 1 & n \end{pmatrix}}{\sqrt{M_1 M_n}} \\ \vdots & \ddots & \vdots \\ \dfrac{\Phi\begin{pmatrix} 0 & l \\ n & 1 \end{pmatrix}}{\sqrt{M_n M_1}} & \cdots & \dfrac{\Phi\begin{pmatrix} 0 & l \\ n & n \end{pmatrix}}{\sqrt{M_n M_n}} \end{pmatrix}, \tag{6.10}$$

for all vectors $\mathbf{k}$ in the first Brillouin zone. The resulting eigenvalues $\lambda_b(k)$ for $b = 1 \ldots 3n$, provide the frequencies of the normal modes through $v_b(k) = \dfrac{1}{2\pi}\sqrt{\lambda_b(k)}$. This information for all k is conveniently summarized by $g(v)$, the phonon density of states (DOS), which specifies the number of modes of oscillation with frequencies in the infinitesimal interval $[v, v + dv]$. The vibrational free energy (per unit cell) F_{vib} is then given by

$$F_{vib} = k_B T \int_0^\infty \ln\left[2\sinh\left(\frac{hv}{2k_B T} \right) \right] g(v)\, dv, \tag{6.11}$$

where h is Planck's constant and k_B is Boltzmann's constant. The associated vibrational entropy S_{vib} of the system can be obtained from the well-known thermodynamic relationship $S_{vib} = -\dfrac{\partial F_{vib}}{\partial T}$. The high temperature limit (which is also the classical limit) of Equation 6.11 is often a good approximation over the range of temperature of interest in solid-state phase diagram calculations

$$F_{\mathrm{vib}} = k_{\mathrm{B}}T \int_0^{\infty} \ln\left(\frac{h\nu}{k_{\mathrm{B}}T}\right) g(\nu)\,\mathrm{d}\nu \qquad (6.12)$$

The high-temperature limit of the vibrational entropy difference between two phases is often used as a convenient measure of the magnitude of the effect of lattice vibrations on phase stability. It has the advantage of being temperature-independent, thus allowing a unique number to be reported as a measure of vibrational effects.

A simple improvement over the harmonic approximation, called the quasiharmonic approximation, is obtained by using volume-dependent force constant tensors. This approach maintains all the computational advantages of the harmonic approximation while permitting the modeling of thermal expansion. The volume dependence of the phonon frequencies induced by the volume dependence of the force constants is traditionally described by the Grüneisen parameter $\gamma_{kb} = -\partial \ln \nu_b(k)/\partial \ln V$. However, for the purpose of modeling thermal expansion, it is more convenient to directly parametrize the volume dependence of the free energy itself. This dependence has two sources: the change in entropy due to the change in the phonon frequencies and the change in elastic energy due to the expansion of the lattice:

$$F(T,V) = E_0(V) + F_{\mathrm{vib}}(T,V), \qquad (6.13)$$

where $E_0(V)$ is the energy of a motionless lattice whose unit cell is constrained to a volume V, while $F_{\mathrm{vib}}(T,V)$ is the vibrational free energy of a harmonic system constrained to remain at a unit cell volume V at temperature T. The equilibrium volume $V^*(T)$ at temperature T is obtained by minimizing $F(T,V)$ with respect to V. The resulting free energy $F(T)$, accounting for thermal expansion, is then given by $F(T,V^*(T))$.

First-principles calculations can be used to provide the necessary input parameters for the above formalism. The so-called direct force method proceeds by calculating, from first principles, the forces experienced by the atoms in response to various imposed displacements. The force constant tensors are then determined by a least-squares fit to the calculated forces. Note that the simultaneous displacements of the periodic images of each displaced atom (due to the periodic boundary conditions used in most *ab initio* methods) typically require the use of a supercell geometry, in order to be able to sample all the displacements needed to determine the force constants. While the number of force constants to be determined is, in principle, infinite, in practice, it can be reduced to a manageable finite number by noting that the force constant tensor associated with two atoms that lie farther than a few nearest-neighbor shells can be accurately neglected for many systems. Alternatively, linear response theory [24–32] can be used to calculate the dynamical matrix $D(k)$ directly using second-order perturbation theory, thus circumventing the need for supercell calculations. Linear response theory is also particularly useful when a system is characterized by nonnegligible long-range force constants, as in the presence of Fermi-surface instabilities or long-range electrostatic contributions.

The previous discussion centered around the application of harmonic (or quasi-harmonic) approximations to the statistical modeling of vibrational contributions to free energies of solids. While harmonic theory is known to be highly accurate for a wide class of materials, important cases exist where this approximation breaks down due to large anharmonic effects. Examples include the modeling of ferroelectric and martensitic phase transformations where the high temperature phases are often dynamically unstable at zero temperature, i.e., their phonon spectra are characterized by unstable modes. In such cases, effective Hamiltonian methods have been developed to model structural phase transitions from first principles [33]. Alternatively, direct application of *ab initio* molecular dynamics offers a general framework for modeling thermodynamic properties of anharmonic solids [16,17].

6.4 THERMODYNAMICS OF DISORDERED ALLOYS

Let us now relax the main assumption made in the previous section by allowing atoms to exit the neighborhood of their local equilibrium position. This is accomplished by considering every possible way to arrange the atoms on a given lattice. As illustrated in Figure 6.1b, the state of order of an alloy can be described by occupation variables σ_i specifying the chemical identity of the atom associated with lattice site i. In the case of a binary alloy, the occupations are traditionally chosen to take the values +1 or −1, depending on the chemical identity of the atom.

Returning to Equation 6.2, all the thermodynamic information of a system is contained in its partition function Z and in the case of a crystalline alloy system, the sum over all possible states of the system can be conveniently factored as follows:

$$Z = \sum_{L} \sum_{\sigma \in L} \sum_{v \in \sigma} \sum_{e \in v} \exp\left[-\beta E\left(L, \sigma, v, e\right)\right], \quad (6.14)$$

where $\beta = (k_B T)^{-1}$ and where

- L denotes a specific underlying lattice type (e.g., bcc, fcc, etc.).
- σ denotes a particular configuration on the above lattice L (i.e., a vector of all occupation variables).
- v denotes a displacement of each atom away from its local equilibrium position.
- e is a particular electronic state when the nuclei are constrained to be in a state described by σ and v.
- $E(L,\sigma,v,e)$ is the energy of the alloy in a state characterized by σ, v, and e.

Each summation defines an increasingly coarse level of hierarchy in the set of microscopic states. For instance, the sum over v includes all displacements such that the atoms remain close to the undistorted configuration σ. Equation 6.14 implies that the free energy of the system can be written as

$$F(T) = -k_B T \ln \sum_\sigma \exp[-\beta F(L,\sigma,T)], \qquad (6.15)$$

where $F(L,\sigma,T)$ is nothing but the free energy of an alloy with a fixed atomic configuration, as obtained in the previous section

$$F(L,\sigma,T) = -k_B T \ln \sum_{\nu \in \sigma} \sum_{e \in \nu} \exp[-\beta E(L,\sigma,\nu,e)] \qquad (6.16)$$

The so-called coarse-graining of the partition function illustrated by Equation 6.15 enables, in principle, an exact mapping of a real alloy onto a simple lattice model characterized by the occupation variables σ and a temperature-dependent Hamiltonian $F(L,\sigma,T)$ [6,23].

6.4.1 CLUSTER EXPANSION FORMALISM

Although the problem of modeling the thermodynamic properties of configurationally disordered solids has been reduced to a more tractable calculation for a lattice model, the above formalism would still require the calculation of the free energy for every possible configuration σ, which is computationally intractable. Fortunately, the configurational dependence of the free energy can often be parametrized using a convenient expansion known as a cluster expansion [5,6,8,34]. This expansion (for a given lattice L, omitted from the notation) takes the form of a polynomial in the occupation variables

$$F(\sigma,T) = J_\varnothing + \sum_i J_i \sigma_i + \sum_{ij} J_{ij} \sigma_i \sigma_j + \sum_{i,j,k} J_{ijk} \sigma_i \sigma_j \sigma_k + \ldots, \qquad (6.17)$$

where the so-called effective cluster interactions (ECI) $J_\varnothing, J_i, J_{ij}, \ldots$ need to be determined. The cluster expansion can be recast into a form that exploits the symmetry of the lattice by regrouping the terms as follows

$$F(\sigma,T) = \sum_\alpha m_\alpha J_\alpha \sigma_\alpha, \qquad (6.18)$$

where α is a cluster (i.e., a set of lattice sites) and where the summation is taken over all clusters that are symmetrically distinct, while

$$\sigma_\alpha = \left\langle \prod_{i \in \alpha} \sigma_i \right\rangle, \qquad (6.19)$$

in which the average $\langle \ldots \rangle$ is taken over all clusters α' that are symmetrically equivalent to α. The multiplicity m_α weight each term by the number of symmetrically equivalent clusters in a given reference volume (e.g., a unit cell). While the cluster expansion is presented here in the context of binary alloys, an extension to multicomponent alloys (where σ_i can take more than two different values) is straightforward [34,35]. Note that the presence of vacancies can also be accounted for within the cluster expansion formalism by merely treating vacancies as an additional atomic species [36,37]. The cluster expansion formalism can also be extended to allow for multiple coupled sublattices [38] (each of which could be occupied by any number of species [35]), thus permitting the treatment of interstitials. The applicability of the cluster expansion extends beyond the representation of the (free) energy: Any configuration-dependent quantity (such as the band gap), including tensor-valued properties (such as elastic constants), admits a cluster expansion [39]. In some systems, the presence of strong long-range elastic [40,41] or electrostatic [42] interactions can be efficiently handled by augmenting the cluster expansion by an analytic form for the long-range interactions (expressed in reciprocal space), so that only the short-range deviation from the asymptotic behavior of the interactions needs to be fitted to *ab initio* data.

Despite a superficial similarity between a lattice model and a spin system, care must be exercised. While it is justified to employ a cluster expansion framework to determine effective interaction between spins, one should refrain from using a lattice model to study the thermodynamics of magnetic systems, because it miscounts the number of states (therefore miscalculating the entropy), and gives an artificial special status to one direction in space. Instead, a Heisenberg model is a more appropriate approach (e.g., Korman et al. [21]).

It can be shown that when *all* clusters α are considered in the sum [18], the cluster expansion is able to represent any function of configuration σ by an appropriate selection of the values of J_α. However, the real advantage of the cluster expansion is that, for many systems, it is found to converge rapidly. An accuracy that is sufficient for phase diagram calculations can often be achieved by keeping only clusters α that are relatively compact (e.g., short-range pairs or small triplets, as illustrated in Figure 6.2b). The unknown parameters of the cluster expansion (the ECI J_α) can then

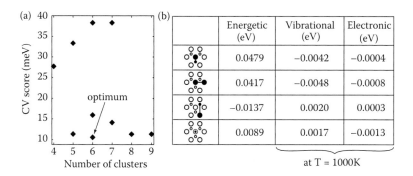

FIGURE 6.2 (a) Optimal selection of clusters via crossvalidation (the optimal number of clusters includes the empty, the point cluster, and four pairs). (b) Values of the effective cluster interactions (ECI) obtained from a least-squares fit to a set of 14 Ti-rich structures. The energetic, vibrational, and electronic components of the free energy at a representative temperature of 1000 K are separately given to illustrate the typical relative magnitude of each contribution.

be determined by fitting them to $F(\sigma,T)$ for a relatively small number of configurations σ obtained from first-principles computations (typically less than 50). Once the ECI have been determined, the free energy of the alloy for any given configuration can be quickly calculated, making it possible to explore a large number of configurations without recalculating the free energy of each of them from first principles.

Within the framework of the cluster expansion formalism, *ab initio* modeling of alloy free energies and phase diagrams proceeds through the following steps: (i) calculation of the free energies $F(\sigma,T)$ for a selected set of alloy configurations to fit the cluster interaction coefficients in Equation 6.17, (ii) determination of the optimal set of clusters to construct a predictive cluster expansion, (iii) determination of the ground-state structures, i.e., those minimum-energy structures representing the stable thermodynamic phases at low temperature, and (iv) MC simulations of configurational free energies for all phases of interest. The remainder of this section describes each of these steps in further detail.

While the methods for performing step (i) were described in Section 6.4, construction of a cluster expansion typically requires calculation of $F(\sigma,T)$ for on the order of 50 structures, which represents a computationally demanding task and thus is commonly performed using more approximate yet computationally more efficient techniques. A popular way to handle this issue is to approximate $F(\sigma,T)$ by the corresponding energy at absolute zero $F(\sigma,0)$, which is readily provided by first-principles total energy methods. However, in recent years, the validity of this approximation has been scrutinized [23,43–49]. It is generally recognized that this approximation is one of the main causes of the widely reported systematic overestimation of transition temperatures in *ab initio* phase diagram calculations [23].

In an effort to alleviate this problem, computationally efficient ways to account for lattice vibrations have been proposed. The simplest scheme is the Moruzzi–Janak–Schwarz (MJS) method [50], in which the vibrational free energy is calculated within the Debye approximation and under reasonable assumptions regarding the elastic tensor in order to obtain an expression that solely depends on a compound's bulk modulus and average density. Agreement with the experimentally measured phase diagrams was substantially improved by including lattice vibrations in this way in a number of alloy systems [51–53].

Another popular scheme [54–56] is the use of a short-range cluster expansion to represent the configuration dependence of $F(\sigma,T) - F(\sigma,0)$, thus necessitating fewer lattice dynamics calculations, while maintaining very accurate cluster expansion for the zero-temperature $F(\sigma,0)$ contribution.

The concept of length-dependent transferable force constants (LDTFC) has recently been suggested [23,57–62] to improve the computational efficiency of lattice dynamics calculations. The basic idea is to rely on the observation that bond length is a good predictor of bond stiffness (for a given lattice and a given type of chemical bond). The bond length–bond stiffness relationships can be determined by calculating, from first-principles, the stretching and bending force constants in a few ordered compounds as a function of volume (see Figure 6.3). Once this relationship is known, the force constants needed for the calculation of the vibrational free energy of a given structure can be predicted solely from the knowledge of its relaxed geometry

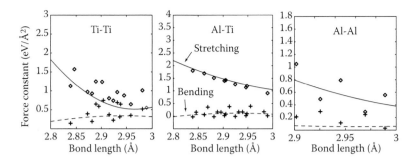

FIGURE 6.3 Determination of the bond-stiffness versus bond-length relationship in the hcp Ti–Al system the for construction of length-dependent transferable force constants (LDTFC).

(which provides the bond lengths). Since it then becomes possible to quickly determine the lattice dynamics for a large number of configurations, the accuracy of the cluster expansion of the vibrational free energy is no longer limited by the number of terms that can be included in the cluster expansion, although it may be limited by the accuracy of the LDTFC approximation, which can fortunately be assessed during the determination of the bond length–bond stiffness relationships.

The second step in the construction of a cluster expansion is the optimal selection of the clusters to be included in the expansion. If too few terms are kept, the predicted energies may be imprecise because the truncated cluster expansion cannot account for all sources of energy fluctuations. If too many terms are kept, the least-squares fit becomes excessively noisy because the number of fitting parameters is too large relative to the amount of available data. The best compromise between those two unwanted effects can be found by minimizing the cross-validation (CV) score [12,63], defined as

$$(CV)^2 = n^{-1} \sum_{i=1}^{n} \left(E_i - \hat{E}_{(i)} \right)^2, \qquad (6.20)$$

where E_i is the calculated energy of structure i, whereas $\hat{E}_{(i)}$ is the predicted value of the energy of structure i obtained from a least-squares fit to the energies of the $(n-1)$ remaining structures. Choosing the number of terms that minimizes the CV score has been shown to be an asymptotically optimal [64] selection rule. In contrast to the well-known mean squared error, the CV score is not monotonically decreasing in the number of fitting parameters. As the number of parameters to be fitted increases, the CV score first decreases because an increasing number of degrees of freedom are available to explain the variations in energy. The CV score then goes through a minimum before increasing, due to a decrease in predictive power caused by an increase in the noise in the fitted ECI. The best compromise between these two effects can then be found. Naturally, as more first-principles data are calculated and included in the fit, the minimum CV score will be achieved by including an increasing number of terms in the cluster expansion. The cluster selection procedure is illustrated in Figure 6.2a, while Figure 6.2b shows the ECI obtained through a least-squares fit to a database of 14 *ab initio* free energies determined using the LDTFC method for the vibrational contributions and Equation 6.3 for the electronic contributions.

In the limiting case when some of the atomic species in a multicomponent alloy are dilute, a so-called local cluster expansion can be used [37] that offers the advantage of reducing the number of ECI that need to be determined, since the interactions between dilute species do not need to be determined. The main difference with the conventional cluster expansion is that the symmetry group used to determine the equivalence between clusters is the local point group associated with a single dilute antisite defect (or vacancy) rather than the full space group of the parent lattice. Local cluster expansions have also found important application in the parametrization of the configurational-dependence of activation barriers for diffusion [65,66].

6.4.2 DETERMINING GROUND-STATE STRUCTURES

The cluster expansion tremendously simplifies the third step of the analysis of phase stability: the search for the lowest energy configuration at each composition of the alloy system. Determining these so-called ground states is important because they determine the general topology of the alloy phase diagram. Each ground state is typically associated with one of the stable phases of the alloy system. There are four main approaches to identifying the ground states of an alloy system.

With the enumeration method, all the configurations whose unit cell contains less than a given number of atoms are enumerated and their energy is quickly calculated using the value of $F(\sigma,0)$ predicted from the cluster expansion. The energy of each structure can then be plotted as a function of its composition (see Figure 6.4a) and the points touching the lower portion of the convex hull of all points indicate the ground states. While this method is approximate, as it ignores ground states with unit cells

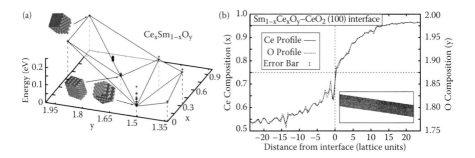

FIGURE 6.4 Computational thermodynamic study of the samarium-doped ceria system [42]. (a) Example of ground-state search via a convex hull construction in energy-composition space. (b) Equilibrium composition profile across (100) interface obtained through a hybrid canonical/grand-canonical multisublattice multicomponent Monte-Carlo simulation. Inset: Snapshot of the 70,000-atom simulation cell used.

larger than the given threshold, it is simple to implement and has been found to be quite reliable, thanks to the fact that most ground states indeed have a small unit cell.

Simulated annealing offers another way to find the ground states. It proceeds by generating random configurations via MC simulations using the Metropolis algorithm [67,68] that mimic the ensemble sampled in thermal equilibrium at a given temperature. As the temperature is lowered, the simulation should converge to the ground state. Thermal fluctuations are used as an effective means of preventing the system from getting trapped in local minima of energy. While the constraints on the unit cell size are considerably relaxed relative to the enumeration method, the main disadvantage of this method is that, whenever the simulation cell size is not an exact multiple of the ground-state unit cell, artificial defects will be introduced in the simulation that need to be manually identified and removed. Also, the risk of obtaining local rather than global minima of energy is not negligible and must be controlled by adjusting the rate of decay of the simulation temperature.

There exists an exact, although computational demanding, algorithm to identify the ground states [5]. This approach relies on the fact that the cluster expansion is linear in the correlations σ_α (see Equations 6.18 and 6.19). Moreover, it can be shown that the set of correlations σ_α that correspond to "real" structures can be defined by a set of linear inequalities. These inequalities are the result of lattice-specific geometric constraints and systematic methods exist to generate them [5]. As an example of such constraints, consider the fact that it is impossible to construct a binary configuration on a triangular lattice where the nearest-neighbor pair correlations take the value −1 (i.e., where all nearest neighbors are between different atomic species). Since both the objective function and the constraints are linear in the correlations, linear programming techniques can be used to determine the ground states. The main difficulty associated with this method is the fact that the resulting linear programming problem involves a number of dimensions and a number of inequalities that both grow exponentially fast with the range of interactions included in the cluster expansion.

Although the three methods described above enable the determination of the ground states that are superstructures of a given parent lattice, a data-mining technique [69,70] has been proposed to carry out ground-state searches among crystal structures that do not share a common parent lattice. Instead of relying on a cluster expansion, this method relies on a large database of existing *ab initio* formation energy calculations in a variety of alloy systems to identify correlations between the formation energies of various crystal structures, through a so-called principal moment analysis. These correlations can then be used as follows to guide the ground-state search for a new alloy system not contained in the database. First, the formation energies of a few simple candidate compounds are calculated. Second, the correlations known from the database are used to predict the most likely ground states, given the known formation energy of the few simple compounds considered. Next, the formation energies of the predicted ground states are calculated and used to further refine the ground-state prediction. The process is iterated until no new ground states are predicted. While this method does not have the ability to discover ground states with a crystal structure that is not included in the database, its predictive capabilities steadily improve over time, as researchers contribute to enlarge the database of available formation energies.

6.4.3 Free Energy Calculations

Once the ground-state structures (i.e., the stable phases at low temperature) have been derived, construction of a solid-state phase diagram from first principles requires calculation of free energies for each of these ordered phases, as well as for the different possible disordered solid solutions, and any other ordered phases that may appear at finite temperatures. Calculation of the required composition and temperature-dependent free energies formally requires summing the configurational partition function given in Equation 6.15. Historically, the infinite summation defining the alloy partition function has been approximated through various mean-field methods [5,6]. However, more recently, the difficulties associated with extending such methods to systems with medium- to long-range interactions and the increase in available computational power, enabling MC simulations to be directly applied, have led to reduced reliance upon these techniques.

MC simulations readily provide thermodynamic quantities such as energy or composition by making use of the fact that averages over an infinite ensemble of microscopic states can be accurately approximated by averages over a finite number of states generated by "importance sampling." Moreover, quantities such as the free energy, which cannot be written as ensemble averages, can nevertheless be obtained via *thermodynamic integration* [15] using standard thermodynamic relationships to rewrite the free energy in terms of integrals of quantities that can be obtained via ensemble averages. For instance, since energy $E(T)$ and free energy $F(T)$ are related through $E(T) = \partial(F(T)/T)/\partial(1/T)$, we have

$$\frac{F(T)}{T} - \frac{F(T_0)}{T_0} = -\int_{T_0}^{T} \frac{E(T)}{T^2} \, dT, \tag{6.21}$$

and free energy differences can therefore be obtained from MC simulations providing $E(T)$. When the cluster expansion is temperature dependent, it is important to note that, although the atomic identity flip probabilities are governed by the effective free energy Hamitonian $F(\sigma, T)$, the quantity $E(T)$ is obtained by averaging the microscopic energy

$$E(\sigma, T) = F(\sigma, T) - T\partial F(\sigma, T)/\partial T \tag{6.22}$$

over the sampled configurations σ. Thermodynamic relationships ideally suited for alloy free energy determination from grand-canonical MC simulations can be found in van de Walle and Asta [71].

Figure 6.5a shows a phase diagram obtained by combining first-principles calculations, the cluster-expansion formalism, and MC simulations, an approach that offers the advantage of handling, in a unified framework, both ordered phases (with potential thermal defects) and disordered phases (with potential short-range order).

Figure 6.5a demonstrates how the inclusion of the effect of lattice vibrations in *ab initio* calculations dramatically improves the agreement between calculated and experimentally determined phase diagrams. Figure 6.5b illustrates that *ab initio*

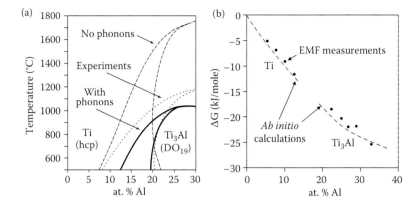

FIGURE 6.5 (a) Comparison between an experimentally assessed phase diagram [85] and the *ab initio* phase diagram calculated including or excluding the effect of lattice vibrations. (b) Comparison between calculated free energies with the ones obtained from electromotive force (EMF) measurements [86].

predictions of free energies that include both configurational and vibrational effects can provide accuracy comparable with the scatter in experimental determinations of the same quantity via electromotive force measurements. A predicted phase diagram with a more complex topology is given in Figure 6.6, along with the pattern of atomic ordering corresponding to the ordered phase.

The usefulness of the cluster-expansion formalism extends beyond the thermodynamics of bulk phases. Figure 6.4b shows a calculated equilibrium composition profile across a coherent interface between two regions of different doping levels in the Samarium-doped Ceria system. Due to the considerable width of the interface and the long-range nature of the electrostatic interactions present in this system, obtaining this result requires a simulation involving about 70,000 atoms, which would clearly be beyond the reach of brute-force *ab initio* calculations.

The applications of the methods described above were facilitated by employing the Alloy Theoretic Automated Toolkit (ATAT), which is a set of open-source programs for performing *ab initio* thermodynamics calculations [12,35,71–73].

6.4.4 FREE ENERGY OF PHASES WITH DILUTE DISORDER

The calculation of configurational free energies for alloy phases with "dilute" compositions is considerably simpler. For phases containing a sufficiently dilute concentration of defects (e.g., solute elements for a solid solution, or antisite defects for an intermetallic compound), such that defect interactions can be safely neglected, the free energy can be written in the following approximate form:

$$F = F_0 + \sum_i \sum_\alpha x_i^\alpha \left(\Delta F_i^\alpha + k_B T \ln x_i^\alpha \right), \tag{6.23}$$

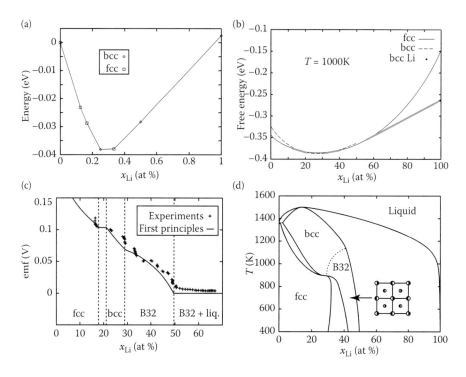

FIGURE 6.6 *Ab initio* thermodynamics of the Cu–Li system [60]. (a) Ground-state convex hull (merging bcc and fcc superstructures). (b) Calculated free energy as a function of composition including both configurational, vibrational, and electronic contributions. Nonconfigurational contributions are essential to accurately predict the numerous phase transitions between bcc- and fcc-based phases. Two-phase equilibria are marked by thick gray tie-lines. (c) Calculated electromotive force (emf) inferred from the calculated free energy in (b) and compared with experimental data. (The solid–liquid Li free energy difference was left as a parameter adjusted to provide the best agreement.) (d) Calculated phase diagram obtained via tangent constuctions as in (b) over a range of temperatures. The liquid phase excess free enegy is taken from the COST database [87]. Note that most of the ground states from (a) disorder at low temperature and are not visible in the phase diagram. Interestingly, neither the magnitude of each ground state's formation energy nor the sharpness of the vertex where a ground state touches the hull bear any correlation with a ground state's disordering temperature. The B32 ground state at 50% (shown in inset) does not have the most negative formation energy and barely breaks the hull, yet it exhibits tremendous stability due to entropy-driven stabilization. The dashed line marks the location of a second-order transition, determined by locating a peak in the heat capacity.

where F_0 is the free energy of the nondefected phase (pure element for a dilute solid solution or stoichiometric compound for an intermetallic phase), the sums are over the sites i of the unit cell and the different defect types α, x_i^α is the site concentration of defect α, and ΔF_i^α is the defect free energy of formation, including vibrational and electronic entropy contributions. When alloy phases are sufficiently dilute, the calculation of configurational free energies and phase diagrams can thus be reduced to the computation of the symmetry-distinct defect formation free energies ΔF_i^α entering Equation 6.23. Such calculations can be accurately performed through the use of supercell geometries within standard first-principles codes. Recent examples illustrating the application of this formalism for the first-principles calculation of the solvus boundaries in Al–Sc and Al–Si can be found in works of Ozoliņš and Asta [74] and Ozoliņš, Sadigh, and Asta [75], respectively. For both systems, vibrational contributions to the calculated solute free energies were found to lead to sizeable corrections to calculated phase-boundary temperatures. The formalism has also been applied recently to investigations of solute partitioning in multicomponent two-phase alloy systems [76,77].

The formalism outlined in the preceding paragraph has also found numerous applications in first-principles calculations of equilibrium vacancy and anti-site-defect concentrations, as well as solute site-selection preferences in multicomponent intermetallic compounds [78–83]. In such applications, the free energy given in Equation 6.23 is more commonly expressed as a grand potential, written in terms of solute chemical potentials rather than the site fractions. In this formalism, equilibrium-defect concentrations can be conveniently derived from the derivative of the grand potential with respect to the chemical potentials.

6.5   CONCLUSION

First-principles thermodynamic modeling of alloys typically proceed by employing a hierarchy of modeling techniques targeted at handling a specific type of thermal fluctuations. At the lowest level, electronic excitations can be modeled within standard finite-temperature DFT, while lattice vibrations can be treated in the quantum or classical regime within the quasiharmonic approximation. At the next level, configurational excitations are described via a lattice model in conjunction with a so-called temperature-dependent cluster expansion that conveniently summarizes the effective interactions between nearby atoms that implicitly account for electronic, vibrational, and relaxation effects via a technique called the "coarse-graining" of the "fast" degrees of freedom of the partition function.

The appeal of this approach is that the cluster expansion can be used to efficiently search for the ground states of the alloy system (which will typically determine the phases present in the phase diagram) and to carry out MC simulations to determine to phase diagram and/or any macroscopic thermodynamic property of interest. This formalism thus encompasses, at the atomic level, dilute or concentrated point defects in ordered phases as well as short-range order in solid solutions.

ACKNOWLEDGMENTS

The author acknowledges support from the NSF under grant DMR-0907669 and wishes to thank Mark Asta, Greg Pomrehn, and Ljubomir Miljacic for helpful input.

REFERENCES

1. P. Hohenberg and W. Kohn.1964. Inhomogeneous electron gas. *Physical Review* 136: B864.
2. W. Kohn and L. J. Sham. 1965. Self-consistent equations including exchange and correlation effects. *Physical Review* 140:A1133.
3. R. M. Dreizler and E. K. U. Gross. 1990. *Density functional theory: An approach to the quantum many-body problem.* Berlin: Springer-Verlag.
4. M. C. Payne, M. P. Teter, D. C. Allan, T. A. Arias, and J. D. Joannopoulos. 1992. Iterative minimization techniques for /it ab initio total-energy calculations: Molecular dynamics and conjugate gradients. *Review of Modern Physics* 64:1045.
5. F. Ducastelle. 1991. *Order and Phase Stability in Alloys.* New York: Elsevier Science.
6. D. de Fontaine. 1994. Cluster approach to order-disorder transformation in alloys. *Solid State Physics* 47:33.
7. G. M. Stocks, D. M. C. Nicholson, W. A. Shelton, B. L. Gyorffy, F. J. Pinski, D. D. Johnson, J. B. Staunton, P. E. A. Turchi, and M. Sluiter. 1994. First principles theory of disordered alloys and alloy phase stability. In *NATO ASI on statics and dynamics of alloy phase transformation*, vol. 319, eds. Turchi and Gonis, p. 305. New York: Plenum Press.
8. A. Zunger. First principles statistical mechanics of semiconductor alloys and intermetallic compounds. 1994. In *NATO ASI on statics and dynamics of alloy phase transformation*, vol. 319, eds. Turchi and Gonis, p. 361. New York: Plenum Press.
9. P. E. A. Turchi. 1995. First principles theory of disordered alloys and alloy phase stability. In *Intermetallic compounds: Principles and practice*, vol. 1, eds. J. H. Westbrook and R. L. Fleischer, p. 21. New York: John Wiley.
10. G. Ceder, A. van der Ven, C. Marianetti, and D. Morgan. 2000. First-principles alloy theory in oxides. *Modelling and Simulation in Materials Science and Engineering* 8:311.
11. M. Asta, V. Ozolins, and C. Woodward. 2001. A first-principles approach to modeling alloy phase equilibria. *JOM—Journal of the Minerals Metals and Materials Society* 53:16.
12. A. van de Walle and G. Ceder. 2002. Automating first-principles phase diagram calculations. *Journal of Phase Equilibria* 23:348.
13. A. V. Ruban and I. A. Abrikosov. 2008. Configurational thermodynamics of alloys from first principles: Effective cluster interactions. *Report on Progress in Physics*, 71:046501.
14. Z. K. Liu. 2009. First-principles calculations and calphad modeling of thermodynamics. *Journal of Phase Equilibrium and Diffusion* 30:517.
15. D. Frenkel and B. Smit. 2002. *Understanding molecular simulation: From algorithms to applications*, 2nd ed. San Diego: Academic Press, San Diego.
16. D. Alfé, G. A. de Wijs, G. Kresse, and M. J. Gillan. 2000. Recent developments in ab initio thermodynamics. *International Journal of Quantum Chemistry* 77:871–879.
17. D. Alfé, M. J. Gillan, and G. D. Price. 2002. Ab initio chemical potentials of solid and liquid solutions and the chemistry of the earth's core. *Journal of Chemical Physics* 116:7127–7136.
18. N. D. Mermin. 1965. Thermal properties of the inhomogeneous electron gas. *Physical Review* 137:A1441.
19. C. Wolverton and A. Zunger. 1995. First-principles theory of short-range order, electronic excitations, and spin polarization in Ni–V and Pd–V alloys. *Physical Review B* 52(12):8813.
20. R. O. Jones and O. Gunnarsson. 1989. The density functional formalism, its applications and prospects. *Review of Modern Physics* 61(3):689.
21. F. Kormann, A. Dick, B. Grabowski, B. Hallstedt, T. Hickel, and J. Neugebauer. 2008. Free energy of bcc iron: Integrated ab initio derivation of vibrational, electronic, and magnetic contributions. *Physical Review B* 78:033102.

22. A. A. Maradudin, E. W. Montroll, and G. H. Weiss. 1971. *Theory of lattice dynamics in the harmonic approximation*, 2nd ed. New York: Academic Press.

23. A. van de Walle and G. Ceder. 2002. The effect of lattice vibrations on substitutional alloy thermodynamics. *Review of Modern Physics* 74:11.

24. S. Baroni, P. Giannozzi, and A. Testa. 1987. Green's-function approach to linear response in solids. *Physical Review Letters* 58(18):1861.

25. P. Giannozzi, S. de Gironcoli, P. Pavone, and S. Baroni. 1991. Ab initio calculation of phonon dispersions in semiconductors. *Physical Review B* 43:7231.

26. X. Gonze, D. C. Allan, and M. P. Teter. 1992. Dielectric tensor, effective charges, and phonons in α-quartz by variational density-functional perturbation theory. *Physical Review Letters* 68:3603.

27. X. Gonze. 1997. First-principles responses of solids to atomic displacements and homogeneous electric fields: Implementation of a conjugate-gradient algorithm. *Physical Review B* 55:10337.

28. X. Gonze and C. Lee. 1997. Dynamical matrices, born effective charges, dielectric permittivity tensors, and interatomic force constants from density-functional perturbation theory. *Physical Review B* 55:10355.

29. J. T. Devreese and P. Van Camp, eds. 1985. *Electronic structure, dynamics and quantum structural properties of condensed matter*. New York: Plenum.

30. S. de Gironcoli. 1995. Lattice dynamics of metals from density-functional perturbation theory. *Physical Review B* 51:6773.

31. U. V. Waghmare. 1996. *Ab initio statistical mechanics of structural phase transitions (lattice dynamics, piezoelectric, ferroelectric)*. PhD thesis, Yale University.

32. V. Ozoliņš. 1996. *Structural and vibrational properties of transition metal systems from ab initio electronic-structure calculations*. PhD thesis, Royal Institute of Technology, Stockholm, Sweden.

33. W. Zhong, D. Vanderbilt, and K. M. Rabe. 1994. Phase transition in $BaTiO_3$ from first principles. *Physical Review Letters* 73(13):1861.

34. J. M. Sanchez, F. Ducastelle, and D. Gratias. 1984. Generalized cluster description of multicomponent systems. *Physica* 128A:334.

35. A. van de Walle. 2009. Multicomponent multisublattice alloys, nonconfigurational entropy and other additions to the Alloy Theoretic Automated Toolkit. *Calphad Journal* 33:266.

36. A. van der Ven, M. K. Aydinol, G. Ceder, G. Kresse, and J. Hafner. 1998. First-principles investigation of phase stability in Li_xCOO_2. *Physical Review B* 58:2975.

37. A. van der Ven and G. Ceder. 2005. Vacancies in ordered and disordered binary alloys treated with the cluster expansion. *Physical Review B* 71:054102.

38. P. D. Tepesch, G. D. Garbulsky, and G. Ceder. 1995. Model for configurational thermodynamics in ionic systems. *Physical Review Letters* 74:2272.

39. A. van de Walle. 2008. A complete representation of structure-property relationships in crystals. *Nature Materials* 7:455.

40. D. B. Laks, L. G. Ferreira, S. Froyen, and A. Zunger. 1992. Efficient cluster expansion for substitutional systems. *Physical Review B* 46:12587.

41. C. Wolverton and A. Zunger. 1995. An Ising-like description of structurally-relaxed ordered and disordered alloys. *Physical Review Letters* 75:3162.

42. A. van de Walle and D. Ellis. 2007. First-principles thermodynamics of coherent interfaces in samarium-doped ceria nanoscale superlattices. *Physical Review Letters* 98:266101.

43. G. D. Garbulsky and G. Ceder. 1994. Effect of lattice vibrations on the ordering tendencies in substitutional binary alloys. *Physical Review B* 49:6327.

44. G. D. Garbulsky. 1996. *Ground-state structures and vibrational free energy in first-principles models of substitutional-alloy thermodynamics*. PhD thesis, Massachusetts Institute of Technology.

45. G. J. Ackland. 1994. Vibrational entropy of ordered and disordered alloys. In *Alloy modelling and design*, eds. G. Stocks and P. Turchi, p. 149. Pittsburgh, PA: The Minerals, Metals and Materials Society.

46. J. D. Althoff, D. Morgan, D. de Fontaine, M. Asta, S. M. Foiles, and D. D. Johnson. 1997. Vibrational spectra in ordered and disordered Ni_3Al. *Physical Review B* 56:R5705.

47. L. Anthony, J. K. Okamoto, and B. Fultz. 1993. Vibrational entropy of ordered and disordered Ni3Al. *Physical Review Letters* 70(8):1128.

48. L. Anthony, L. J. Nagel, J. K. Okamoto, and B. Fultz. 1994. Magnitude and origin of the difference in vibrational entropy between ordered and disordered Fe_3Al. *Physical Review Letters* 73(22):3034.

49. L. J. Nagel. 1996. *Vibrational entropy differences in materials*. PhD thesis, California Institute of Technology.

50. V. L. Moruzzi, J. F. Janak, and K. Schwarz. 1988. Calculated thermal properties of metals. *Physical Review B* 37:790.

51. M. Asta, R. McCormack, and D. de Fontaine. 1993. Theoretical study of alloy stability in the Cd–Mg system. *Physical Review B* 48(2):748.

52. J. M. Sanchez, J. P. Stark, and V. L. Moruzzi. 1991. First-principles calculation of the Ag–Cu phase diagram. *Physical Review B* 44(11):5411.

53. C. Colinet, J. Eymery, A. Pasturel, A. T. Paxton, et al. 1994. A first-principles phase stability study on the Au–Ni system. *Journal of Physics: Condensed Matter* 6:L47.

54. G. D. Garbulsky and G. Ceder. 1996. Contribution of the vibrational free energy to phases stability in substitutional alloys: Methods and trends. *Physical Review B* 53:8993.

55. P. D. Tepesch, A. F. Kohan, G. D. Garbulsky, and G. Ceder, et al. 1996. A model to compute phase diagrams in oxides with empirical or first-principles energy methods and application to the solubility limits in the CaO–MgO system. *Journal of the American Ceramic Society* 49:2033.

56. V. Ozoliņš, C. Wolverton, and A. Zunger. 1998. First-principles theory of vibrational effects on the phase stability of Cu–Au compounds and alloys. *Physical Review B* 58(10):R5897.

57. A. van de Walle and G. Ceder. 2000. First-principles computation of the vibrational entropy of ordered and disordered Pd_3V. *Physical Review B* 61:5972l.

58. E. Wu, G. Ceder, and A. van de Walle. 2003. Using bond-length-dependent transferable force constants to predict vibrational entropies in Au–Cu, Au–Pd, and Cu–Pd alloys. *Physical Review B* 67:134103.

59. B. P. Burton and A. van de Walle. 2006. First principles phase diagram calculations for the system NaCl-KCl: The role of excess vibrational entropy. *Chemical Geology* 225:222.

60. A. van de Walle, Z. Moser, and W. Gasior. 2004. First-principles calculation of the Cu–Li phase diagram. *Archives of Metallurgy and Materials* 49:535.

61. O. Adjaoud, G. Steinle-Neumann, B. P. Burton, and A. van de Walle. 2009. First-principles phase diagram calculations for the HfC–TiC, ZrC–TiC, and HfC–ZrC solid solutions. *Physical Review B* 80:134112.

62. B. Burton, A. van de Walle, and U. Kattner. 2006. First principles phase diagram calculations for the wurtzite-structure systems AlN–GaN, AlN–InN and GaN–InN. *Journal of Applied Physics* 100:113528.

63. M. Stone. 1974. Cross-validatory choice and assessment of statistical predictions. *Journal of the Royal Statistical Society Series B Methodological* 36:111.

64. K.-C. Li. 1987. Asymptotics optimality for c_p, c_l, cross-validation and generalized cross-validation: Discrete index set. *Annals of Statistics* 15:956.

65. A. van der Ven and G. Ceder. 2005. First principles calculation of the interdiffusion coefficient in binary alloys. *Physical Review Letters* 94:045901.

66. A. van der Ven, G. Ceder, M. Asta, and P. D. Tepesch. 2001. First principles theory of ionic diffusion with non-dilute carriers. *Physical Review B* 64:184307.
67. K. Binder and D. W. Heermann. 1988. *Monte Carlo simulation in statistical physics.* New York: Springer-Verlag.
68. M. E. J. Newman and G. T. Barkema. 1999. *Monte Carlo methods in statistical physics.* Oxford: Clarendon Press.
69. S. Curtarolo, D. Morgan, K. Persson, J. Rodgers, and G. Ceder. 2003. Predicting crystal structures with data mining of quantum calculations. *Physical Review Letters* 91:135503.
70. D. Morgan, J. Rodgers, and G. Ceder. 2003. Automatic construction, implementation and assessment of Pettifor maps. *Journal of Physics: Condensed Matter* 15:4361.
71. A. van de Walle and M. Asta. 2002. Self-driven lattice-model Monte Carlo simulations of alloy thermodynamic properties and phase diagrams. *Modelling and Simulation in Materials Science and Engineering* 10:521.
72. A. van de Walle. *The alloy theoretic automated toolkit*, http://www.its.caltech.edu/~avdw/atat/.
73. A. van de Walle, M. Asta, and G. Ceder. 2002. The Alloy Theoretic Automated Toolkit: A user guide. *CALPHAD Journal* 26:539.
74. V. Ozoliņš and M. Asta. 2001. Large vibrational effects upon calculated phase boundaries in Al–Sc. *Physical Review Letters* 86:448.
75. V. Ozoliņš, B. Sadigh, and M. Asta. 2005. Effects of vibrational entropy on the Al–Si phase diagram. *Journal of Physics: Condensed Matter* 17:1–14.
76. E. A. Marquis, D. N. Seidman, M. Asta, C. Woodward, and V. Ozoliņš. 2003. Equilibrium Mg segregation at Al/Al$_3$Sc heterophase interfaces on an atomic scale: Experiments and computations. *Physical Review Letters* 91:036101.
77. R. Benedek, A. van de Walle, S. S. A. Gerstl, M. Asta, D. N. Seidman, and C. Woodward. 2005. Partitioning of impurities in multi-phase Ti–Al alloys. *Physical Review B* 71: 094201.
78. C. L. Fu, Y. Y. Ye, M. H. Yoo, and K. M. Ho. 1993. Equilibrium point defects in intermetallics with the B2 structure—NiAl and FeAl. *Physical Review B.* 48:6712–6715.
79. C. L. Fu and G. S. Painter. 1997. Point defects and the binding energies of boron near defect sites in Ni$_3$Al: A first-principles study. *Acta Materialia* 45:481–488.
80. C. Woodward and S. Kajihara. 1999. Density of thermal vacancies in γ-TiAl-M, M=Si, Cr, Nb, Mo, Ta or W. *Acta Materialia* 47:3793–3798.
81. C. Woodward, M. Asta, G. Kresse, and J. Hafner. 2001. Density of constitutional and thermal defects in Li$_2$ Al$_3$Sc. *Physical Review B* 63:094103.
82. M. Fahnle. 2002. Atomic defects and diffusion in intermetallic compounds: The impact of the ab initio electron theory. *Defects and Diffusion Forum* 203-2:37–46.
83. R. Drautz, I. Schultz, F. Lechermann, and M. Fahnle. 2003. Ab-initio statistical mechanics for ordered compounds: Single-defect theory vs. cluster-expansion techniques. *Physica Status Solidi B* 240:37–44.
84. P. E. Turchi and A. Gonis, editors. 1994. *NATO ASI on statics and dynamics of alloy phase transformation*, vol. 319. New York: Plenum Press.
85. I. Ohnuma, Y. Fujita, H. Mitsui, K. Ishikawa, R. Kainuma, and K. Ishida. 2000. Phase equilibria in the Ti–Al binary system. *Acta Materialia* 48:3113.
86. V. V. Samokhva, P. A. Poleshchuk, and A. A. Vecher. 1971. Thermodynamic properties of aluminum–titanium and aluminum–vanadium alloys. *Russ. J. Phys. Chem.*, 45:1174.
87. N. Saunders. 1998. COST 537 Thermochemical database for light metal alloys. In *COST 537 Thermochemical database for light metal alloys*, volume 2, eds. I. Ansara, A. T. Dindail, and M. H. Rand, p. 168. European Communities.

7 Nonlinear Finite Element Model for the Determination of Elastic and Thermal Properties of Nanocomposites

Paul Elsbernd and Pol Spanos

CONTENTS

7.1   INTRODUCTION

As nanotechnology has blossomed over the past two decades, countless applications for a variety of nanoparticles have been demonstrated or proposed. However, the most commonly proposed application of nanoparticles is their use in composite materials. While the choice of particle highly depends on the functionality sought for the composite, the most widely studied one is the carbon nanotube reinforced composite.

Carbon nanotubes offer exceptional mechanical, thermal, and electrical properties at an almost insignificant weight on the macroscale. Theoretically, single-walled carbon nanotubes (SWCNTs) exhibit elastic moduli of the order of 1 TPa and fracture strains of 10% to 30% [1–4]. These values indicate that SWCNTs have elastic moduli that are three times that of carbon fibers and five times that of steel at one-sixth of the weight [5]. In addition to their outstanding mechanical properties, carbon nanotubes exhibit exceptional thermal and electrical properties. The theoretical thermal conductivity of SWCNTs is usually reported as 6000 W/mK at room temperature [6], approximately three times the thermal conductivity of diamond and approximately 10^4 that of most polymers [7]. This value is in fact highly temperature-dependent, as reported by Grujicic et al. [8], ranging from values as high as 12,000 W/mK at 100K to about 2000 W/mK at 300K. Carbon nanotubes are also found to have exceptional electrical conductivity, on the order of 10^4~10^7 S/m, approximately 20 decades higher than that of most polymers [7,9]. These properties, along with an aspect ratio of approximately 1000, make SWCNTs ideal candidates for composite reinforcement.

Their mechanical properties make them ideal candidates for use as reinforcing agents, either alone or as filler with other reinforcements, such as carbon or glass fibers. Their thermal properties make them desirable for thermal management applications and their electrical properties make them ideal for countless other composite applications such as electromagnetic shielding, preventing electrostatic charging, and damage sensing [10,11]. Collectively these properties make nanotube composites of extreme interest to diverse fields including the aerospace, automotive, medical, and military industries.

Due to the interest in these new composites, a need to predict their physical properties has developed as well. Several researchers have sought to characterize these new composites experimentally. Skakalova et al. [12] measured the electrical conductivity, stress–strain characteristics, and Raman spectra of SWCNT composites of various volume fractions. Zhu et al. [13,14], Li et al. [15], and Sun et al. [16] measured the elastic moduli and stress–strain characteristics. In addition, Huxtable et al. [17], Xu et al. [18], Wang et al. [19], Moisala et al. [9], and Cai and Song [20] all focused on the thermal properties of SWCNT-composites.

While experiments have proven the utility of SWCNTs in composite materials, the ability to accurately predict properties of nanocomposites using a versatile theoretical model is crucial since the sheer number of different possible composites limits the experimental options. Not only are there countless different polymers to be used as the matrix material of the composite, but the effect of different volume fractions of nanotubes also needs to be studied for each composite. Because of this need, many researchers have sought to model these new nanocomposites. Tserpes et al. [21] developed a multiscaled modeling approach to test the tensile behavior of composites in which nanotubes are modeled independently beam elements and integrated into a representative volume element (RVE), which can then be solved using finite element analysis (FEA). Li and Chou [22] presented another multiscale approach to test compressive behavior. Xio and Gillespie, Jr. [23] reported a nanomechanics model for predicting elastic properties of SWCNTs as composite reinforcement and then used a micromechanics model to calculate the elastic properties of the composite as a whole. Shi et al. [24] used a micromechanics model as well and include waviness and agglomeration in their calculation of elastic properties. Song and Youn [25] calculated the effective thermal conductivity of nanotube/polymer composites using the control volume finite element method (FEM). All of these models use a single CNT that traverses the entire RVE. They assume that the effective properties of the composite as a whole can be calculated based on a tiny RVE with a single nanotube due to computational simplicity. This assumption is not made by the current model and this attribute is one of the major advantages of the proposed model.

In addition to the models that calculate effective properties based upon a single nanotube RVE, there are several models that do not make this assumption. Xue [26] presents a numerical method based on Maxwell's theory to calculate the thermal conductivity of nanotube/oil and nanotube/decene composites. Bagchi and Nomura [27] approximate nanotubes as spheroidal inclusions and use effective medium theory to calculate the thermal conductivity of aligned nanotube/polymer composites. One of the most versatile models was presented by Zhang and Tanaka [28]. Using the hybrid boundary node method in conjunction with the fast mulitpole method, they calculate the thermal properties of CNT composites while including waviness and alignment for a variety of different nanotube configurations. Odegard et al. [29] model the nanotubes as an effective continuum fiber using the equivalent-continuum modeling method along with a micromechanics approach to calculate mechanical properties of CNT/polymer composites. Seidel and Lagoudas [30] model nanotubes using a composite cylinders approach and use the Mori-Tanaka methods to obtain effective elastic properties for the composite as a whole. They consider both aligned and randomly oriented nanotubes and also consider nanotube bundling. Finally, Li and Chou present a model for damage sensing by calculating electrical properties of CNT/glass-fiber composites using the FEM.

These efforts indicate many of the great leaps made in understanding these new nanocomposites. However, to date limited attention has been devoted to model the nonlinear properties of either CNTs or the polymer matrix for a realistic composite size. Elastic nonlinearity must be considered to accurately model the stress–strain relationship of these composites. Further, the effect the nonlinear thermal conductivity of SWCNTs on the effective thermal conductivity of the composite is of interest. The proposed model studies the effects of these two types of nonlinearity on the physical properties of SWCNT reinforced polymer composites using a novel method described in Spanos and Esteva [31].

7.2 METHODOLOGY

The proposed nonlinear FEM model is based upon the embedded fiber finite element method (EFFEM), first applied to nanocomposites in Spanos and Esteva [31]. In this method, a RVE of a user-specified volume fraction of SWCNTs is developed in three distinct steps. First, the nanotube geometry throughout the RVE is generated in a random manner, incorporating nanotube waviness, entanglement, and CNT diameter/length distributions. The RVE thickness is assumed to be much smaller than the width or diameter, and thus, the geometry is only generated in two dimensions. By making this assumption, the problem becomes a two-dimensional plane-stress problem, greatly reducing the computational effort ultimately required. Once the geometry is generated, a square finite element mesh size is chosen and the nanotubes are divided into two new, shorter nanotubes when they cross the boundary from one element into another. This process is known as partitioning and is necessary to perform the final step of the method, fiber embedment. After partitioning, all nanotubes lie completely inside individual elements. This fact allows the direct addition of their elastic or thermal properties to those of the parent element in a finite element stiffness matrix sense. This direct combination of properties is known as fiber embedment. These three steps form the foundation upon which a nonlinear finite element solution approach can be used to calculate a nonlinear stress–strain curve and thermal conductivity for any arbitrary SWCNT/polymer composite.

7.3   RVE GEOMETRY GENERATION

Unlike the majority of computational models that currently exist for SWCNT/polymer composites, the EFFEM does not build up the composite for analysis based upon smaller RVEs with a single nanotube that traverses the entire RVE. Rather, by directly incorporating nanotube properties into the matrix, a complex nanotube geometry where some nanotubes overlap and curve can be modeled. With this in mind, a scheme for generating complex nanotube geometries, including nanotube morphology and waviness, is developed as in Spanos and Esteva [31]. In this scheme, each nanotube is generated individually. The length and diameter of each nanotube are randomly generated from Weibull and lognormal distributions, respectively. In order to generate a random value from a distribution, the inverse transformation method (ITM) is employed to isolate the random variable from the cumulative distribution function (CDF) of the distribution. To generate a random length, the CDF for a Weibull distribution, given by the equation

$$F(x) = 1 - e^{-(x/\alpha)\gamma} \tag{7.1}$$

is solved for x using the ITM to isolate a random number from this distribution, given by the equation

$$x = -\alpha[\ln(1 - u)]^{1/\gamma} \tag{7.2}$$

To generate a random diameter, the same approach is taken using a lognormal distribution with the CDF given by the equation

$$F(x) = \frac{1}{2} + \frac{1}{2}\,\mathrm{erf}\left[\frac{\ln x - u}{\sigma\sqrt{2}}\right] \tag{7.3}$$

The variable x is then isolated using the ITM to yield the equation

$$x = e^{\left(\sigma\sqrt{2}\,\mathrm{erfinv}[2u-1]+u\right)}, \tag{7.4}$$

where x is a random number from a lognormal distribution.

The Weibull distribution parameters in Equation 7.1 are taken from the experimental observations of Wang et al. [32]. They are related to the shape and scale parameters reported by Wang et al. [32] in the following manner:

$$\alpha = e^{\ln a/b}, \quad \gamma = b \tag{7.5}$$

The lognormal parameters are calculated from the data reported by Ziegler et al. [33]. Additionally, nanotube waviness is incorporated as a linear function of nanotube length based on SEM and TEM images of nanotubes in polymer matrices [13,34]. Each nanotube is generated in ten segments and the angle between each segment varies uniformly between $-\dfrac{\Theta_{max}}{2}$ and $\dfrac{\Theta_{max}}{2}$, where Θ_{max} is some angular upper limit. The upper limit of the angle is determined by the nanotube length, l. Due to insufficient experimental data on the subject, a linear model of waviness is assumed and thus the function for the maximum angle of deviation between any two segments for a particular nanotube is given as

$$\Theta_{max} = \frac{180°}{\text{upper length limit}} \times l \tag{7.6}$$

Using this method, complex RVE geometries can be generated. One particular such geometry, called a *microstructure*, is shown in Figure 7.1. For greater detail on the method of geometry generation, see Spanos and Esteva [31].

7.3.1 FIBER PARTITIONING

After generating the geometry of a microstructure, the next step is to partition the nanotubes that cross from one element to another. This step is necessary because, in order to combine the stiffness matrices of the fibers within an element with the stiffness matrix of the polymer matrix for that element, the reinforcing fiber must lie completely within the element. To accomplish this partitioning, the coordinate pairs for every nanotube segment are examined to see whether they lie in the same element. If they do not, a line equation is generated for the segment. This equation is compared with the equations for the lines of the RVE mesh grid and the points of intersection are found. The nanotube is then divided at these points. For a more detailed description of the partitioning process, see Spanos and Esteva [31]. Visualizations of the partitioning process along with a partitioned view of an actual microstructure section are presented in Figures 7.2 and 7.3, respectively.

7.3.2 EMBEDDED FIBER METHOD

Once the nanotube geometry has been generated and the fibers are partitioned, the final step to prepare the RVE for traditional FEA is to embed the nanotubes in the polymer matrix. The process of embedding, as described above, is to directly add the nanotubes' stiffness properties to that of the surrounding matrix. This is

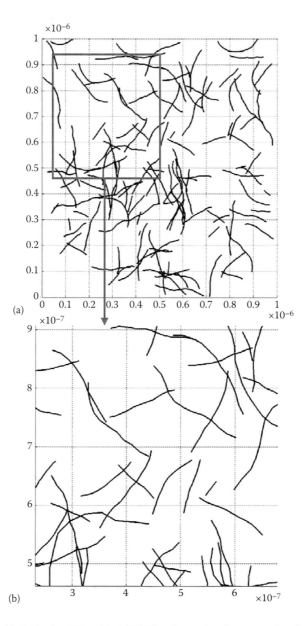

FIGURE 7.1 (a) A single RVE with 0.0889% volume fraction nanotubes is depicted. (b) A close-up view of one section of the RVE is shown. The waviness of the nanotubes is easily observed. Longer nanotubes are wavier than shorter nanotubes to simulate the physical phenomenon.

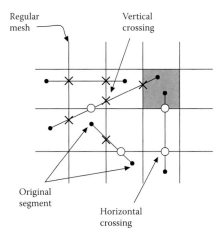

FIGURE 7.2   Visualization of partitioning process.

accomplished for each element in the RVE finite element mesh. Regardless of the kind of problem to be solved, be it elastic, thermal, electrical, etc., the same embedding process is applied to the RVE.

In general there are three steps to the embedding process. The first step is to use the traditional finite element formulation to derive the stiffness matrix of the

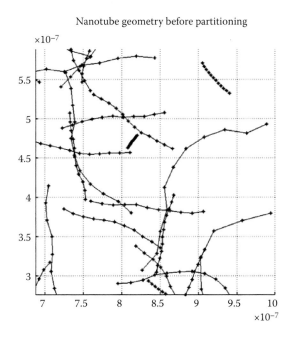

FIGURE 7.3   Section of partitioned microstructure.

polymer element. This formulation can be found in any basic finite element text and is given by

$$[K]^e_{mat} = h \int_{A^e} [B]^e[D][B]^{Te} \, da = h \iint_{\Omega^e} [B]^e[D][B]^{Te} |J| \, ds \, dt, \tag{7.7}$$

where the strain matrix B, the material matrix D, and the Jacobian matrix J are given, respectively, by

$$B^e = \begin{bmatrix} du/dx \\ dv/dy \\ du/dy + dv/dx \end{bmatrix} \tag{7.8}$$

$$= \begin{bmatrix} dN_1/dx & 0 & dN_2/dx & 0 & dN_3/dx & 0 & dN_4/dx & 0 \\ 0 & dN_1/dy & 0 & dN_2/dy & 0 & dN_3/dy & 0 & dN_4/dy \\ dN_1/dy & dN_1/dx & dN_2/dy & dN_2/dx & dN_3/dy & dN_3/dx & dN_4/dy & dN_4/dx \end{bmatrix}$$

$$D = \frac{E}{1-v^2} \begin{bmatrix} 1 & v & 0 \\ v & 1 & 0 \\ 0 & 0 & \dfrac{1-2v}{2} \end{bmatrix} \tag{7.9}$$

$$J = \begin{bmatrix} dx/dr & dy/dr \\ dx/ds & dy/ds \end{bmatrix} \tag{7.10}$$

Equations 7.9 and 7.10 represent the matrices for the two-dimensional plane-stress FEA problem, which is what the current model is solving. In Equation 7.9, E is the modulus of elasticity of the matrix material and v is Poisson's ratio. For a four-noded quadrilateral element with two degrees of freedom per node, there are four shape functions used to map between the local parametric coordinates r and s and the global coordinates x and y. These shape functions are defined by

$$[N] = [N_1 N_2 N_3 N_4]^T \tag{7.11}$$

$$N_1 = \frac{1}{4}(1-r)(1-s) \tag{7.12}$$

$$N_2 = \frac{1}{4}(1+r)(1-s) \tag{7.13}$$

$$N_3 = \frac{1}{4}(1+r)(1+s) \tag{7.14}$$

$$N_4 = \frac{1}{4}(1-r)(1-s) \tag{7.15}$$

These shape functions allow the mapping of local coordinates to global coordinates using the equation

$$x(r,s) = [N][u] \text{ and } y(r,s) = [N][v], \tag{7.16}$$

where u and v are the x and y coordinates of the element nodes, respectively. In other words, by using the shape functions in Equations 7.12 through 7.15, the x and y positions corresponding to any parametric point (r,s) can be found readily using Equation 7.16.

Furthermore, the local derivatives are mapped to global derivatives using the Jacobian from Equation 7.16 and the derivates of the shape functions. This mapping is given by the equation

$$\begin{bmatrix} dN/dx \\ dN/dy \end{bmatrix} = [J]^{-1} \begin{bmatrix} dN/dr \\ dN/ds \end{bmatrix} \tag{7.17}$$

With these derivatives and coordinates in hand, the strain matrix and stiffness matrix for an element are easily calculated. Due to the discrete nature of FEA, the integral of Equation 7.7 is determined numerically using Gaussian quadrature. For any particular element, the shape functions and the Jacobian are evaluated at each of the Gaussian quadrature points within the element and summed. For a two-dimensional square element, as is the case in the present model, these points are given as

$$\xi = \eta = \pm 1/\sqrt{3} \tag{7.18}$$

The stiffness matrix can now be written as

$$[K]_{\text{mat}} = h \sum_{i=1}^{n} \sum_{j=1}^{n} w_i w_j \left[B(s_j, r_j) \right] [D] \left[B(s_j, r_j) \right]^{\text{T}} |J| \tag{7.19}$$

These finite element formulations can be found in any standard FEM book, such as those by Akin [35], Desai and Abel [36], and Zienkiewicz et al. [37].

The second step is to calculate the stiffness matrix of each nanotube within the element. In the current model, the nanotubes are approximated as line elements. This approximation is accurate and computationally efficient as shown by Konrad and

Graovac [38]. With this approximation, the stiffness matrix of a one-dimensional fiber, expanded using zeros to two dimensions, is given as

$$K_{fib_L} = \frac{EA}{L} \begin{bmatrix} 1 & 0 & -1 & 0 \\ 0 & 0 & 0 & 0 \\ -1 & 0 & 1 & 0 \\ 0 & 0 & 0 & 0 \end{bmatrix} \tag{7.20}$$

The stiffness matrix is then transformed from local to global coordinates using a direction cosine matrix given by the equation

$$K_{fibG} = [C][K_{fibL}][C]^{\mathrm{T}}, \tag{7.21}$$

with

$$[C] = \begin{bmatrix} \cos t & \sin t & 0 & 0 \\ -\sin t & \cos t & 0 & 0 \\ 0 & 0 & \cos t & \sin t \\ 0 & 0 & -\sin t & \cos t \end{bmatrix}, \tag{7.22}$$

and then interpolated to the element nodes using a transformation matrix of quadrilateral shape functions, given by the equations

$$K_{fib} = [T][K_{fibG}][T]^{\mathrm{T}} \tag{7.23}$$

$$[T] = \begin{bmatrix} N_1(r_1,s_1) & 0 & N_2(r_1,s_1) & 0 & N_3(r_1,s_1) \\ 0 & N_1(r_1,s_1) & 0 & N_2(r_1,s_1) & 0 \\ N_1(r_2,s_2) & 0 & N_2(r_2,s_2) & 0 & N_3(r_2,s_2) \\ 0 & N_1(r_2,s_2) & 0 & N_2(r_2,s_2) & 0 \\ N_3(r_1,s_1) & 0 & N_4(r_2,s_1) & 0 \\ 0 & N_3(r_1,s_1) & 0 & N_4(r_1,s_1) \\ N_3(r_2,s_2) & 0 & N_4(r_2,s_2) & 0 \\ 0 & N_3(r_2,s_2) & 0 & N_4(r_2,s_2) \end{bmatrix} \tag{7.24}$$

The final step in the embedding process is to add the nanotube stiffness properties to those of the surrounding polymer for the element. This process is simple, accomplished by direct addition, and is described by the equation

$$[K]^e_{eqv} = [k]^e_{elem} + \sum_{i=1}^{n} [k]^i_{nt},\qquad(7.25)$$

where $[K]^e_{eqv}$ is the equivalent stiffness matrix, $[k]^e_{elem}$ is the stiffness matrix for the polymer for the element, and $\sum_{i=1}^{n} [k]^i_{nt}$ is the sum of the stiffness matrices of all the nanotubes in the element.

By approaching the problem in this manner, the need for complex, manual meshing around the nanotube-polymer matrix is eliminated. A simple, square finite element mesh is all that is necessary for the EFFEM and this automation of the model, without human intervention, directly follows from the process. This is the primary advantage of this approach. However, an important assumption is made in this regard. The EFFEM assumes perfect bonding between the nanotubes and the polymer matrix. While initial experiments indicated that this was not a valid assumption, recent findings have found this assumption justified in two ways. First, dramatic advancements have been achieved in interfacial bonding through chemical functionalization of the nanotube sidewalls and endcaps. Garg et al. report that even a high degree of sidewall functionalization will degrade the mechanical strength of SWCNTs by only 15% [39]. Demonstrating the effectiveness of functionalization, Zhu et al. [13] report that by utilizing acid treatment and fluorination of the nanotubes, they were able to achieve a 30% increase in elastic modulus of an epoxy composite with the addition of only 1% weight fraction (wt.) of SWCNTs. Second, conventional observation of reinforcing inclusions indicates that interfacial weakening often occurs, causing a detrimental effect on the load transfer properties of the composite. Esteva and Spanos [40] showed that for small volume fractions, $\leq 30\%$, the interface weakening effect of elliptical inclusions is not relevant for nanotube reinforced composites. Based on these two justifications, the EFFEM is deemed as a valid approach to modeling nanocomposites and this fact is exemplified in the results of the work reported herein.

7.4   NONLINEAR MODEL

After establishing the EFFEM as the framework in which to approach the problem of modeling a SWCNT reinforced polymer composite while incorporating many realistic and essential features of such a material, the next step is to determine how to model nonlinearity. The nonlinear mechanical properties of SWCNT reinforced composites for small strains are well observed in several experiments. Sun et al. [16], Zhu et al. [13], Skalova et al. [12], and Gojny et al. [34] all report clearly nonlinear stress–strain relationships for nanotube reinforced composites at various volume fractions. Since strain levels are small, the nonlinearity is dominated by material properties rather than geometric nonlinearity. The nonlinear thermal properties of composite are a less explored area. Xu et al. [18], Hong and Cai [41], Cai and Song [20], and Biercuk et al. [42] have all investigated this theme with mixed results.

However, for this nonlinear analysis, the deformation of the finite element mesh is not an issue; thus only material nonlinearity is important.

Since material nonlinearity is the sole concern, two different approaches are considered to incorporate the nonlinear properties both of the carbon nanotubes and the polymer matrix. The first is the incremental approach. By solving a linear problem in a series of small increments, the nonlinearity of the materials can be incorporated in the model through a recursive updating procedure. The second approach is iterative. In this approach, the full linear problem is solved repeatedly, again updating the properties of the nanotubes or the polymer after each step. The incremental approach is more versatile yet more computationally costly. It allows the user to recover intermediate solution values useful for calculations such as a stress–strain curve. The iterative approach is more stable and usually less computationally costly. However, the iterative approach provides only final solution values and therefore has limited applicability for certain applications. Both approaches are described in detail by Desai and Abel [36]. To decide which approach to use for the elastic analysis and which to use for the thermal analysis, an understanding of the mechanisms of nonlinearity in both carbon nanotubes and epoxies is needed.

7.4.1 NONLINEAR PROPERTIES OF CARBON NANOTUBES

The nonlinear elastic and thermal properties of carbon nanotubes are well-documented in the current literature, both theoretically and experimentally. For the nonlinear elastic properties, Tserpes et al. [22] presented a detailed finite element model of lone nanotubes using a structural mechanics approach. They identify the stress–strain relationship for armchair and zigzag SWCNTs. Additionally, Tserpes et al. [6] presented a progressive fracture model to calculate the stress–strain relationship for SWCNT's with perfect structures and for those with imperfections. Belytschko et al. [7] developed a molecular mechanics model to simulate nanotube fracture. They report the stress–strain curve for perfect and imperfect zigzag nanotubes as well as a curve for various different perfect nanotube structures. Natsuki et al. [43] describe a structural mechanics approach to modeling nanotube elastic properties. Finally, Meo and Rossi [5] presented a molecular-mechanics-based FEM for the prediction SWCNT elastic properties. Each of these models takes a slightly different approach, but they all identify ≈10% to 15% strain as the region where significant nonlinear behavior is first exhibited by SWCNTs. The stress–strain curve of Tserpes et al. [3] is shown in Figure 7.4.

The nonlinear thermal properties of SWCNTs have also been modeled, though less frequently than the elastic properties. Grujicic et al. [15] presented a molecular-dynamics-based model of the thermal conductivity of SWCNTs in which they investigate the nonlinear dependence of thermal conductivity on temperature. The strong dependence they report is depicted in Figure 7.5 and is the motivation for the nonlinear thermal analysis performed in this paper.

Kawamura et al. [44] also reported the temperature dependence of SWCNT thermal conductivity using a molecular dynamics simulation. Additionally, Xu et al. [18] reported this effect indirectly by discussing the temperature dependence of the thermal conductivity of composites reinforced w/ SWCNTs.

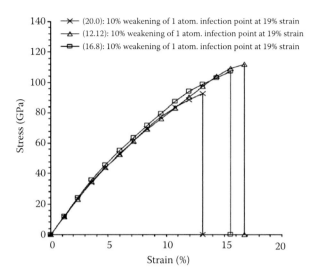

FIGURE 7.4 Stress–strain curves for SWCNTs. (Reprinted from Tserpes, K., Papanikos P., and Tsirkas, S., *Compos. B: Eng.*, vol. 37, 2006, pp. 662–669.)

Along with the theoretical modeling evidence presented above, experimental evidence corroborates the nonlinear thermal and elastic properties of SWCNTS. Yu et al. [45,46] performed tensile load testing using an AFM tip to document the nonlinear stress–strain relationship of SWCNTs. Their results show the onset of nonlinear behavior to occur at greater than 10% strain. Walters et al. [47] also report the lower bound of the yield strain of SWCNTs to be approximately 6%. For thermal

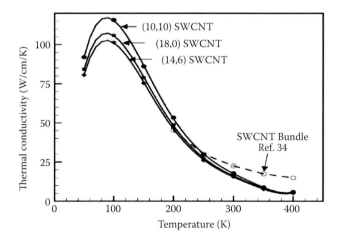

FIGURE 7.5 The nonlinear thermal properties of SWCNTs are shown in the analysis of Grujicic et al. (Reproduced from Grujicic, M., Cao, G., and Gersten, B., *Mater. Sci. Eng. B*, vol. 107, Mar. 2004, pp. 204–216.)

properties, Shaikh et al. [48] report a range of thermal conductivities for CNT ropes to range from 0.1–2000 W/mK.

7.4.2 NONLINEAR PROPERTIES OF EPOXIES

Epoxy composites exhibit nonlinear elastic properties at a much lower strain level. Zhu et al. [13,14] present stress–strain curves for neat epoxy in which nonlinearity is shown at 3% strain. Li et al. [15] present similar effects with a load-displacement curve of neat epoxy. Sun et al. [16] and Gojny et al. [34] also indicate 3% strain as the onset of nonlinearity in neat epoxy. Each author indicates epoxy failure at around 6% without reinforcement. To illustrate, the curve reported by Zhu et al. [13] is shown in Figure 7.6.

The significantly lower onset of nonlinear behavior and failure in the epoxy than in the lone nanotubes led to the conclusion that the nonlinear behavior of the nanotubes may be ignored and that of the epoxy must be taken into account for an accurate elastic model at small strains less than 6%.

In general, polymers are reported to have constant values of thermal conductivity. Xu et al. [18] report a constant thermal conductivity of ~0.2 W/mK for their PVDF polymer. Wang et al. [19] report a constant thermal conductivity of ~0.18 W/mK for the neat epoxy resin used in their experiments. Moisala et al. [16] report a constant value of ~0.255 W/mK for epoxy consisting of a bisphenol-A resin and an aromatic amine hardener. Du et al. [14] also report a constant thermal conductivity of ~0.18 W/mK for their PMMA polymer. Finally, Cai and Song [20] used a semicrystalline PU dispersion with a constant thermal conductivity of ~0.15 W/mK. Because of the lack of any indication of a nonlinear temperature dependence of the thermal conductivity

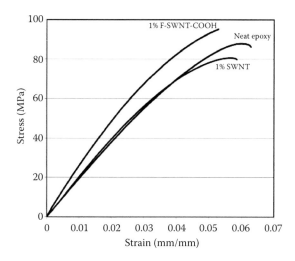

FIGURE 7.6 Stress–strain curves reported by Zhu et al. (Reproduced from Zhu, J., Kim, J., Peng, H., Margrave, J., Khabashesku, V., and Barrera, E. *Nano Letters*, vol. 3, Aug. 2003, pp. 1107–1113.)

of the neat epoxy, only the nonlinear thermal conductivity of the nanotubes should be considered for a thermal model of CNT reinforced polymer composites.

Based upon this review of the nonlinear thermal and elastic properties of both carbon nanotubes and polymer matrix materials, the elastic properties of the epoxy matrix are included in the elastic analysis of the model presented in this paper and the nonlinear thermal properties of the CNTS are included in the thermal analysis. With these decisions made, the next step is to choose the nonlinear approach to take in each case.

7.4.3   CHOICE AND IMPLEMENTATION OF NONLINEAR APPROACHES

After a careful consideration of the incremental and iterative methods, the incremental technique is chosen for the elastic analysis to be performed and both techniques are studied for the thermal analysis. The incremental technique is chosen both for its versatility and because it most closely resembles the physical phenomenon of a tensile strain test. The iterative technique is not applicable in the elastic analysis of the current model because it does not provide any information about the stresses and strains at intermediate load increments. Thus, a stress–strain relationship cannot be obtained from an iterative approach in the elastic case. Also, the availability of experimental data for comparison contributed to this choice.

For the thermal analysis, both an incremental approach and an iterative approach are considered for comparison. An iterative approach is more readily employed than an incremental approach and is often more quickly convergent than the incremental approach. For this reason, it is utilized to calculate an effective thermal conductivity for the composite. However, there is no guarantee that an iterative technique will converge and it provides no conductivity information at intermediate temperature increments, so an incremental approach was also used.

Upon deciding upon the approaches to take to the thermal and elastic analyses, a nonlinear FEM is developed, using the embedded fiber method as a foundation. After generating the nanotube geometry, partitioning the fibers, and embedding the fibers, the stiffness matrix of the entire RVE is assembled using the standard finite element procedures found in any finite element text. The standard force balance equation for FEA is given by

$$[K]\vec{u} = \vec{F}, \tag{7.26}$$

where $[K]$ is the system stiffness matrix assembled from the element stiffness matrices defined by Equation 7.25, $\vec{u}$ is the displacement or temperature vector for all of the mesh nodes in the RVE, and $\vec{F}$ is the reaction force or temperature gradient vector at the nodes. For the elastic analysis case, the general finite element problem is solved using a displacement-controlled approach, rather than force-controlled because, for a uniform displacement, an entire edge is needed to generate a stress–strain curve for the RVE as a whole.

To employ the incremental solution technique, displacement increments are applied to the elastic problem and temperature increments were applied to the thermal problem. After each increment is applied, the resulting displacements or temperatures of every node in the mesh are calculated.

For the elastic case, there are two degrees of freedom per node, x and y displacements. With the displacements at each node in hand after an incremental displacement is applied, the strain at each node can be calculated and interpolated to the Gaussian quadrature points using the strain matrix B from Equation 7.8 and the equation

$$\vec{\varepsilon} = [B]\,\vec{u}. \tag{7.27}$$

In two-dimensions, the strain vector ε is a three-component vector including the strain in the x and y directions, ε_x and ε_y, as well as the shear stress, γ_{xy}. With the strain, the stress at each Gauss point can be calculated using the relation

$$\vec{\sigma} = [D]\,\vec{\varepsilon}. \tag{7.28}$$

Recalling the earlier description of the basic FEM, $[D]$ is the material matrix defined by Equation 7.9.

It is worth discussing the modulus of elasticity E that appears in the material matrix and how it is calculated in the current model. In any linear model, the modulus of elasticity of a solid material is assumed to be constant. For a perfectly elastic material, this modulus can be calculated as the slope of the stress–strain curve for that material. Though no material is perfectly elastic, this assumption is valid for most "nearly" elastic materials. For materials that exhibit nonlinearity, the slope of the stress–strain curve changes as stress levels change. For materials that exhibit plasticity, the curve decreases at the onset of plastic deformation. For work-hardening materials, the slope of the curve may even increase at higher stress or strain levels. The slope of the curve at any given strain level is known as the tangent modulus. In this report, the tangent modulus of a polymer material is a function of strain, not assumed to be constant, and will simply be referred to as the modulus or elastic modulus of the material.

In general plasticity theory, there are two competing ways to assess whether a material has begun to undergo plastic deformation. The Tresca and von Mises criteria are two different methods that take into account the full stress state at one specific point of a material to come up with a single value that describes the effective stress at that point. The von Mises criterion is the more conservative of the two and is thus employed in the current model. The von Mises equivalent stress in two dimensions is given by [36,49]

$$\sigma_{\text{VMS}} = \sqrt{1/2\left[\left(\sigma_x - \sigma_y\right)^2 + \sigma_y^2 + \sigma_x^2 + 6\tau_{xy}^2\right]}. \tag{7.29}$$

After finding the strain vector at a node using Equation 7.27 and using it to find its corresponding stress vector by Equation 7.28, the von Mises stress at each Gauss point in each element can be calculated using Equation 7.27. The von Mises stress is then used to refer to the stress–strain curve of the neat epoxy to calculate the tangent modulus at any particular Gauss point. In this way, the modulus at each Gauss point throughout the RVE can be updated after each displacement increment. This procedure allows the nonlinearity of the material to be effectively incorporated into the model. A synopsis of the incremental procedure is presented in Figure 7.7.

For the thermal problem, the incremental approach follows essentially the same approach as the elastic problem, with only a few differences. The first difference between the thermal and elastic problem is that the thermal case requires only one degree of freedom per node, temperature. This in turn greatly reduces the computational cost to solve the problem because the number of equations to solve is halved. The second difference is that instead of updating the elastic modulus of the material, the thermal conductivity of the nanotubes is updated after each incremental temperature increase. This process is simply performed with the knowledge of the temperature at each of the nodes throughout the RVE. Using the shape functions of a quadrilateral element, the temperature is interpolated from the nodal temperatures to define a temperature at each of the nanotube endpoints. Because the nanotube thermal conductivity is quite high, approximately four orders of magnitude greater than that of the polymer matrix, an average of the two nanotube endpoint temperature is

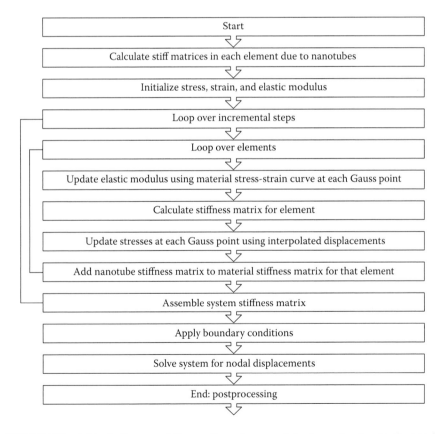

FIGURE 7.7 A block diagram of the nonlinear incremental scheme for the elastic case is presented. Notice the two major loops that are the driving force behind the computational time required by the model.

assumed as the constant temperature along that nanotube. The thermal conductivity of that particular nanotube is then updated based upon the curve of Grujicic et al. [15] shown in Figure 7.9. The equation used to update the thermal conductivity is given by

$$K_{nt} = 3E^{-8}T^5 + 7E^{-9}T^4 - 0.0011T^3 + 0.1568T^2 - 26.503T + 2471.7, \quad (7.30)$$

where K_{nt} is the thermal conductivity of the nanotube and T is the temperature of the nanotube in degrees celsius. This process is repeated for every nanotube in each element and is repeated after each incremental temperature step.

In addition to the incremental approach, the iterative technique is also applied to the thermal problem. In this process, the finite element solution is solved with a single, large temperature difference. Using the temperature data recovered from this solution, the thermal conductivities are updated using the same procedure as the incremental approach. The problem is then solved again using this data. This process is repeated until the solution converges to a solution. The number of iterations required for convergence is one of the results presented in the discussion section of the paper.

Overall, there are two objectives to the nonlinear analysis that is performed by the proposed model. The first objective is to create a stress–strain curve for any CNT reinforced polymer composite. The second objective is to assess whether or not the nonlinear thermal properties of carbon nanotubes need to be included in thermal models of CNT/polymer composites. The results of incremental and iterative nonlinear approaches are compared to determine if one is superior to the other. However, whatever the results of the model may be, the success of these objectives is ultimately determined by whether or not the model accurately represents experimental results. Due to this fact, the present model is compared to several different experiments by incorporating the parameters reported in those studies into the details of the model.

7.5   RESULTS AND DISCUSSION

Having discussed the specific details and assumptions of the current model, the elastic and thermal results are now presented. In order to calculate an effective stress–strain curve of the composite, the RVE, a tensile displacement is applied for the elastic problem and transverse movement of the structure was unrestricted. By adopting these boundary conditions, transverse stresses at the bottom nodes are eliminated and a direct correlation can be made between the applied displacement and the stresses recovered at the bottom nodes that are held in the tensile direction.

For the thermal problem, the bottom edge of the RVE is held at 0°C and the top edge is held at 100°C. The sides of the RVE are insulated and the effective thermal conductivity of the RVE are calculated using the sum of the heat fluxes through the bottom nodes held at 0°C. These boundary conditions are shown in Figure 7.8.

To qualitatively verify the EFFEM method, the preliminary elastic and thermal analysis was performed. To verify the elastic results, the 1 μm × 1 μm RVE was displaced to a strain of 3.5% for an arbitrary nanotube geometry of 1% wt. SWCNTs. By visualizing the displacement contours in the tensile stretching direction along

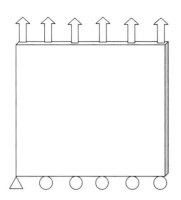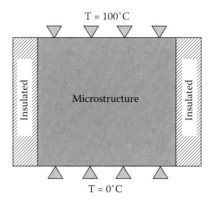

FIGURE 7.8   A diagram of the boundary conditions for the elastic and thermal test conditions is presented. The bottom left corner is pinned and the bottom boundary is on rollers for the elastic case. The top edge is held at 100°C and the bottom is held at 0°C while the sides are insulated in the thermal case.

with the nanotube geometry, the reinforcing effect of the nanotubes can clearly be seen by noticing the areas of the plot in which color remains constant. The significance of the reinforcement is seen by examining the displacements at a high-strain value, after the polymer has begun to exhibit a large degradation of elastic modulus. These two contours are shown in Figure 17.9 for comparison.

For the thermal problem, the preceding boundary conditions are utilized and a thermal contour plot are produced. The temperature contour for one particular nanotube geometry of 1% wt. SWCNTs is presented in Figure 7.10. Again, the enhancement of thermal conductivity can be seen by identifying the areas where color remains constant.

After confirming the general validity of the EFFEM for both the elastic and thermal nonlinear problems, the model is compared to several experiments by utilizing the same model parameters that are reported in the corresponding experimental reports.

7.5.1   ELASTIC RESULTS

Before comparing to experiment, a Monte Carlo Mesh Convergence Analysis (MCMCA) is performed to determine the mesh refinement and incremental step size necessary for accurate results. Esteva [31] performed MCMCA for his linear EFFEM model and varied the number of random microstructures used. He determined that 500 microstructures were sufficient to report statistically accurate values of effective moduli of elasticity and thermal conductivities. This same number of microstructures is employed in the proposed model. For this part of the analysis, the epoxy properties used were from Sun et al. [16]. First, MCMCA was performed for 500 microstructures of 1% wt. SWCNTs with only one element to determine the number of incremental steps necessary for convergence. This analysis is shown in Figure 7.11 and one hundred steps are found to be necessary to capture the nonlinearity of the polymer.

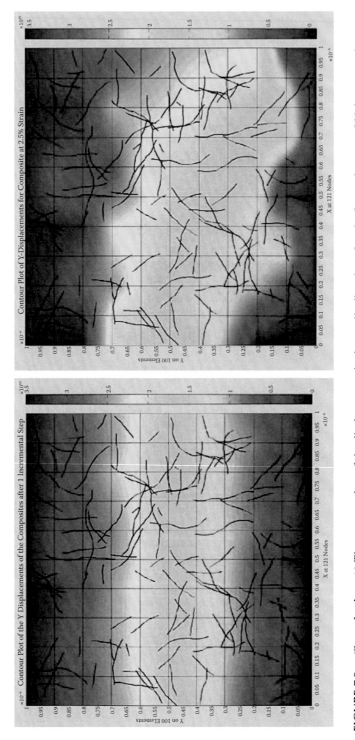

FIGURE 7.9 **(See color insert.)** The contour plots of the displacements in the tensile direction after the first and seventy-fifth incremental steps are presented. By examining the deformation before and after the onset of nonlinear polymer behavior, the role of the nanotubes in mechanical reinforcement is evident.

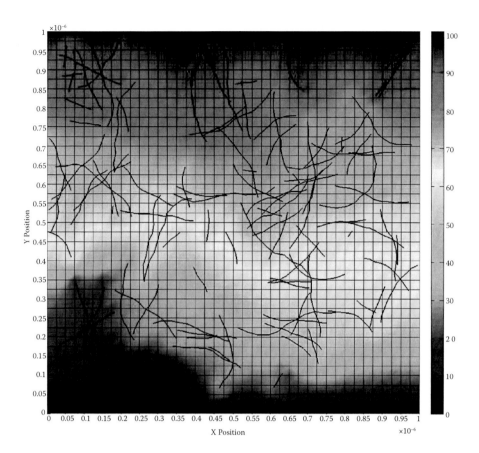

FIGURE 7.10 **(See color insert.)** A contour plot of temperatures from the nonlinear thermal model is presented.

MCMCA is then performed on 500 microstructures for 100 incremental steps over a range of finite element mesh refinements. While an increase in mesh refinement would always produce more accurate results, it was found that 60 divisions were sufficient for the aims of the proposed model.

With 60 divisions and 100 incremental steps selected as the parameters needed for an accurate finite element model that captures the nonlinear elastic properties of the polymer, the final step is to compare model results to experiment (Figure 7.12).

The first experiment chosen for comparison is that of Sun et al. [16]. They used 1% wt. SWCNTs in an epoxy resin and performed a tensile test to 3.5% strain. Fitting a polynomial to the neat epoxy stress–strain curve they reported and taking the derivative, the equation used to update the modulus of elasticity of the epoxy is given by

$$\frac{d\sigma}{d\varepsilon} = E_{\tan} = -1.193E^{12}\varepsilon^2 - 3.148E^{10}\varepsilon + 2.864E^9 \tag{7.31}$$

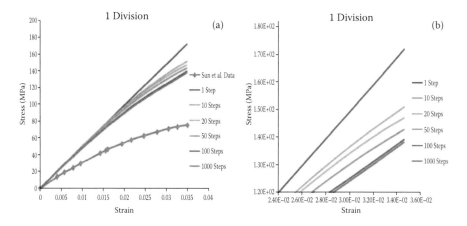

FIGURE 7.11 **(See color insert.)** (a) The mesh convergence analysis for the model with 100 incremental steps is shown for 1, 10, 20, 40, 60, and 80 divisions of the RVE. (b) A closer view of the model results at high strains identifies 60 divisions as sufficient mesh refinement.

The standard deviation for each model point is calculated using the standard equation [50]

$$\sigma = \sqrt{1\big/N \sum_{i=1}^{N} \left(x_i - \mathbf{x}\right)^2} \qquad (7.32)$$

The stress-curve calculated using the proposed model is compared to the stress–strain curve reported by Sun et al. [16] in Figure 7.13.

The disagreement between the two curves is no more than 5% at any point along the curve. The disagreement is explained by two factors. First, an increase in mesh refinement of the model would bring the model curve closer to the experimental

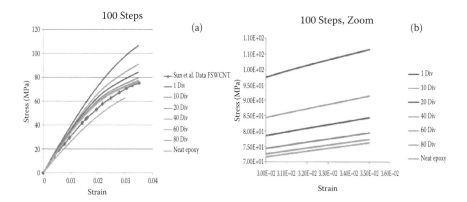

FIGURE 7.12 **(See color insert.)** (a) The model results for 60 divisions, 100 steps are compared to the results of Sun et al. (b) A close-up view of the stress–strain curve high strain values is shown. (Reproduced from Sun, L. et al., *Carbon*, vol. 46, Feb. 2008, pp. 320–328.)

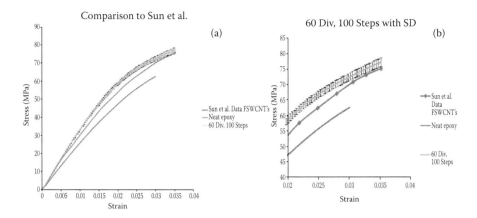

FIGURE 7.13 **(See color insert.)** The maximum and minimum RVEs out of the 500 RVEs sampled in the MCMCA are compared to the experimental results of Sun et al. to highlight the span of possible outcomes for random nanotube geometries. (Reproduced from Yu, M., Lourie, O., Dyer, M. J., Moloni, K., Kelly, T.F., and Ruoff, R.S., *Science*, vol. 287, Jan. 2000, pp. 637–640.)

curve. Because the objective is to simply show that the model is accurate, not exact, the mesh refinement was not changed. Second, the model curve is an average over 500 random microstructures whereas the experimental curve is a single realization of one particular nanotube geometry. Any single geometry curve strongly depends on the particular orientation of nanotubes in that sample as is illustrated by comparing the maximum and minimum curves of the 500 microstructures to the same experimental curve in Figure 7.14.

The second experiment chosen for comparison was that of Zhu et al. [13]. They also used 1% wt. SWCNTs in an epoxy matrix. They performed a tensile test to 5% strain and reported a different stress–strain curve for their neat epoxy. The equation for the modulus of elasticity of the neat epoxy from Zhu et al. is given by

$$\frac{d\sigma}{d\varepsilon} = E_{tan} = -3.325E^{11}\varepsilon^2 - 8.0E^{18}\varepsilon + 1.93E^9 \tag{7.33}$$

With these parameters, a comparison of the stress–strain curves of the model and the experiment was performed and is shown in Figure 7.15.

Akin to the first experiment, the model is accurate to within 5% for the entire curve. Again, this could be improved with an increase in mesh refinement and Zhu et al.'s data is based on one particular sample.

These two comparisons show that the proposed nonlinear model captures the material nonlinearity of polymer-based nanotube composites. Utilizing a nonlinear finite element procedure with only 60 divisions in a simple square mesh, an accurate stress–strain curve can be generated for an arbitrary nanotube-reinforced polymer composite if the polymer matrix's stress–strain curve is known.

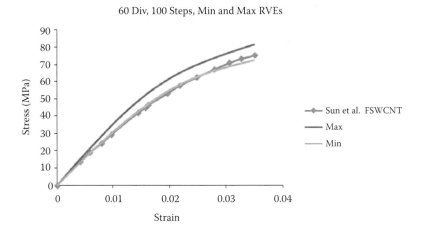

FIGURE 7.14   (See color insert.) The model results for 60 divisions, 100 steps are compared to the results of Zhu et al. The center blue F-SWCNT line represents the experimental data. (Reproduced from Zhu, J., Kim, J., Peng, H., Margrave, J., Khabashesku, V., and Barrera, E. *Nano Letters*, vol. 3, Aug. 2003, pp. 1107–1113.)

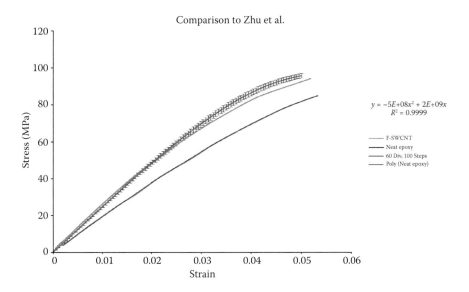

FIGURE 7.15   (a) The MCMCA for the thermal conductivity of 500 RVEs is presented for the incremental approach. A mesh refinement of 130 divisions is needed for accurate results. (b) By zooming in, it is also evident that ten incremental steps are sufficient to capture the nonlinearity of the thermal conductivity of the nanotubes. (c) The MCMCA for the thermal conductivity of 500 RVEs is presented for the iterative approach. A mesh refinement of 130 divisions is needed for accurate results. (d) By zooming in, it is also evident that only two iterative steps are needed to capture the nonlinearity of the thermal conductivity of the nanotubes.

7.5.2 THERMAL RESULTS

The effect of nonlinear thermal properties on the effective thermal conductivity of the composite is also examined herein. To model the nonlinear thermal conductivity of SWCNTs, both incremental and iterative approaches are examined. In both approaches, the thermal conductivity of the composite are calculated using the heat fluxes at the bottom nodes after the final step. The equation used to calculate the effective thermal conductivity is given by

$$K_{\text{eff}} = \frac{\sum q}{At\Delta T}, \tag{7.34}$$

where $\sum q$ is the sum of the heat fluxes across the bottom nodes, A is the area of the RVE, t is the thickness of the RVE, and ΔT is the temperature difference between the bottom and top of the RVE. To determine which nonlinear approach is superior, as well as how many divisions are necessary in the mesh, MCMMA is again performed. These results are shown in Figure 7.16.

Based on these results, the iterative approach is selected because it converged to a single value after only two steps. A mesh refinement of 130 divisions is also chosen since there is only degree of freedom per node in the thermal analysis and so few iterative steps is needed.

Having selected these parameters, the thermal results are compared to several different experiments. As many of the experimental results are reported in weight fractions, a conversion from weight fraction to volume fraction was performed using the equation [51]

$$vf_{\text{nt}} = \frac{wf_{\text{nt}}\rho_{\text{matrix}}}{wf_{\text{nt}}\rho_{\text{matrix}} + \left(1 - wf_{\text{nt}}\right)\rho_{\text{nt}}} \tag{7.35}$$

A matrix density of 1.2 g/cm³ is assumed since most journals did not detail the density of their polymer and this value is fairly typical for epoxies. Additionally, considerations are made to account for the high thermal resistance between the nanotube matrix and nanotube–nanotube interfaces. Xue [52] found that the effects of interfacial resistance, or Kapitza resistance, between the nanotubes and the surrounding matrix can effectively reduce the thermal conductivity of the nanotubes by several orders of magnitude. Using the work of Huxtable et al. [17], Wilson et al. [53], and Xue [52] as justification, a phenomenological degradation of nanotube thermal conductivity by a factor of twenty was incorporated into the model.

The data of Wang et al. [19], Moisala et al. [16], Biercuk et al. [42], Hong and Tai [41], Cai and Song [20], and Xu et al. [18] were selected to provide a broad spectrum of polymers and volume fractions for comparison.

Utilizing Equation 7.35 along with the reported matrix thermal conductivity for each experiment, a comparison to these six experiments is presented in Figure 7.17.

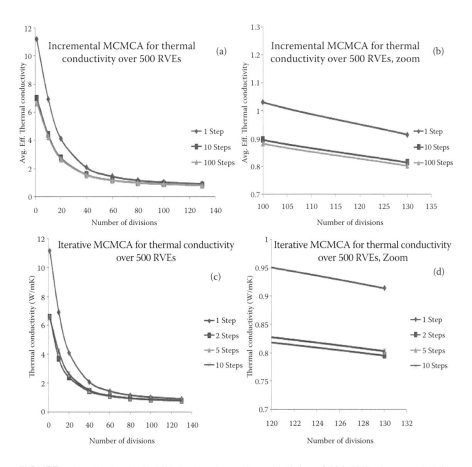

FIGURE 7.16 (a) The MCMCA for the thermal conductivity of 500 RVEs is presented for the incremental approach. A mesh refinement of 130 divisions is needed for accurate results. (b) By zooming in, it is also evident that ten incremental steps are sufficient to capture the nonlinearity of the thermal conductivity of the nanotubes. (c) The MCMCA for the thermal conductivity of 500 RVEs is presented for the iterative approach. A mesh refinement of 130 divisions is needed for accurate results. (d) By zooming in, it is also evident that only two iterative steps are needed to capture the nonlinearity of the thermal conductivity of the nanotubes.

For every comparison except that to Xu et al., the thermal conductivity predicted by the model is within ~33% of the value reported by the corresponding experiment. While this value may seem large at first, considering the poorly understood and complex mechanisms that inhibit efficient heat conduction, this agreement is good.

These comparisons show that the thermal conductivity calculated using a two-step iterative finite element procedure using the embedded fiber method produces results that give a good approximation of the thermal properties of a SWCNT reinforced polymer composite for volume fractions less than about 3%. Comparing model results to six different experiments and finding good agreement demonstrates

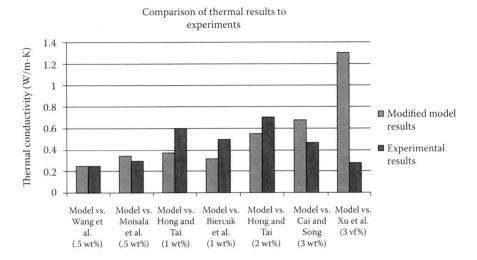

FIGURE 7.17 **(See color insert.)** A comparison of the modified model's thermal conductivity results to various experiments is presented. Substantially better agreement with experimental results is clearly evident.

the versatility of the model for simulating a variety of different polymers over a large range of SWCNT volume fractions. However, the accuracy of the model could be improved with a better understanding of the mechanisms inhibiting efficient heat transfer between the polymer and the nanotubes and between nanotube junctions.

7.6   CONCLUDING REMARKS

Utilizing the EFFEM, a novel model to predict nonlinear elastic and thermal properties of SWCNT reinforced composites has been presented. By incorporating the nonlinear modulus of elasticity of the polymer matrix using an incremental approach, an accurate stress–strain curve for these nanocomposites can be generated in a short amount of time. The accuracy of numerical results has been compared to experimental results in the literature and found to be accurate within 5% for small strains less than 5%. The effective conductivity of the composite has been also investigated while incorporating the nonlinear thermal conductivity of SWCNTs. Using an iterative approach, the numerical results has been compared to several experiments in the literature. The model has proven quite accurate for small volume fractions, less than 3%, but somewhat inaccurate for large volume fractions. This is attributed to complex mechanisms inhibiting heat conduction at the nanotube-matrix interface. Utilizing the EFFEM provides a way to overcome the human intervention required in other FEMs and establishes the described method as a new means to characterize nanocomposites using a simple, automated routine. Additional discussions on the discritization and the Monte Carlo aspects of the method can be found in Esteva [54] and Spanos and Kontsos [55].

REFERENCES

1. T. Natsuki, K. Tantrakarn, and M. Endo. 2004. Prediction of elastic properties for single-walled carbon nanotubes. *Carbon* 42:39–45.
2. M. Meo and M. Rossi. 2007. A molecular-mechanics based finite element model for strength prediction of single wall carbon nanotubes. *Materials Science and Engineering: A* 454–455:170–177.
3. K. Tserpes, P. Papanikos, and S. Tsirkas. 2006. A progressive fracture model for carbon nanotubes. *Composites Part B: Engineering* 37:662–669.
4. T. Belytschko, S. P. Xiao, G. C. Schatz, and R. S. Ruoff. 2002. Atomistic simulations of nanotube fracture. *Physical Review B* 65:235430.
5. A. Desai and M. Haque. 2005. Mechanics of the interface for carbon nanotube-polymer composites. *Thin-Walled Structures* 43:1787–1803.
6. S. Berber, Y. Kwon, and D. Tomanek. 2000. Unusually high thermal conductivity of carbon nanotubes. *Physical Review Letters* 84:4613.
7. F. Du, C. Guthy, T. Kashiwagi, J. E. Fischer, and K. I. Winey. 2006. An infiltration method for preparing single-wall nanotube/epoxy composites with improved thermal conductivity. *Journal of Polymer Science Part B: Polymer Physics* 44:1513–1519.
8. M. Grujicic, G. Cao, and B. Gersten. 2004. Atomic-scale computations of the lattice contribution to thermal conductivity of single-walled carbon nanotubes. *Materials Science and Engineering B* 107:204–216.
9. A. Moisala, Q. Li, I. Kinloch, and A. Windle. 2006. Thermal and electrical conductivity of single- and multi-walled carbon nanotube-epoxy composites. *Composites Science and Technology* 66:1285–1288.
10. C. Li and T. Chou. 2008. Modeling of damage sensing in fiber composites using carbon nanotube networks. *Composites Science and Technology* 68(15–16):3373–3379.
11. Q. Wang, J. Dai, W. Li, Z. Wei, and J. Jiang. 2008. The effects of CNT alignment on electrical conductivity and mechanical properties of SWNT/epoxy nanocomposites. *Composites Science and Technology* 68:1644–1648.
12. V. Skakalova, U. Dettlaff-Weglikowska, and S. Roth. Electrical and mechanical properties of nanocomposites of single wall carbon nanotubes with PMMA. *Synthetic Metals* 152:349–352.
13. J. Zhu, J. Kim, H. Peng, J. Margrave, V. Khabashesku, and E. Barrera. 2003. Improving the dispersion and integration of single-walled carbon nanotubes in epoxy composites through functionalization. *Nano Letters* 3:1107–1113.
14. J. Zhu, H. Peng, F. Rodriguez-Macias, J. Margrave, V. Khabashesku, A. Imam, K. Lozano, and E. Barrera. 2004. Reinforcing epoxy polymer composites through covalent integration of functionalized nanotubes. *Advanced Functional Materials* 14:643–648.
15. X. Li, H. Gao, W. A. Scrivens, D. Fei, X. Xu, M. A. Sutton, A. P. Reynolds, and M. L. Myrick. 2004. Nanomechanical characterization of single-walled carbon nanotube reinforced epoxy composites. *Nanotechnology* 15:1416–1423.
16. L. Sun, G. Warren, J. O'Reilly, W. Everett, S. Lee, D. Davis, D. Lagoudas, and H. Sue. 2008. Mechanical properties of surface-functionalized SWCNT/epoxy composites. *Carbon* 46:320–328.
17. S. T. Huxtable, D. G. Cahill, S. Shenogin, L. Xue, R. Ozisik, P. Barone, M. Usrey, M. S. Strano, G. Siddons, M. Shim, and P. Keblinski. 2003. Interfacial heat flow in carbon nanotube suspensions. *Nature Materials* 2:731.
18. Y. Xu, G. Ray, and B. Abdel-Magid. 2006. Thermal behavior of single-walled carbon nanotube polymer-matrix composites. *Composites Part A: Applied Science and Manufacturing* 37:114–121.
19. S. Wang, R. Liang, B. Wang, and C. Zhang. 2009. Dispersion and thermal conductivity of carbon nanotube composites. *Carbon* 47:53–57.

20. D. Cai and M. Song. 2008. Latex technology as a simple route to improve the thermal conductivity of a carbon nanotube/polymer composite. *Carbon* 46:2107–2112.
21. K. Tserpes, P. Papanikos, G. Labeas, and S. Pantelakis. Multi-scale modeling of tensile behavior of carbon nanotube-reinforced composites. *Theoretical and Applied Fracture Mechanics* 49:51–60.
22. C. Li and T. Chou. 2006. Multiscale modeling of compressive behavior of carbon nanotube/polymer composites. *Composites Science and Technology* 66:2409–2414.
23. J. Xiao and J. W. Gillespie Jr. 2006. Nanomechanics of single-walled carbon nanotubes as composite reinforcement. *Polymer Engineering and Science* 46:1051.
24. D. Shi, X. Feng, Y. Y. Huang, K. Hwang, and H. Gao. 2004. The effect of nanotube waviness and agglomeration on the elastic property of carbon nanotube-reinforced composites. *Journal of Engineering Materials and Technology* 126:251.
25. Y. S. Song and J. R. Youn. 2006. Evaluation of effective thermal conductivity for carbon nanotube/polymer composites using control volume finite element method. *Carbon* 44:710–717.
26. Q. Xue. 2005. Model for thermal conductivity of carbon nanotube-based composites. *Physica B: Condensed Matter* 368:302–307.
27. A. Bagchi and S. Nomura. 2006. On the effective thermal conductivity of carbon nanotube reinforced polymer composites. *Composites Science and Technology*, vol. 66, Sep. 2006, pp. 1703–1712.
28. J. Zhang and M. Tanaka. 2007. Systematic study of thermal properties of CNT composites by the fast multipole hybrid boundary node method. *Engineering Analysis with Boundary Elements* 31:388–401.
29. G. M. Odegard, T. S. Gates, K. E. Wise, C. Park, and E. J. Siochi. 2003. Constitutive modeling of nanotube-reinforced polymer composites. *Composites Science and Technology* 63:1671–1687.
30. G. D. Seidel and D. C. Lagoudas. 2006. Micromechanical analysis of the effective elastic properties of carbon nanotube reinforced composites. *Mechanics of Materials* 38:884–907.
31. P. D. Spanos and M. Esteva. 2009. Effect of stochastic nanotube waviness on the elastic and thermal properties of nanocomposites by fiber embedment in finite elements, *Journal of Computational and Theoretical Nanoscience* 6:2317–2333.
32. S. Wang, Z. Liang, B. Wang, and C. Zhang. 2006. Statistical characterization of single-wall carbon nanotube length distribution. *Nanotechnology* 17:634–639.
33. K. J. Ziegler, U. Rauwald, Z. Gu, F. Liang, W. Billups, R. H. Hauge, and R. E. Smalley. 2007. Statistically accurate length measurements of single-walled carbon nanotubes. *Journal of Nanoscience and Nanotechnology* 7:2917–2921.
34. F. H. Gojny, M. H. Wichmann, B. Fiedler, and K. Schulte. 2005. Influence of different carbon nanotubes on the mechanical properties of epoxy matrix composites—A comparative study. *Composites Science and Technology* 65:2300–2313.
35. J. Akin. 2005. *Finite element analysis with error estimators: An introduction to the FEM and adaptive error analysis for engineering students.* Oxford: Elsevier Butterworth-Heinemann.
36. C. S. Desai and J. F. Abel. 1972. *Introduction to the finite element method.* New York: Litton Educational Publishing.
37. O. Zienkiewicz, R. Taylor, and J. Zhu. 2005. *The finite element method: Its basis and fundamentals.* Oxford: Elsevier Butterworth-Heinemann.
38. A. Konrad and M. Graovac. 1996. An application for line elements embedded in a 2D or 3D finite element mesh. *IEEE Transactions on Magnetics* 32:647–650.
39. A. Garg and S. B. Sinnott. 1998. Effect of chemical functionalization on the mechanical properties of carbon nanotubes. *Chemical Physics Letters* 295:273–278.
40. M. Esteva and P. D. Spanos. 2009. Effective elastic properties of nanotube reinforced composites with slightly weakened interfaces. *Journal of Mechanics of Materials and Structures* 4:887–990.

41. W. Hong and N. Tai. 2008. Investigations on the thermal conductivity of composites reinforced with carbon nanotubes. *Diamond and Related Materials* 17:1577–1581.

42. M. J. Biercuk, M. C. Llaguno, M. Radosavljevic, J. K. Hyun, A. T. Johnson, and J. E. Fischer. 2002. Carbon nanotube composites for thermal management. *Applied Physics Letters* 80:2767–2769.

43. T. Natsuki, K. Tantrakarn, and M. Endo. 2004. Effects of carbon nanotube structures on mechanical properties. *Applied Physics A: Materials Science & Processing* 79:117–124.

44. T. Kawamura, Y. Kangawa, and K. Kakimoto. 2008. Investigation of the thermal conductivity of a fullerene peapod by molecular dynamics simulation. *Journal of Crystal Growth* 310:2301–2305.

45. M. Yu, O. Lourie, M. J. Dyer, K. Moloni, T. F. Kelly, and R. S. Ruoff. 2000. Strength and breaking mechanism of multiwalled carbon nanotubes under tensile load. *Science* 287:637–640.

46. M. Yu, B. S. Files, S. Arepalli, and R. S. Ruoff. Tensile loading of ropes of single wall carbon nanotubes and their mechanical properties. *Physical Review Letters* 84:5552.

47. D. A. Walters, L. M. Ericson, M. J. Casavant, J. Liu, D. T. Colbert, K. A. Smith, and R. E. Smalley. 1999. Elastic strain of freely suspended single-wall carbon nanotube ropes. *Applied Physics Letters* 74:3803–3805.

48. S. Shaikh, L. Li, K. Lafdi, and J. Huie. 2007. Thermal conductivity of an aligned carbon nanotube array. *Carbon* 45:2608–2613.

49. M. Meyers and K. Chawla. 1999. *Mechanical behavior of materials.* Upper Saddle River, NJ: Prentice-Hall, Inc.

50. J. R. Taylor. 1997. *An introduction to error analysis: The study of uncertainties in physical measurements.* Sausalito, CA: University Science Books.

51. R. B. Pipes, S. J. V. Frankland, P. Hubert, and E. Saether. 2003. Self-consistent properties of carbon nanotubes and hexagonal arrays as composite reinforcements. *Composites Science and Technology* 63:1349–1358.

52. Q. Z. Xue. Model for the effective thermal conductivity of carbon nanotube composites. *Nanotechnology* 17:1655–1660.

53. O. M. Wilson, X. Hu, D. G. Cahill, and P. V. Braun. Colloidal metal particles as probes of nanoscale thermal transport in fluids. *Physical Review B* 66:224301.

54. M. Esteva. 2008. *Hybrid finite elements nanocomposite characterization by stochastic microstructuring.* PhD dissertation, Rice University.

55. P. D. Spanos and A. Kontsos. 2008. A multiscale Monte Carlo finite element method for the determination at the mechanical properties of polymer nanocomposites. *Journal of Probabilistic Engineering Mechanics* 66:456–470.

8 Ensemble Monte Carlo Device Modeling: High-Field Transport in Nitrides

Cem Sevik

CONTENTS

8.1 INTRODUCTION

Semiconductor device transport is, in general, a tough problem from both the mathematical and the physical points of view. In fact, the Boltzmann transport equation (BTE), assumed to be the fundamental description of carrier transport, does not offer simple analytical solutions due to the carrier density evolving in time and space according to a complicated nonlinear integro-differential equation. Furthermore, numerical solution techniques, such as iterative, path integral, and orthogonal polynomial expansion methods, usually are not applicable to real device systems. Another alternative method that has been widely used in studies of the physics of hot-carrier transport, allows simulation of the trajectories of individual carriers as they move through a device under the applied fields and experience various scattering or collision events. In order to implement such a model we need to directly incorporate a statistical method of at least the scattering processes, and this is done by the use of Monte Carlo (MC) techniques.

An MC technique is a stochastic method that is based on the solution of mathematical or physical systems by simulation of random quantities. The American mathematicians J. Neyman and S. Ulam are considered to be its originators due to their article published in 1949, "The Monte Carlo Method." This powerful method is used in many different fields such as mathematics, physical sciences, design and visualization, finance and business, telecommunications, games, and others.

Kurosawa first used the MC method in the field of computational electronics in 1966 [1]. Subsequently, this method was developed to a high degree of refinement by Price [2], Rees [3], and Fawcett et al. [4], and since then they have been widely used to obtain results for various situations in practically all materials of interest. It was, in fact, clear that with the aid of large and fast modern computers, the MC method would become popular and it would become possible to obtain exact numerical solutions of the Boltzmann equation for microscopic physical models of considerable complexity. In many cases, MC simulation is the most accurate method in the field of computational electronics related to the simulation of the state of the art devices because of its straightforward implementation and direct interpretation from a physical point of view.

8.2 BOLTZMANN TRANSPORT EQUATION

The Boltzmann transport equation (BTE), introduced by Austrian physicist, Ludwig Boltzmann (1844–1906) describes the statistical distribution function of carriers under an external perturbation for given appropriate boundary conditions. Here, the critical quantity, the distribution function, $f(k, r, t)$, is defined as the probability of finding a carrier with crystal momentum k at location r at time t. Under equilibrium conditions, this function coincides with the Fermi–Dirac distribution function.

Possible reasons such as carrier motion (diffusion), external forces $\left(F_{ext} = \dfrac{dr}{dt} \right)$, and scattering processes might be responsible for change in the equilibrium distribution function in momentum, k-space, and real, r-space. Considering the scattering processes as a balancing mechanism to the change in $f(k, r, t)$ due to an external force, BTE can be written as

$$\frac{df(k,r,t)}{dt} = \frac{\partial f(k,r,t)}{\partial t} + v\nabla_r f(k,r,t) + \frac{1}{\hbar} F_{ext}\nabla_K f(k,r,t) = \frac{\partial f(k,r,t)}{\partial t}\Big|_{sc}, \quad (8.1)$$

where $v = \dfrac{d\boldsymbol{r}}{dt}$, electron group velocity, $\nabla_r = \dfrac{\partial}{\partial \boldsymbol{r}}$, and $\nabla_r = \dfrac{\partial}{\partial \boldsymbol{k}}$. The analytical solution of Equation 8.1 is only possible in simple cases; the particular advantage of the MC method is that it provides a first-principles transport formulation based on the exact solution of this equation, limited only by the extent to which the underlying physics of the system is included.

There has been considerable discussion in the literature about the connection between the BTE and the MC technique. Most of this discussion relates to whether or not they yield the same results and, if so, upon what time scale. In fact, it was easily pointed out many years ago that the MC procedure only approached the BTE result in the long time limit [3,4]. Yet, there are still efforts to put more significance into the BTE on the short time scale. The problem is that the BTE is Markovian in its scattering integrals; a retarded, or non-Markovian, form of the BTE is required for the short time scale. For this reason, it should be mentioned that the ensemble Monte Carlo (EMC) technique supersedes the BTE, even if it could be solved exactly in short time scales and, therefore, the EMC technique is intrinsically much more suited to tackle the high-field transport phenomena [2].

8.3 ENSEMBLE MONTE CARLO CHARGE TRANSPORT SIMULATION

The EMC method is based on the successive and simultaneous simulation of the dynamics of many particle systems with a small time increment, Δt. The method is basically dynamic, and it is therefore appropriate to analyze carrier diffusion, carrier transport in an inhomogeneous field, nonstationary behavior of carriers, and so on. Accordingly, the EMC method is one of the most used simulation techniques to analyze high-field transport in bulk semiconductors and devices. One can find excellent resources on the MC technique [5–8]; here, we refer only to essential parts of EMC device simulation.

8.3.1 EMC FLOW CHART

The EMC method is based on the simulation of a many particle system with a small time increment, Δt, as shown schematically in Figure 8.1. Each point labeled with *sc* shows the scattering time, and the time interval between two successive *sc* points determines the free-flight time, t_f. The simulation starts with n particles with given initial conditions with a wave vector, $\boldsymbol{k}_0$, and a real space position, $\boldsymbol{r}_0$; then, t_f is chosen by a random number, generated with probability determined by the total scattering rate. If t_f is larger than Δt, the particle merely drifts during Δt. If t_f is less than Δt, the particle drifts up to t_f at which time a particle is scattered with the scattering process chosen according to the relative probabilities of all scattering rates. Depending on the type of chosen scattering event, a new $\boldsymbol{k}$ is chosen as an initial state of the new free flight. The new scattering time has to then be determined by another random number, and the same hierarchy is repeated (see the EMC flowchart in Figure 8.2). Over the entire course of simulation, the quantity of interests, velocity, energy, etc. are recorded at each specified time interval, Δt.

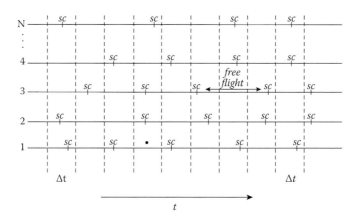

FIGURE 8.1 Time evaluation of N particle during the EMC simulation. The points labeled with *sc* show scattering time.

8.3.1.1 Free Flight

The probability, $P(t_f)$, of a carrier being scattered after traveling for a time t_f, is given by

$$P(t_f) = \Gamma(E_k)\exp- \int_0^{tf} \Gamma(E_k)\,dt, \tag{8.2}$$

where $\Gamma(E_k)$ is the total scattering rate given by

$$\Gamma(E_k) = \sum_{j=1}^{N} \Gamma_j(E_k) \tag{8.3}$$

Here, N is equal to the total scattering events taken into account. Because a carrier's energy, $E(k)$, changes with time as it moves under external force and scattering events, $\Gamma(E_k)$ is a function of time. Hence, the solution of the integral in Equation 8.2 is very complicated. The most straightforward way to avoid this problem is to define a time-independent (constant) total scattering rate, Γ, by introducing a virtual scattering mechanism self-scattering that makes no change to k. The rate of the self-scattering event, Γ_0, can be defined as

$$\Gamma = \Gamma_0 - \sum_{j=1}^{N} \Gamma_j(E_k) \tag{8.4}$$

Now, Equation 8.2 can be written simply as

$$P(t_f) = \Gamma \exp(-\Gamma t_s), \tag{8.5}$$

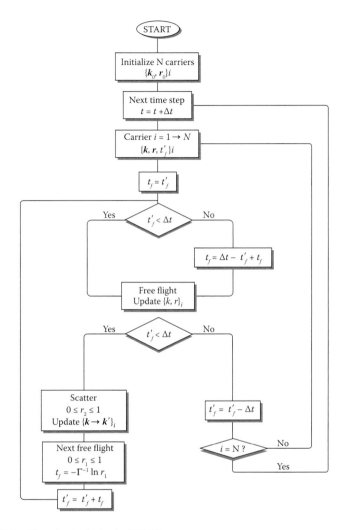

FIGURE 8.2   Flowchart of a typical EMC program.

and the free-flight time can be determined by a random number, r_1 distributed uniformly between 0 and 1 with the following equation,

$$t_f = -\frac{\ln r_1}{\Gamma} \tag{8.6}$$

Here, the constant total scattering rate, Γ_0, is equalized to the largest value of $\Gamma(E_k)$. During a single free flight, a carrier is treated as a classical particle with an effective mass m^*, and thus, carrier (electron) momentum is determined by the relation,

$$\hbar\boldsymbol{k} = -e\varepsilon,\ \Delta\boldsymbol{k} = -\frac{e\varepsilon}{\hbar}t_f, \tag{8.7}$$

where ε is the applied electric field. Therefore, the mean carrier velocity $\langle \boldsymbol{v} \rangle_{t_f}$ and increment of the position vector, $\Delta\boldsymbol{r}$, during the t_f is determined by,

$$\langle \boldsymbol{v} \rangle_{t_f} = -\frac{\Delta E_k}{e\varepsilon}t_f,\ \Delta\boldsymbol{r} = \langle \boldsymbol{v} \rangle_{t_f} t_f \tag{8.8}$$

8.3.1.2 Scattering Event

At the end of the free flight, a random scattering event is chosen and carrier momentum is updated, depending on the type of the selected scattering mechanism. Selection of a mechanism is performed by matching a random number, r_2, distributed uniformly between 0 and 1, with a normalized scattering table (successive summation of all scattering rates normalized with Γ) as shown in Figure 8.3.

Once the scattering mechanism that caused the end of the carrier's free flight is determined, both the magnitude and orientation of $\boldsymbol{k}$ must be updated, considering the selected scattering event. The magnitude of the $\boldsymbol{k}$ is easily calculated by the energy conservation law,

$$E(\boldsymbol{k}) = E(\boldsymbol{k}') + \Delta E, \tag{8.9}$$

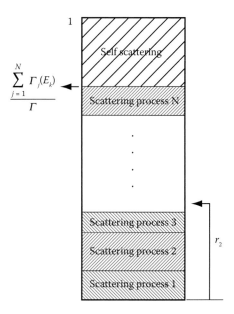

FIGURE 8.3   Illustration of the procedure for identifying a scattering event, R.

where change in energy, ΔE, is equal to zero for elastic scattering and equal to the phonon energy for inelastic scattering. In the case of isotropic scattering (if the scattered carrier has the same probability of being in any direction after scattering), the polar θ and azimuthal Φ angle can be determined by a couple of uniform random numbers, r_3 and r_4, generated between 0 and 1,

$$\varphi = 2\pi r_3, \cos\theta = 1 - 2r_4, \tag{8.10}$$

and the new orientation can be determined by transforming new directions to the simulation frame. For anisotropic scattering, determination of the orientation of a carrier's momentum is dependent on the type of scattering event. Further details can be found in the works of Lundstrom [7] and Tomizawa [8].

8.3.1.3 Carrier Energy

Characterization of the electronic band structure of carriers during EMC simulation is the key issue to increase its accuracy. The most common models might be listed, from simple to complicated, as follows: parabolic, nonparabolic, and full-band approximations. The simplest one, parabolic band approximation, generally is used to imitate carrier transport under low-field conditions in simple systems such as bulk crystals. In this model, carriers behave just like the ones in free space, moving with effective mass m^*. Thus, their energy can be written as a quadratic function of k,

$$E_k = \frac{\hbar^2 \mathbf{k}^2}{2m^*}, \tag{8.11}$$

where $\dfrac{1}{m^*} = \dfrac{1}{\hbar^2}\dfrac{\partial^2 E_k}{\partial \mathbf{k}^2}$ is the inverse of the carrier effective mass. Free-energy band approximation is simple, but too crude to provide meaningful results.

The nonparabolic band approximation is one of the most used models to simulate moderate and high-field conditions. In this model, the nonparabolic behavior of E_k is approximated by

$$E_k(1 + \alpha E_k) = \frac{\hbar^2 \mathbf{k}^2}{2m^*} = \gamma(\mathbf{k}), \tag{8.12}$$

where m^* *is* the effective mass defined at band edge and is a nonparabolic factor simply given by $\alpha = \dfrac{1}{E_g}1 - \dfrac{m^*}{m_0}$. However, m^* and α of each band can be obtained by using first-principles band structure and density of states to obtain much better accuracy. The other physical quantities, group velocity of carriers and energy independent carrier momentum trajectory, can be calculated by the solution of Equations 8.7 and 8.8, respectively.

The most accurate method for high-field and avalanche breakdown conditions is full-band numerical description of $E(k)$ throughout the whole Brillouin zone. In that case, the energy of a carrier is easily evaluated from the numerical table $E(k)$ by using the final momentum calculated by the band structure independent equation, Equation 8.7. The real space position of a carrier can still be evaluated using Equation 8.8 with numerically calculated group velocity $v(k) = \Delta_k E(k)$ of carriers. However, selection of the final state after a scattering event is a bit more complicated within this approximation. The final state, which conserves both momentum and energy, must be chosen from the numerical band structure tabulated $E(k)$ [7].

8.3.2 SCATTERING

In a semiconductor, a carrier moves under the influence of a periodic crystal potential associated with the array of ions at the lattice points. Various mechanisms such as quasiparticles of the lattice vibrations (phonon), disorders in composites, ionized dopants, crystal defects, and collisions with other carriers produce particularly instantaneous changes in the energy of carriers. When a carrier encounters such a perturbation induced by a Hamiltonian H', it scatters from a state $\psi(k, r)$ to $\psi(k', r')$. The probability of such a transition can be calculated by Fermi's golden rule, which is derived from the time-dependent perturbation theory of the first order [9]

$$P(k,k) = \frac{2\pi}{\hbar} \left\langle \psi(k,r) \mid H \mid \psi(k,r) \delta \left[E(k) - E(k) + \Delta E \right] \right\rangle \qquad (8.13)$$

Here, $E(k)$ and $E(k')$ are the initial and final carrier energies, respectively, and ΔE is the energy difference between the initial and final states.

Scattering processes in semiconductors can be classified as intravalley, when initial and final states lie in the same valley, or intervalley, when they lie in different valleys. The remarkable scattering mechanisms that determine these transitions in bulk or homogeneous semiconductors might be listed as shown in Figure 8.4. The interaction of phonons with carriers is due to the malformation of the perfect crystal

Phonons	Defects	Carriers
• Acoustic / optical: deformation potential	• Neutral impurities	• Binary:
• Acoustic / optical: polar	• Dislocations	electron – electron
	• Alloy scattering	electron – hole
	• Ionized impurities	radiative rec.
		non radiative rec.
		impact ionization
		• Collective plasmons

FIGURE 8.4 Various scattering mechanisms.

potential by the deformation potential in covalent semiconductors and by both the deformation potential and electrostatic forces produced by the polarization of the phonons in covalent semiconductors. Defect scattering includes scattering by both ionized and neutral impurities and by crystal defects such as dislocations. In semiconductor alloys such as AlGaAs, fluctuations of compositions also produce scattering. Carrier–carrier scattering includes both binary collisions and interactions with the carrier plasma.

Equation 8.13 is the basic result of scattering theory, which can be applied to the scattering rate calculation of carriers in semiconductors. To apply Fermi's golden rule, the perturbation potential H' for each individual mechanism must be identified so that the matrix elements $<\psi(k, r)|H'|\psi(k', r')>$, and so scattering rates can be evaluated. Thus, the total scattering rate is obtained by the sum of the rates from each of the individual processes, as indicated in Equation 8.3.

8.4   DEVICE APPLICATIONS

In this section, we would like to turn your attention to EMC applications by considering our previous study on the high-field transport phenomena in wide-band gap semiconductors such as GaN, AlGaN, and AlN. As mentioned several times before, for the high-field transport phenomena, the EMC technique is currently the most reliable choice, free from major simplifications [10]. With this method, the dynamics of three different types of devices, operating under high-field conditions, namely, unipolar n-type structures, avalanche photodiodes (APDs), and, finally, the Gunn diodes have been analyzed. The applied EMC treatment includes all standard scattering mechanisms other than dislocation, neutral impurity, and the piezoacoustic scattering, as they only become significant at low temperatures and fields [11]. Impact ionization parameters for bulk GaN are extracted from a recent experiment of Kunihiro et al. [12]. As for the case of AlN, due to a lack of any published results, we had to resort to a Keldysh approach, while Bloch overlaps were taken into account via the f-sum rule [5] (see Bulutay [13] for details). Furthermore, the polar optical phonon and ionized impurity potentials are screened by using random-phase-approximation-based dielectric function [14].

The band structures for GaN and AlN are obtained using the empirical pseudopotential technique fitted to available experimental results and first-principles computations [15,16]. For the alloy, $Al_xGa_{1-x}N$, we resort to linear interpolation (Vegard's law) between the pseudopotential form factors of the constituent binaries. The necessary band edge energy, effective mass, and nonparabolicity parameters of all valleys in the lowest two conduction bands and valence bands located at high symmetry points are extracted through the computed bands of GaN and $Al_xGa_{1-x}N$. To account for the remaining excited conduction and valence bands, we further append additional higher lying parabolic free electron and hole bands. At this point, it is important to stress that we use the actual density of states computed using the Lehmann–Taut approach [17], rather than the valley-based nonparabolic band approximation, in calculating the scattering rates [18]. This assures perfect agreement with rigorous full-band EMC simulations [19] even for the hole drift velocities at a field of 1 MV/cm.

8.4.1 Hot-Electron Effects in *N*-Type Structures

GaN, AlN, and their ternary alloys are technologically important semiconductors with application in high-power microelectronic devices such as GaN/AlGaN HEMTs as well as in optoelectronic devices like visible- and solar-blind AlGaN photodiodes. The impact ionization (II) scattering is an important process for all those devices subject to extreme electric fields. Despite being considered an undesired mechanism leading to breakdown in high-power devices, II is the key process for the sole operation of APDs.

A surprisingly limited number of experimental and theoretical studies have been published related to the II coefficient for the $Al_xGa_{1-x}N$ system. From the computational side, II in bulk AlGaN alloys has been analyzed extensively [13]. However, in this study, we focus on the device-related aspects of II, and transient and steady-state hot-electron phenomena taking place within this structure. A useful model system for understanding hot-electron effects is the unipolar n^+-n-n^+ homojunction channel (see Figure 8.5), which is, to some extent, impractical as it gives rise to an excessive amount of current density.

8.4.1.1 Computational Details

In the course of the EMC calculations, the energy loss of the impacting electrons is accurately modeled as described by Bulutay [13]. More than 20,000 super particles are employed within the ensemble, for a total simulation time of about 7.5 ps. The time interval of invoking the Poisson solver is taken to be 0.1 fs. To take into account degeneracy effects, the Lugli–Ferry recipe [20] is chosen in several methods [21–24].

8.4.1.2 Alloy Scattering

The subject of alloy scattering has caused substantial controversy over the years, which is still unsettled [25–28]. In the case of group-III nitrides, Farahmand et al.

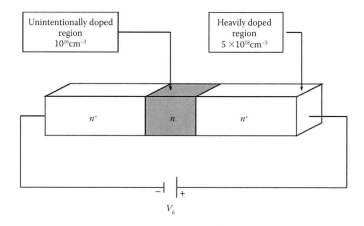

FIGURE 8.5   Structural schematics.

[28] have dealt with this issue and reported that using the conduction band offset between the binary constituents as the alloy potential leads to an upper bound for alloy scattering. Being more conservative, for this value we prefer to use half of the corresponding GaN/AlN conduction band offset, 0.91 eV. The particular implementation of alloy scattering within the MC simulation is also another source of concern. Following Fischetti and Laux [29], we treat the alloy scattering as an intravalley process with the distribution of the final scattering angles assumed to be isotropic, even though at higher energies it attains a forward directional character that should presumably weaken the effect of this mechanism on the momentum relaxation. Therefore, we are led to think that the effect of alloy scattering may still be overestimated.

8.4.1.3   Results

The high-field transport phenomena in GaN-, AlN-, or $Al_xGa_{1-x}N$-based submicron-sized dimensions is analyzed considering a simple n^+-n-n^+ homojunction channel device [31] having 0.1 μm-thick unintentionally doped (10^{16} cm^{-3}) n region sandwiched between two heavily doped (5×10^{18} cm^{-3}) n^+ regions at least 0.2 μm thick (see Figure 8.5). As seen in velocity profiles for these materials in Figure 8.6, the $Al_{0.4}Ga_{0.6}N$-based structure suffers severely from alloy scattering and has a substantially reduced velocity. If we turn off the alloy scattering, then the curve for $Al_{0.4}Ga_{0.6}N$ (not shown) almost coincides with that of GaN. In fact, previous analysis of Bulutay for bulk AlGaN alloys has identified the alloy scattering to modify the high energy electron distribution and lead to an increased II threshold [13].

The electric field along this device is distributed highly nonuniformly, reaching a few MV/cm values, which peak at the right nn^+ interface, as shown in Figures 8.7a and 8.8a. Also note the penetration of the electric field into the heavily doped anode n^+ region with increasing applied bias, which amounts to widening of the unintentionally

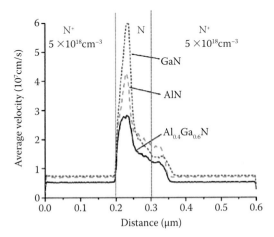

FIGURE 8.6   Velocity distribution over the n^+-n-n^+ channel under an applied bias of 20 V. (Reprinted with permission from IET. Sevik, C., and Bulutay, C. *Proc. Optoelectron.* Vol. 150, p. 87, 2003. EBSCO Publishing.)

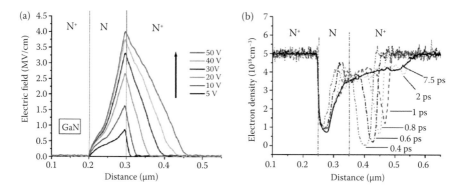

FIGURE 8.7 (a) Electric field distribution over the n^+–n–n^+ GaN channel at applied biases ranging from 5 V to 50 V. (b) Time evolution of the transient electron density profile over the n^+–n–n^+ GaN channel under an applied bias of 50 V; the steady-state result is also shown, evaluated at 7.5 ps. (Reprinted with permission from IET. Sevik, C., and Bulutay, C. *Proc. Optoelectron.* Vol. 150, p. 87, 2003. EBSCO Publishing.)

n-doped base region, as in the Kirk effect; this viewpoint is further supported by the same figures. The time evolution of the electron density profile is depicted in Figure 8.7b, starting from 0.4 ps. Oscillations around the unintentionally doped n region are clearly visible until steady state is established (7.5 ps curve in Figure 8.7b).

The fermionic degeneracy effects are seen to be operational at high fields and at high-concentration spots. However, if degeneracy is ignored, the electron distribution is observed to develop a dip in the n^+ anode region, as shown in Figure 8.8a.

At a higher applied bias (80 V in GaN), the effect of II becomes dominant. As illustrated in Figure 8.8b, this mechanism introduces a substantial energy loss mechanism for the energetic carriers that have just traversed the unintentionally doped n region.

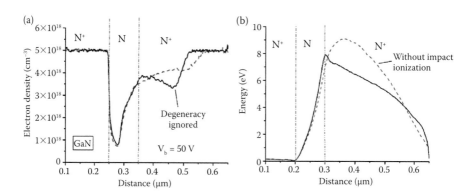

FIGURE 8.8 (a) Steady-state density profile at a bias of 50 V, with and without the degeneracy effects included. (b) Energy distribution over the n^+–n–n^+ GaN channel under an applied bias of 80 V, with and without impact ionization being included. (Reprinted with permission from IET. Sevik, C., and Bulutay, C. *Proc. Optoelectron.* Vol. 150, pp. 87–88, 2003. EBSCO Publishing.)

8.4.2   ALGAN SOLAR-BLIND AVALANCHE PHOTODIODES

Achieving ultraviolet solid-state photodiodes having internal gain due to avalanche multiplication is a major objective with a potential to replace photomultiplier tube-based systems for low-background applications [31,32]. The $Al_xGa_{1-x}N$ material with the aluminum mole fraction $x \geq 0.38$ becomes a natural candidate for the solar-blind APD applications that can also meet high-temperature and high-power requirements. Unfortunately, due to growth-related problems such as high defect and dislocation densities causing premature microplasma breakdown, there have been a limited number of experimental demonstrations of an APD with the $Al_xGa_{1-x}N$ material [33–37]. Even for the relatively mature GaN-based technology, few reports of observations of avalanche gain exist [38–41].

Within the past decade, several techniques have been reported that model the gain and time response of APDs [44]. Most, however, approximate the carriers as always being at their saturated drift velocity and impact ionization rates are usually assumed to depend only on the local electric field; see Dunn et al. [45] for references. While nonlocal effects have recently been incorporated [44], the dubious assumption on carrier drift velocity remains. Among all possible techniques, the EMC method is potentially the most powerful as it provides a full description of the particle dynamics. However, only a small number of such simulations have been reported, predominantly on GaAs-based APDs [45,46].

8.4.2.1   Computational Details

During the simulation, the Schottky barrier height is neglected in comparison to the applied very high reverse bias, whereas it should be included in the case of a forward bias. Similarly, this eliminates the subtle complications regarding the choice of a suitable boundary condition. Hence, we use the standard neutral-contact model, which keeps the charge density constant at the boundary regions via injecting or removing majority/minority carriers. More than 60,000 superparticles are employed within the ensemble and use the higher order triangular-shaped-cloud representation of the superparticle charge densities [57] to decrease the statistical noise on the current. The Poisson solver is invoked in 0.25-fs time intervals so as not to cause an artificial plasma oscillation. All computations are done for a temperature of 300 K. To avoid prolonged transients following the sudden application of a high field, the reverse DC bias is gradually applied across the APD over a linear ramp within the first 1.25 ps.

8.4.2.2   Results

The results and discussions of our simulations related to the gain and temporal response of APDs can be presented in two different sections: the GAN and AlGaN APDs.

8.4.2.2.1   GaN APDs

We first test the performance of our methodology on GaN-based visible-blind APDs where a few experimental results have recently been reported [38–41]. Among these, we choose the structure (Figure 8.9) reported by Carrano et al. [40] having a 0.1-µm-

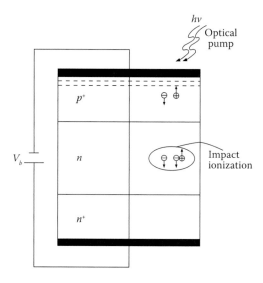

FIGURE 8.9 Structural details.

thick unintentionally doped (10^{16} cm^{-3}) n (multiplication) region sandwiched between a 0.2-μm-thick heavily doped (10^{18} cm^{-3}) p^+ region and a heavily doped (10^{19} cm^{-3}) n^+ region. Figure 8.10a shows that the current gain of this structure where, following Carrano et al. [40], the current value at 1 V is chosen as the unity gain reference point. The overall agreement between EMC and the measurements [40] is reasonable. Notably, EMC simulation yields somewhat higher values over the gain region, and the breakdown at 51 V cannot be observed with the simulations. Nevertheless,

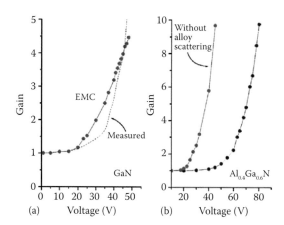

FIGURE 8.10 (a) Current gain of the GaN APD; EMC simulation (symbols) compared with measurements (dotted). (b) Current gain of the Al$_{0.4}$Ga$_{0.6}$N APD simulated using EMC with and without alloy scattering. Full lines in EMC curves are used to guide the eye. (Reprinted with permission from Sevik, C. and Bulutay, C., *Applied Physics Letters*, vol. 83, pp. 1382–1384, 2003. Copyright 2003, American Institute of Physics.)

given the fact that there is no fitting parameter used in our simulation, we find this agreement quite satisfactory.

An important characteristic of the APDs is their time response under an optical pulse. For this purpose, an optical pump is turned on at 6.25 ps creating electron-hole pairs at random positions consistent with the absorption profile of the electromagnetic radiation with a skin-depth value of 10^{-5} cm for GaN. The photon flux is assumed to be such that an electron-hole pair is created in 0.5-fs time intervals. The optical pump is kept on for 25 ps to assure that steady state is attained and afterward it is turned off at 31.25 ps to observe the fall of the current. As we are assuming a p-side illumination, it is mainly the electrons that travel through the multiplication region, even though the impact ionization of both electrons and holes are included in the simulation.

The falling edge of the optical pulse response can be fitted with a Gaussian profile $exp\left(-t^2/\tau_f^2\right)$ (see the parameters in Table 8.1). As seen in Figure 8.11 and Table 8.1, the width of the Gaussian profile increases with the applied bias. Hence, the temporal response of the device degrades in the high-gain region where a substantial amount of secondary carriers exist, as expected. The rising edge of the pulse shows an under-damped behavior, becoming even more pronounced toward the gain region; this can approximately be fitted by the function $1 - exp\left(-t^2/\tau_f^2\right)\cos(\omega_r t)$, with the parameters listed in Table 8.1.

Figure 8.12a demonstrates the electric field profile of the GaN APD. Observe that, as the applied bias increases, the moderately doped p^+ region becomes vulnerable to the penetration of the electric field, hence preventing further build-up in the multiplication region and increasing the impact ionization events. In this regard, it should be mentioned that achieving very high p doping persists as a major technological

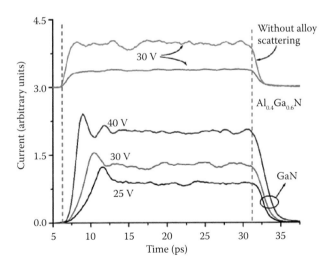

FIGURE 8.11   Temporal response of the GaN and $Al_{0.4}Ga_{0.6}N$ (vertically shifted for clarity) APD to a 25-ps optical pulse, applied between the dashed lines. (Reprinted with permission from Sevik, C. and Bulutay, C., *Applied Physics Letters*, vol. 83, p. 1383, 2003. Copyright 2003, American Institute of Physics.)

TABLE 8.1

Fitted Temporal Response Functions $exp(-t^2/\tau_f^2)$ **and** $1 - exp(-t^2/\tau_f^2)\cos(\omega rt)$
for the GaN APD

Bias (V)	τ_f (ps)	τ_r (ps)	ω_f (ps)
25	1.72	3.06	0.35
30	1.78	2.57	0.38
40	2.04	1.75	0.51

Source: Reprinted with permission from Sevik, C. and Bulutay, C., *Applied Physics Letters*, vol. 83, p.
1383, 2003. Copyright 2003, American Institute of Physics.

challenge. Therefore, in our considerations of the AlGaN APDs, we replace the problematic p^+ region with a Schottky contact.

8.4.2.2.2   Schottky contact-AlGaN APDs

With this insight, we analyze an $Al_{0.4}Ga_{0.6}N$ APD of 0.1-μm-thick unintentionally
doped (10^{16} cm^{-3}) n region sandwiched between a Schottky contact and a heavily
doped (10^{19} cm^{-3}) n^+ region. Previously, in unipolar AlGaN structures, we observed
the alloy scattering to be substantial [47], whereas the actual significance of this
mechanism has always been controversial [5]. For this reason, we provide in Figure
8.10b the gain characteristics of this $Al_{0.4}Ga_{0.6}N$ APD, both with and without alloy
scattering. The presence of alloy scattering almost doubles the breakdown voltage
with respect to the case when there is no alloy scattering. The time response of
the $Al_{0.4}Ga_{0.6}N$ APD is shown in Figure 8.11 in the low-gain region (30 V), under

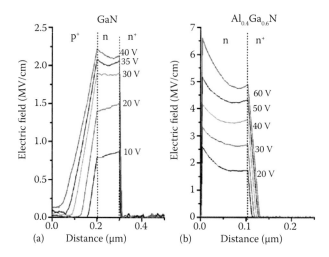

FIGURE 8.12   Electric field distribution over (a) GaN and (b) $Al_{0.4}Ga_{0.6}N$ APDs at several bias levels. (Reprinted with permission from Sevik, C. and Bulutay, C., *Applied Physics
Letters*, vol. 83, p. 1383, 2003. Copyright 2003, American Institute of Physics.)

TABLE 8.2

Fitted Temporal Response Functions $exp(-t/\tau_f)$ and $1-exp\left(-t^2/\tau_f^2\right)$ for Al$_{0.4}$Ga$_{0.6}$N APD Under a Reverse Bias of 30 V

Alloy Scattering	τ_f (ps)	τ_r (ps)
No	0.75	0.67
Yes	1.06	2.14

Source: Reprinted with permission from Sevik, C. and Bulutay, C., *Applied Physics Letters*, vol. 83, p. 1384, 2003. Copyright 2003, American Institute of Physics.

the same optical illuminations discussed in the GaN case. The falling edge of the response can be fitted by an exponential $exp(-t\,/\,\tau_f)$, whereas the rising edge by a Gaussian function, $1-exp\left(-t^2/\tau_f^2\right)$ (see Table 8.2 for the parameters).

Figure 8.12b demonstrates the electric field profile of this Al$_{0.4}$Ga$_{0.6}$N structure. It is observed that for all values of the applied bias, the electric field is confined in the intrinsic (multiplication) region, which is very desirable for the APD operation. Finally, we would like to check the standard assumption made in other theoretical APD treatments that assume the carriers travel at their saturated drift velocities. It is seen in Figure 8.13 that this assumption may be acceptable for the Schottky structure, which has a uniform field distribution within the multiplication region, whereas it is not appropriate in the p^+-n-n^+ case with our doping values. Also it should be noted that some of the wild oscillations in the n-region of Figure 8.13a are possibly due to poor statistical averaging of our EMC simulation. Any experimental support that is currently impeded by the poor AlGaN material quality will be extremely valuable in further refining our models. In other words, our simulations await verification or contradiction by other researchers.

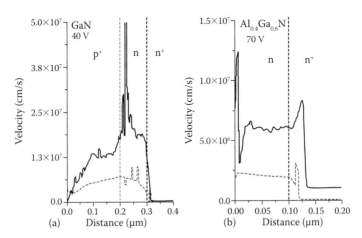

FIGURE 8.13   Average velocity distribution over (a) GaN and (b) Al$_{0.4}$Ga$_{0.6}$N APDs for electron (solid) and holes (dotted). (Reprinted with permission from Sevik, C. and Bulutay, C., *Applied Physics Letters*, vol. 83, p. 1384, 2003. Copyright 2003, American Institute of Physics.)

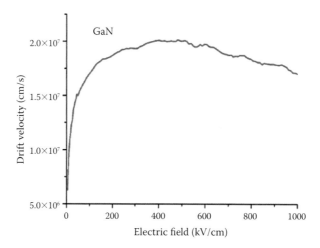

FIGURE 8.14 EMC simulation of the drift velocity versus field for bulk GaN at 300 K.

8.4.3 GUNN OSCILLATIONS IN GaN CHANNELS

At high electric fields, the electron velocity, v, in GaAs, GaN, AlN, and some other compound semiconductors decreases with an increase in the electric field, F, so that the differential mobility, μ_d = dv/dF becomes negative (see Figure 8.14). Ridley and Watkins in 1961 and Hilsum in 1962 were first to suggest that such a negative differential mobility in high electric fields is related to an electron transfer between different valleys of the conduction band (intervalley transfer). When the electric field is low, electrons are primarily located in the central valley of the conduction band. As the electric field increases, many electrons gain enough energy for the intervalley transition into higher satellite valleys. The electron effective mass in the satellite valleys is much greater than in the central valley. Also, the intervalley transition is accompanied by an increased electron scattering. These factors result in a decrease of the electron velocity in high electric fields. There are mechanisms to achieve negative differential mobility other than the intervalley transfer as well [48]. From the technological point of view, this effect is exploited to build oscillators up to terahertz frequencies.

8.4.3.1 Basics

A simplified equivalent circuit [49] of a uniformly doped semiconductor may be presented as a parallel combination of the differential resistance (see Figure 8.15)

$$R_d = \frac{L}{q\mu_d n_0 S},$$

(8.14)

and the differential capacitance:

$$C_d = \frac{\varepsilon S}{L'},$$

(8.15)

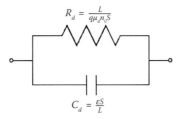

FIGURE 8.15   Equivalent circuit of a uniform piece of semiconductor.

where S is the cross-section of the sample, L is the sample length, and n_0 is the electron concentration. The equivalent RC time constant determining the evolution of the space charge is given by

$$\tau_{md} = R_d C_d = \frac{\varepsilon}{q\mu_d n_0},\tag{8.16}$$

where τ_{md} is called the differential dielectric relaxation time or Maxwell dielectric relaxation time. In a material with a positive differential conductivity, a space charge fluctuation decays exponentially with this time constant. However, if the differential conductivity is negative, the space charge fluctuation may grow with time. What actually happens depends on the relationship between τ_{md} and the electron transit time, $\tau_{rd} = L/v$. If $(-\tau_{md}) \gg \tau_{rd}$, a fluctuation of the electron concentration occurring near the negatively biased terminal (cathode) grows very little during its transit time toward the positively biased terminal (anode). However, when $(-\tau_{md}) \ll \tau_{rd}$, a space charge fluctuation grows tremendously during a small fraction of the transit time. In this case, it develops into a high-field region (called a high-field domain), which propagates from the cathode toward the anode with a velocity that is approximately equal to the electron saturation velocity, v_s.

The condition $(-\tau_{md}) \ll \tau_{rd}$, leads to the following criterion of a high-field domain formation:

$$n_0 L \gg \frac{\varepsilon v_s}{q|\mu_d|}\tag{8.17}$$

For $v = 10^5$ m/s, $|\mu_d| = 0.15$ m^2/Vs, $\varepsilon = 1.14 \times 10^{-10}$ F/m, we obtain $n_0 L \gg 1.5 \times 10^{11}$ cm^{-2}. This condition (first introduced by Professor Herbert Krömer in 1965) is called the Krömer criterion. On the experimental side, Ian Gunn was first to observe high-field domains in GaAs in 1963. Ever since, these GaAs two-terminal devices are often called Gunn diodes

8.4.3.2 Motivation
The negative differential mobility threshold field due to intervalley carrier transfer for GaN is quite high, above 200 kV/cm, which becomes appealing for building very

high-power millimeter-wave oscillators (see Figure 8.14). In addition to their techno-logical importance, these Gunn diodes still pose a number of physical puzzles, such as the detailed understanding of the domain nucleation process in different doping profiles [50]. Also, the onset of chaotic behavior [51] in these structures is another intriguing subject. As a matter of fact, the presence of impact ionization has been reported to give rise to chaotic multidomain formation [52]. This result was based on a numerical solution of a set of partial differential equations under simplifying assumptions. The EMC approach is believed to be significantly better-suited for this task [53–55], and, for instance, it has been successfully tested in the analysis of InP Gunn diodes [56].

An ever-present objective is to increase the operating frequency of the Gunn diodes. This can be achieved in several ways. Our approach is to operate the Gunn diodes at their higher harmonic frequencies rather than the fundamental. However, the drawback here is the very low efficiency associated with these high harmonic modes. Therefore, we devote much of this chapter to the harmonic RF conversion efficiency enhancement.

8.4.3.3 Computational Details

Here we employ the EMC method to shed light on the dynamics of millimeter-wave Gunn domain oscillations with large amplitudes in GaN channels. The same GaN material was the subject of another recent study with an emphasis on multiple-transit region effects on the output power [56]. Unfortunately, their analysis utilized unre-alistic values for the two important satellite valleys, chosen as 2.27 eV and 2.4 eV above the conduction band edge. These are about 1 eV higher than the experimental and theoretical values. In this work, as mentioned, the necessary band structure data is extracted from our empirical pseudopotential calculations [15] fitted to available experimental and theoretical data.

The basic structure we investigate is of the form, $n^+-n^--n-n^+$, with the active region being formed by the n^- notch with a doping of 10^{16} cm^{-3} and the main n-doped channel having 3×10^{17} cm^{-3} doping; the n^+ contact regions are assumed to have 2×10^{18} cm^{-3} dopings (see Figure 8.16). The length of the notch region is varied to investigate its effect on the harmonic operation, while keeping the total length of the active region (n^--n) constant at 1.2 μm. Our EMC simulations all start from a neutral charge distribution, and unless otherwise stated, are at 300 K. As a standard prac-tice in modeling Gunn diodes (see Dunn and Kearney [50] and references therein), a single-tone sinusoidal potential of the form $V_{DC} + V_{AC} \sin(2\pi ft)$ is imposed across the structure; in our work $V_{DC} = 60$ V and $V_{AC} = 15$ V (if not stated). This choice sig-nificantly simplifies our frequency performance analysis; its validity will be checked later on. The oscillator efficiency is defined as $\eta = P_{AC} / P_{DC}$, where P_{AC} is the time-average generated AC power and P_{DC} is the dissipated DC power by the Gunn diode. Therefore, a negative efficiency corresponds to a resistive (dissipative) device and a positive value designates an RF conversion from DC.

8.4.3.4 Results

Figure 8.17a displays Gunn domains for operations at the fundamental, second, third, and fourth harmonic frequencies for a 250-nm-notch device. As usual, the domains

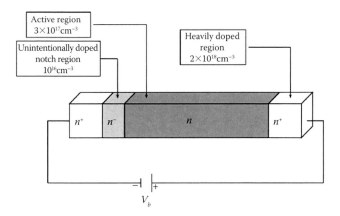

FIGURE 8.16   Structural details.

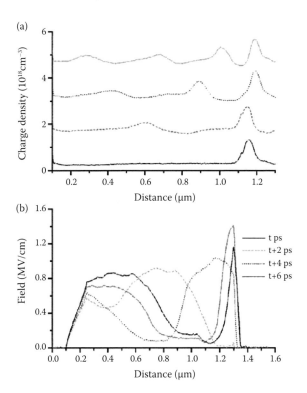

FIGURE 8.17   (a) Typical charge density profiles for a 250-nm-notch device operating at the fundamental, second, third, and fourth harmonic modes, each respectively vertically up-shifted for clarity. (b) Time evolution of the electric field profile within one period of the Gunn oscillation at 2-ps intervals for the 250-nm-notch device. (Reproduced from Sevik, C. and Bulutay, C., *Semiconductor Science and Technology*, vol. 19, page 189, 2004, Institute of Physics Publishing.)

build up as they approach the anode side. Figure 8.17b illustrates the evolution of the electric field in one period for the fundamental frequency operation (122.5 GHz). It can be noted that due to the relatively wide notch width, a significant amount of the electric field accumulates around this region, with a value that can exceed 1.2 MV/ cm (under a DC bias of 60 V), reaching the impact ionization threshold [13]. To analyze this further, we increased the DC bias to 90 V and the operating temperature to 500 K; the effect of turning off the impact ionization mechanism was observed to be marginal even at these extreme conditions for all notch widths considered.

8.4.3.4.1   Effect of Notch Width

In Figure 8.18a, different notch widths are compared in terms of their frequency performance. Our main finding is that, by increasing the notch width, GaN Gunn diodes can be operated with more efficiency at their second harmonic frequency than the fundamental, as seen for the 250-nm-notch width curve. However, we observed that increasing the notch width above 400 nm gives rise to total loss of the Gunn oscillations. These results are extracted from long simulations, up to 500 ps, to capture the steady-state characteristics at each frequency, which becomes quite demanding. Hence, the Pauli degeneracy effects requiring extensive memory storage are not included. Figure 8.18b illustrates the effect of including the Pauli exclusion principle using the Lugli–Ferry recipe [20,47]. Note that for Gunn diodes, this effect is quite negligible, slightly lowering the resonance frequencies at higher harmonics.

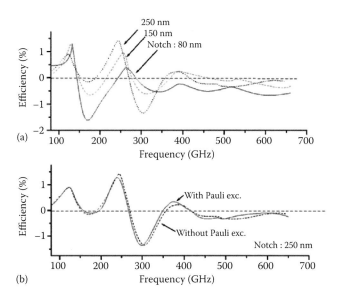

FIGURE 8.18   Gunn diode efficiency versus frequency. (a) Effect of different doping-notch widths, while keeping the total active channel length fixed at 1.2 μm. (b) Effect of including the Pauli exclusion principle for the 250-nm-notch device. (Reproduced from Sevik, C. and Bulutay, C., *Semiconductor Science and Technology*, vol. 19, page 189, 2004, Institute of Physics Publishing.)

8.4.3.4.2   Effect of Lattice Temperature

Being unipolar devices, Gunn diodes operate under high current levels which lead to excessive heating of the lattice. Therefore, we consider the effects of temperature on our results, using a 250-nm-notch device (being the most promising one in harmonic enhancement). As seen in Figure 8.19, in response to an increase in the lattice temperature, the second harmonic efficiency increases, whereas the fundamental and third harmonic efficiencies decrease. Also, as expected, the curves shift to lower frequencies due to increased phonon scattering, which reduces the carrier-saturation velocity and, hence, the Gunn oscillation frequency. The important finding is that, at even higher lattice temperatures, the second harmonic efficiency is further reinforced.

8.4.3.4.3   Effect of Channel Doping

Next, we consider the effect of channel doping again, using the same 250-nm-notch device for comparison purposes. One can easily observe from Figure 8.20 that channel doping is highly critical in the operation of the Gunn diodes. This doping is mainly instrumental in the differential dielectric relaxation time (see Equation 8.17), which decreases as the doping increases. For relatively low channel dopings, the $(-\tau_{md}) \ll \tau_{rd}$ condition gets violated, which explains the behavior displayed in Figure 8.20. Furthermore, at lower channel dopings, not only is the AC current magnitude decreased, but the phase-angle difference with respect to AC voltage also shifts from active toward lossless and dissipative regimes. If the channel doping is increased beyond 3×10^{17} cm^{-3}, then the part of the active channel on the anode side begins

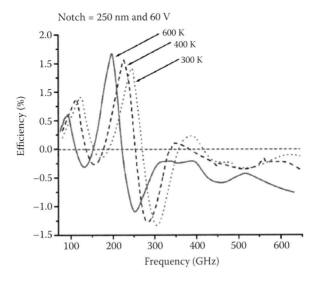

FIGURE 8.19   RF conversion efficiency versus frequency for several lattice temperatures; a 250-nm-notch device at 60-V bias is used.

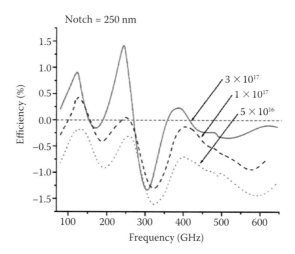

FIGURE 8.20 RF conversion efficiency versus frequency for several channel dopings; a 250-nm-notch device at 60-V bias is used.

to act like a contact, reducing the dimension of the active channel length, hence increasing the Gunn oscillation frequency.

8.4.3.4.4 Effect of the DC Operating Point

Another parameter that can affect the achievement of harmonic enhancement is the DC operating point. In order not to lose the grounds for comparison, we need to scale the RF amplitude while changing the applied DC voltage across the Gunn diode. Hence, we simulate the same 250-nm-notch Gunn diodes operating under three

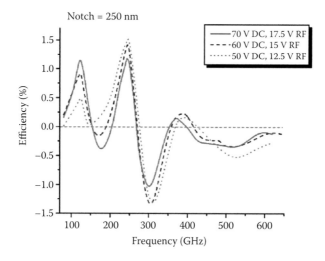

FIGURE 8.21 RF conversion efficiency versus frequency for several DC bias voltages; a 250-nm-notch device is used.

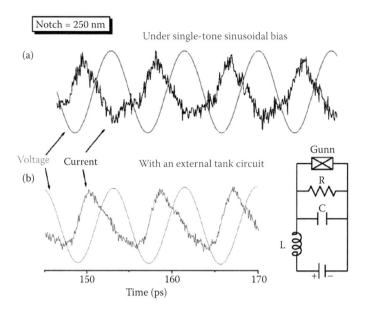

FIGURE 8.22   Current and voltage waveforms for a 150-nm-notch Gunn diode (a) under an imposed single-tone sinusoidal voltage, and (b) connected to an external tank circuit shown in the inset. (Reproduced from Sevik, C. and Bulutay, C., *Semiconductor Science and Technology*, vol. 19, page 189, 2004, Institute of Physics Publishing.)

different DC and RF bias levels, which are 50-V DC bias with 12.5-V RF bias, 60-V DC bias with 15-V RF bias, and 70-V DC bias with 17.5-V RF bias. The important observation is that the harmonic enhancement is favored toward the critical field, at which point negative differential mobility sets in (50 V case in Figure 8.21).

8.4.3.4.5   Connecting to a Tank Circuit

Finally, we relax the imposed single-tone sinusoidal voltage across the Gunn diode and connect it to an external tank circuit with the voltage across the device being self-consistently updated at each simulation step (0.4 fs) through solving in time-domain a Gunn diode in parallel with a capacitor and a resistor, all in series with an inductor, and with a DC source (see Figure 8.22b inset). The AC voltage and current of the Gunn diode are shown in Figures 8.22a and 8.22b, comparing respectively the imposed single-tone bias with the tank circuit tuned to the fundamental frequency of the 150-nm-notch device. For both cases, the current and voltage are in antiphase with each leading to RF generation as intended; the main discrepancy is higher harmonic content in the tank circuit case as governed by the quality factor of the resonator.

8.5   CONCLUDING REMARKS

In this chapter, we have described the EMC method and demonstrated its use in the simulation and modeling of charge transport in various devices such as hot-electron effects in n-type structures, AlGaN solar-blind APDs, and Gunn oscillations in GaN channels.

REFERENCES

1. T. Kurosawa. 1966. Monte Carlo calculation of hot electron problems, *J. Phys. Soc. Jpn.* 21, 424–426.
2. P. J. Price, in *Proceedings of the 9th International Conference on the Physics of Semiconductors*, R. E. Burgess (eds.), p. 355, New York: Academic.
3. H. D. Rees. 1969. *J. Phys. Chem.* Solids 30, 643.
4. W. Fawcett, D. A. Boardman, and S. Swain. 1970. *J. Phys. Chem. Solids* 31, 1963.
5. D. K. Ferry. 2000. *Semiconductor transport.* Boca Raton: Taylor & Francis.
6. C. Jacoboni and L. Reggiani. 1983. *Reviews of Modern Physics* 55:645.
7. M. Lundstrom. 2000. *Fundamentals of carrier transport*, 2nd ed. Cambridge: Cambridge University Press.
8. K. Tomizawa. 1993. Numerical simulation of submicron semiconductor devices. Boston: Artech House.
9. L. I. Schiff. 1968. *Quantum mechanics.* New York: McGraw-Hill.
10. D. Vasileska and S. M. Goodnick. 2002. *Materials Science and Engineering R* 38:181.
11. B. K. Ridley. 1999. *Quantum processes in semiconductors*, 4th ed. Oxford: Oxford University Press.
12. K. Kunihiro, K. Kasahara, Y. Takahashi, and Y. Ohno. 1999. *IEEE Electron Device Letters* 20:608.
13. C. Bulutay. 2002. *Semiconductor Science and Technology* 17:L59.
14. M. A. Osman and D. K. Ferry. 1987. *Physical Review B* 36:6018.
15. C. Bulutay, B. K. Ridley, and N. A. Zakhleniuk. 2000. *Physical Review B* 62:15754.
16. C. Bulutay, B. K. Ridley, and N. A. Zakhleniuk. 2002. *Physica B* 314:63.
17. G. Lehmann and M. Taut. 1972. *Physica Status Solidi B* 54:469.
18. M. V. Fischetti and J. M. Higman. 1991. *Monte Carlo device simulation: Full band and beyond*, ed. K. Hess. Norwell, MA: Kluwer.
19. H. Ouzman, J. Kolnik, and K. F. Brennan. 1996. *Journal of Applied Physics* 80:8.
20. P. Lugli and D. K. Ferry. 1985. *IEEE Transactions on Electron Devices* 32:2431.
21. M. Zebarjadi, C. Buluay, K. Esfarjani, and A. Shakouri. 2007. *Applied Physics Letters* 90:092111.
22. S. Bosi and C. Jacoboni. 1976. *Journal of Physics C* 9:315.
23. P. Borowik and J. L. Thobel. 1998. *Journal of Applied Physics* 84:3706.
24. M. V. Fischetti and S. E. Laux. 1998. *Physical Review B* 38:9721.
25. E. Bellotti, F. Bertazzi, and M. Goano. 2007. *Journal of Applied Physics* 101:123706.
26. D. Jena, S. Heikman, J. S. Speck, A. Gossard, and U. K. Mishra. 2003. *Physical Review B* 67:153306.
27. J. Simon, A. K. Wang, H. Xing, S. Rajan, and D. Jena. 2006. *Applied Physics Letters* 88:042109.
28. M. Farahmand, C. Garetto, E. Bellotti, K. F. Brennan, M. Goano, E. Ghillino, G. Ghione, J. D. Albrecht, and P. P. Ruden. 2001. *IEEE Transactions on Electron Devices* 48:535.
29. M. V. Fischetti and S. E. Laux. 1996. *Journal of Applied Physics* 80:2234.
30. K. Tomizawa, Y. Awano, N. Hashizume, and M. Kawashima. 1982. *IEEE Proceedings* 129:131.
31. C. Sevik and C. Bulutay. 2003. *Applied Physics Letters* 83:1382.
32. M. Razeghi. 2002. *Proceedings of the IEEE* 90:1006.
33. T. Tut, M. Gokkavas, B. Butun, S. Butun, E. Ulker, and E. Ozbay. 2006. *Applied Physics Letters* 89:183524.

34. T. Tut, M. Gokkavas, A. Inal, and E. Ozbay. 2007. *Applied Physics Letters* 90:163506.
35. T. Tut, M. Gokkavas, and E. Ozbay. 2008. *Physica Status Solidi C* 5:2316.
36. S. K. Zhang, W. B. Wang, A. M. Dabiran, A. Osinsky, A. M. Wowchak, B. Hertog, C. Plaut, P. P. Chow, S. Gundry, E. O. Troudt, and R. R. Alfano. 2005. *Applied Physics Letters* 87:262113.
37. R. McClintock, A. Yasan, K. Minder, P. Kung, and M. Razeghi. 2005. *Applied Physics Letters* 87:241123.
38. A. Osinsky, M. S. Shur, R. Gaska, and Q. Chen. 1998. *Electronics Letters* 34:691.
39. K. A. McIntosh, R. J. Molnar, L. J. Mahoney, A. Lightfoot, M. W. Geis, K. M. Molvar, I. Melngailis, R. L. Aggarwal, W. D. Goodhue, S. S. Choi, D. L. Spears, S. Verghese. 1999. *Applied Physics Letters* 75:3485.
40. J. C. Carrano, D. J. H. Lambert, C. J. Eiting, C. J. Collins, T. Li, S. Wang, B. Yang, A. L. Beck, R. D. Dupuis, J. C. Campbell. 2000. *Applied Physics Letters* 88:924.
41. B. Yang, T. Li, K. Heng, C. Collins, S. Wang, J. C. Carrano, R. D. Dupuis, R. C. Campbell, M. J. Schurman, and I. T. Ferguson. 2000. *IEEE Journal of Quantum Electronics* 36:1389.
42. M. Moresco, F. Bertazzi, and E. Bellotti. 2009. *Journal of Applied Physics* 106:063719.
43. P. J. Hambleton, S. A. Plimmer, J. P. R. David, G. J. Rees, and G. M. Dunn. 2002. *Journal of Applied Physics* 91:2107.
44. K. F. Li, D. S. Ong, J. P. R. David, G. J. Rees, R. C. Tozer, P. N. Robson, and R. Grey. 1998. *IEEE Transactions on Electron Devices* 45:2102.
45. G. M. Dunn, G. J. Rees, J. P. R. David, S. A. Plimmer, and D. C. Herbert. 1997. *Semiconductor Science and Technology* 12:111.
46. F. Ma, S. Wang, X. Li, K. A. Anselm, X. G. Zheng, A. L. Holmes, and J. C. Campbell. 2002. *Journal of Applied Physics* 92:4791.
47. C. Sevik and C. Bulutay. 2003. *IEE Proc.-J: Optoelectronics* 150:86.
48. N. Balkan, B. K. Ridley, and A. J. Vickers, eds. 1993. *Negative differential resistance and instabilities in 2-D semiconductors.* New York: Plenum Publishing.
49. M. Shur. 1996. *Introduction to electronic devices*, 1st ed. Hoboken, NJ: John Wiley & Sons.
50. G. M. Dunn and M. J. Kearney. 2003. *Semiconductor Science and Technology* 18:794.
51. E. Schll. 2001. *Nonlinear spatio-temporal dynamics and chaos in semiconductors.* Cambridge: Cambridge University Press.
52. K. Oshio and H. Yahata. 1995. *Journal of the Physical Society of Japan* 64:1823.
53. C. Sevik and C. Bulutay. 2004. *Applied Physics Letters* 85:3908.
54. C. Sevik and C. Bulutay. 2004. *Semiconductor Science and Technology* 19:188.
55. K. Hess. 1998. *Advanced theory of semiconductor devices.* Englewood Cliffs, NJ: Prentice Hall.
56. V. Gruzinskis, E. Starikov, P. Shiktorov, L. Reggiani, and L. Varani. 1994. *Journal of Applied Physics* 76:5260.
57. R. W. Hockney and J. W. Eastwood. 1988. *Computer simulation using particles.* New York: Institute of Physics.

9 Modeling Two-Dimensional Charge Devices

Afif Siddiki

CONTENTS

9.1   INTRODUCTION

The enormously fast improvements achieved in solid-state-based technologies are triggered by the unusual transport properties of semiconductors. Likewise, in the "metal age" of humanity, the technological applications of semiconductors enabled us to develop tools that (mostly) increased our quality of life. Solid-crystal materials are classified as metals, semiconductors, and insulators when comparing their electrical transport properties at room temperature. In general, metals are good conductors because there are many available states at the Fermi level that contribute to electronic transport [1]. In contrast, insulators have a large energy gap between their valance and conduction bands, and there are no available states at Fermi energy, E_F. Semiconductors, as can be understood from their name, present a smaller band gap compared to insulators. However, they still have no states at E_F under thermal and electrochemical equilibrium conditions. These classifications are illustrated in Figure 9.1. How do we calculate the energy band diagrams of a given material, starting from crystal (growth) parameters? It is a quite complicated procedure that is discussed in other chapters of this book. However, in this chapter, knowing only the energy gap of the given semiconductor device, we perform reasonably reliable transport calculations [2,3]. Here, we will focus on the electronic states and their influence on the transport properties of two-dimensional (2D) charge systems. The first section is devoted to introducing the fundamental concepts of the calculation scheme to obtain the electrostatic potential, charge distribution, and energy dispersion of a semiconductor heterojunction in one dimension (1D), namely, in the growth direction z. A particular emphasis is placed on the GaAs/AlGaAs material, which we will focus on throughout this chapter, because the main objective is to discuss the intriguing physics observed in high-mobility 2D electron systems, particularly when subject to a perpendicular **B** field. In the first section, we generalize our simple 1D calculation scheme to three dimension (3D) and obtain the electrostatic properties of the crystal by solving the equation the Poisson equation (PE) numerically, utilizing iterative methods [4]. We discuss different edge profiles of the system. Section 9.2.4 deals with the solution of the PE and the Schrödinger equation (SE) self-consistently in the growth direction to obtain the energy dispersion, where the conditions to obtain a 2D electron system (2DES) or 2D hole system will be also discussed. Next, in a short subsection, we consider multiple layers of 2D systems. In the last part of this section, we confine ourselves mainly to a single-layer 2D system and investigate

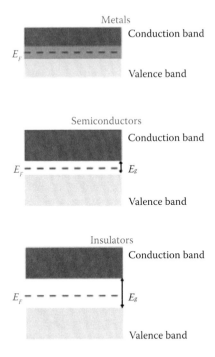

FIGURE 9.1 Energy band classifications of solid-state crystal materials. The E_F is indicated by the broken horizontal line and the energy gap is E_g. Metals are good conductors at room temperature due to the high density of states at the Fermi level, whereas semiconductors and insulators are not.

the transport properties with and without a magnetic (**B**) field. We tackle the impurity problem in a superficial manner and the narrow (quantizing) constrictions are studied in some detail within the Thomas–Fermi approximation. The next section is reserved for specific application of the previous sections, namely, the quantized Hall effect (QHE). The QHE section starts with an introductory subsection, where the mainstream theories are presented, followed by a long discussion on the interactions that are important to describe integer QHEs (IQHEs). The last section is devoted to a specific topic that concerns the many-body effects in a single layer, i.e., the fractional QHEs (FQHEs) and double-layer system, namely, the Bose–Einstein condensation (BEC) of excitonic particles and related interesting systems. We summarize our chapter in a conclusion section where we emphasize the fundamental concepts and open questions of the related field, s.

9.2 THE ELECTROSTATIC MODELING OF SEMICONDUCTOR CRYSTALS: SOLVING DIFFERENTIAL EQUATIONS

9.2.1 SOLVING THE POISSON EQUATION IN 1D

Most physics problems deal with at least one differential equation and, therefore, with the boundary conditions that determine the solution [5]. Throughout this chapter, we deal with an equilibrium state and leave out time-dependency, thereby seeking steady-state solutions. The first simple but fundamental challenge is to obtain the potential distribution for a given set of boundary conditions, i.e., to solve the Laplace equation,

$$\vec{\nabla}^2 V(\vec{r}) = 0,\tag{9.1}$$

which has no charge sources within the region that we are interested in. The solution is given by the Green's function for a test particle in a vacuum as

$$G(\vec{r}, \vec{r}') \propto \frac{e^2}{|r - r'|},\tag{9.2}$$

which resides at $r = (x_o, y_o, z_o)$ and generates an electrostatic potential at $r' = (x', y', z')$, with the boundary condition $V(\pm\infty)$. The solutions to Equation 9.1 differ by the boundary conditions dictated by the physical system. However, if there exists a charge density (distribution) in the real space, one has to deal with the PE:

$$\nabla^2 V(\vec{r}) = 4\pi\rho(\vec{r}),\tag{9.3}$$

where $\rho(\vec{r}) = \rho(\vec{r}) - \rho(\vec{r})$ is composed of positive $\rho^+(\vec{r})$ and negative $\rho^-(\vec{r})$ charge densities. As with the Laplace equation, closed formed solutions for a given system can be obtained, provided that the boundary conditions are given explicitly. Although performing self-consistent calculations, we will only consider mobile electrons and assume that the ionized donors are fixed in space. Now, we turn our attention to a semiconductor heterojunction.

By a heterojunction, we mean that two different semiconductors (i.e., with different band structures) are grown on top of each other. Such a band diagram (conduction and valance bands) is shown in Figure 9.2, which depicts an AlGaAs/GaAs/AlGaAs undoped sandwich. We confine ourselves to 1D not only for simplicity but also because it is appropriate for our self-consistent calculations concerning a heterojunction. The different heights of the potential landscape are due to the different band gaps of the materials considered. We assume implicitly the boundary conditions $V(z = 0) = V_0 = 0$ and $E(z \rightarrow \infty) = 0$. Now the challenge is to solve the Laplace equation, either analytically or numerically. After the solution of the homogeneous part is obtained, as a "good" starting point, we assume a potential distribution and add the positively charged particles (the donors) at $z = z_D$ to solve the PE:

$$\nabla^2 V(\vec{r}) = 4\pi\rho^+(z_D).\tag{9.4}$$

Up to this step, the solution of the problem does not require self-consistency, i.e., for a fixed potential charge distribution under certain boundary conditions one has to solve the above equation only once. However, if one also includes mobile (negative) charges, the solution becomes complicated and requires self-consistency, in the sense that we have to distribute the negative changes $\rho^+(z)$ such that the boundary conditions

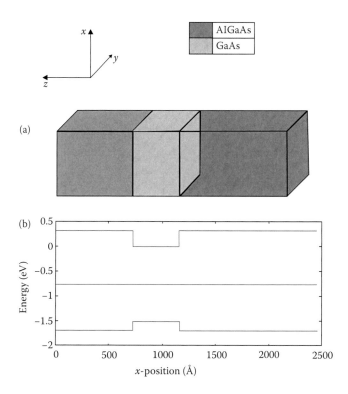

FIGURE 9.2 (a) Schematic representation of a semiconductor heterojunction. (b) The conduction (upper dashed line) and the valance (lower dashed line) bands calculated considering the heterojunction at zero temperature without doping, together with the Fermi level (straight line) pinned to the mid-gap at the surface.

are satisfied as well as the charge neutrality condition, i.e., $\int dz[\rho^+(z) - \rho^-(z)]$. The following are the explicit calculation steps of the solution procedure. Start with the given potential and obtain the Green's function for the given boundary conditions, and then calculate the potential profile generated by the positive charges while preserving the boundary conditions. Next, choose an initial distribution function for the electrons and recalculate the potential distribution, while keeping an eye on the charge neutrality condition. After the potential distribution is obtained, redistribute the negative charges (i.e., the electrons). In each iteration step, the overall electrochemical potential is kept fixed (i.e., the number of particles is constant) and the difference between two successive iterations is checked. It is important to note that direct iteration is not usually the best iterative procedure. One may use different schemes, such as Newton–Raphson, successive overrelaxation (SOR), or Gauss–Seidel, depending on the convergence criteria [6]. Figure 9.3 presents numerical results obtained by Aquilla [7]. The material properties are depicted in Figure 9.3a and the boundary conditions $V(0) = 0, \vec{E}(z = \infty) = 0$ are imposed implicitly.

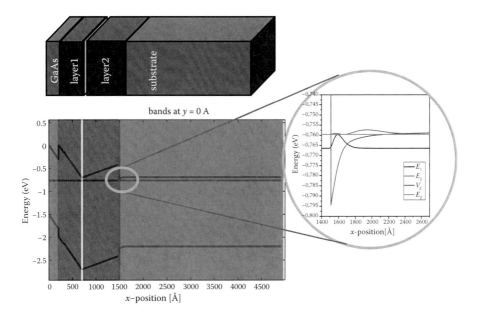

FIGURE 9.3 Self-consistently calculated energy dispersion in the growth direction considering the heterojunction shown on top; the vertical light thin line indicates the doping layer. The inset focuses on the interface; the two lowest subbands and corresponding wave functions are shown. The first excited state is unoccupied since it lies above the Fermi level.

One observes a couple of interesting differences when comparing the potential distribution with and without doping. The stepwise behavior is altered to triangular-shaped regions. The most dramatic change is observed at the interfaces: The potential at the doped side is bent, whereas the GaAs side presents a quantum well in the z direction. This behavior can be understood if one looks at the self-consistently obtained charge distribution in Figure 9.3. We see that most of the negative charges reside at the surface, $z = 0$; however, some of the electrons are confined at the boundary of the interface. We leave quantum mechanical treatment of this region to Section 9.2.3.

The above described self-consistent scheme is relatively simple when compared to realistic calculations done in 3D. In 1D, one has to perform numerical differentiation on an array defining the potential $V(z_i) = [V(z_1), V(z_2),\ldots, V(z_N)]$ to obtain the charge distribution as an array $\rho(z_i) = [\rho(z_1), \rho(z_2),\ldots, \rho(z_N)]$ when considering N (mesh) points to span the space. It is not necessary to have equidistant mesh points to perform differentiation. A better numerical method is to use finite element methods (FEMs) to span the space. Such an FEM treatment is more accurate if the potential and/or the density distribution presents irregularities, similar to the interface and doped region (for charge distribution). Such FEM codes are available in the literature [8]. As mentioned, solving the PE self-consistently in 3D is a great challenge; one has to span the 3D space by matrices corresponding to each layer while preserving the boundary conditions. We discuss such a case in the next section.

9.2.2 GENERALIZATION TO 3D

Increasing the dimensions from one to three increases the number of mesh points, N to N^3, and therefore the computational time required for the iteration process becomes enormously large. However, choosing the mesh points in a clever way (FEM) or using an accurate iteration process reduces the computational requirements considerably (both memory- and time-wise). In this subsection, we present results obtained by the latter optimization, namely, the fourth-order grid approach. Our method is based on the minimization of the points used in iteration and instead of direct iteration we use SOR. The details of the calculation scheme and its successful applications can be found in the literature [4–11]. Here, we present the results of our model calculations considering a generic sample that consists of two types of semiconductor materials, namely, GaAs/AlGaAs, and essentially vacuums surrounding the crystal. We assume open-boundary conditions, i.e., $V(\pm x, \pm y, \pm z \rightarrow \infty) = 0$, whereas the dielectric constants of the material (and vacuum) and the band gap of the semiconductors are given as input parameters, together with the geometrical constraints of the growth process. Figure 9.4 depicts the schematic drawing of the material under consideration. A GaAs undoped crystal is grown as a substrate followed by an (again) undoped 280-nm-thick AlGaAs and δ-doped silicon layer 120 nm above the 2DES. To prevent unintentional oxidation, the top-most undoped AlGaAs layer is capped by a 10-nm-thick GaAs layer. Later we will define our quantum constrictions on top of this structure either physically (by chemical etching or

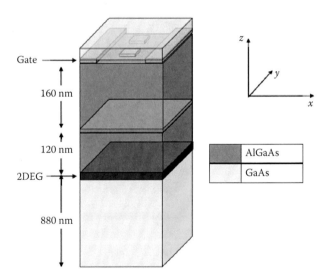

FIGURE 9.4 **(See color insert.)** The heterojunction composed of GaAs/AlGaAs, where the edges are defined by gates (gray planes on surface). Silicon doping is depicted by the thin layer and electron gas that lies 280 nm below the surface. The dielectric constants of the materials are assumed to be the same, $\kappa = 12{:}4$, and the crystal is surrounded by air in five directions, except in the substrate direction.

oxidation) or electrostatically (by gating). Figure 9.5 presents the essential outcome of the calculations, namely, the electrostatic potential (right) and the charge distribution (left) for characteristic layers: Figures 9.5a and 9.5b depict the surface; Figures 9.5c and 9.5d depict the electron layer; Figures 9.5e and 9.5f depict the donor layer; and, finally, the substrate is depicted by Figures 9.5g and 9.5h. Starting from the substrate, we see that no free charges exist and the potential is approximately constant, as expected. At the interface, a 2D electron gas is formed and the potential is complete, pointing to a perfect screening. The donor layer accumulates "all" the positive charges, while electrons at the surface pin the Fermi energy to the mid-gap of the GaAs, also at the edges of the inner contacts.

At this point, we should emphasize once more that our calculations are purely "semiclassical" in the sense that, in both one and three dimensions, we do not solve the Schrödinger equation. The "semi" prefix arises from the fact that we consider the Fermi–Dirac distribution function, albeit with a continuous energy dispersion in calculating the occupation and assuming that the electronic wave functions are of the Dirac-delta type. In the next subsections, we will treat the SE on an equal footing

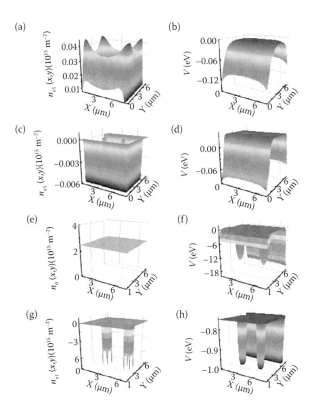

FIGURE 9.5 **(See color insert.)** 3D density (left) and potential (right) profiles calculated at zero temperature considering the structure shown in Figure 1.4. The inner (square) contacts deplete electrons at the center. Interestingly, one also observes charge fluctuations at the substrate.

in 1D, after we discuss the effects of edges on the potential profile. As mentioned before, there are mainly two different ways to define quantum constrictions on the 2D electron system, and we will now discuss those cases.

9.2.2.1 Chemical Etching or Surface Oxidation

Some of the crystal is removed by chemical etching from the top, following the predefined pattern. In Figure 9.6, we schematically present a Hall bar defined on the crystal in this manner. The details of the "real" etching process can be found in many experimental works; however, for our modeling purposes, we replace the semiconductor material for a given pattern with another dielectric material, i.e., air. The etching processes are classified as deep-etching (as deep as the 2DES or more) and shallow-etching (less than the depth of 2DES), which generate different edge profiles as shown in Figure 9.7. The difference arises from the side surface charges generated by the etching procedure, similar to the "top" surface charges that pin the Fermi energy to the mid-gap of the material. It is common to define micrometer-sized devices on the crystal by etching; however, the chemical process itself induces disorder on the system. The numerical calculation is based on a fourth-order grid approach, which minimizes the number of mesh points used at the iteration procedure. Open boundary conditions are imposed by placing a dielectric material at the boundaries of the matrix that defines our system.

At the surfaces, charges are accumulated due to the surface states pinned to the mid-gap of the material. Our sample is defined by removing the crystal, starting from the top with the dielectric; hence, the electrons are depleted from the edges. The density and potential profiles for such an etching-defined sample are shown in Figure 9.7. One can see that, for different etching depths, both the density and the potential distribution (not shown here) become steeper.

The steepness is due to the side surface charges that repel electrons from the edges. As we will see with the gate-defined samples, the edge profile will be smoother

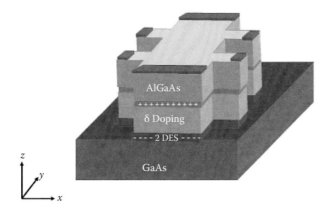

FIGURE 9.6 Chemically etched sample. The semiconductor material is removed from the edges following the lithographically defined mask as deep as the depth of the 2DES. The contacts are illustrated by metallic planes on the surface. The doping layer is depicted with "+" signs and the electron layer is depicted by "−".

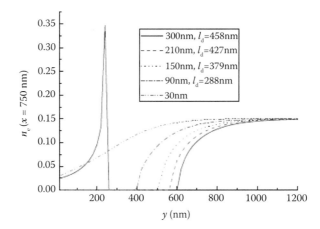

FIGURE 9.7 Electron density as a function of a lateral coordinate considering various etching depths. The deep-etched sample (black solid line) presents a peak at the edge of the sample corresponding to the side-surface charges, and the shallowest (dashed magenta line, 30 nm) etching does not deplete electron gas from the edge.

because the charges residing at the metallic (top) gates are at a distance from the 2DEG. It is important to note that the accuracy of the pattern is limited by the resolution of the mask, by which we define our geometry. If one needs to define finer structures, a fairly recent technique is used, specifically, surface oxidation. This method utilizes a charged scanning force microscope tip to create sparks between the surface and the tip, where the surface is covered by water. Hence, negative charges are accumulated on the surface within a narrow region that depletes electrons beneath it. The oxidation is a somewhat superposition of etching and gating, i.e., one removes a thin layer from the surface and generates a narrow pattern where negative charges can reside. Such an oxidation-defined pattern geometry, together with the potential profile, are shown in Figure 9.8.

Both the etching- and oxidation-defined samples have disadvantages in controlling the edge profiles, which become important when considering high perpendicular magnetic fields or charge transport. Namely, if one defines the sample by etching, the heights and the widths of the barriers that electrons have to tunnel through are fixed. However, a full control of the charge transport requires small changes at the barriers [like quantum dots, quantum-point contacts (QPCs), etc.], where even single-particle tunneling can be observed. Such systems are usually defied by metallic gates deposited on the surface, which we discuss next.

9.2.2.2 Gating

The commonly accepted method of defining structures on a 2DES is done by metallic gates deposited on the surface of the crystal. These metallic gates can be well controlled by applying appropriate potentials; hence, very precise experiments can be performed, such as single-particle transport measurements [12] and interference experiments [13]. Another advantage of gate-defined samples is the smooth potential

(a)

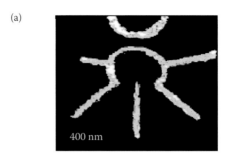

(b)

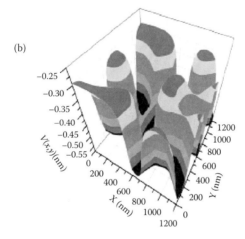

FIGURE 9.8 **(See color insert.)** (a) Scanning electron microscopy picture of the oxidation-defined quantum dot (R. Haug's group, Hannover University). (b) The self-consistently calculated potential profile.

profile generated at the edges of the pattern. As a minor illustration, Figure 9.9 shows the sample structure, the potential, and the density profiles where the applied potential is increased in relatively small steps. The drastic change happens after the gate(s) completely deplete the electrons beneath the pinch-off regime. In such a situation, the potential varies rapidly because there are no electrons to screen the external potential. At this point we would like to mention a couple of interesting but underestimated effects of the confinement potential or, in general, gate patterns. In most textbooks, quantum dots are discussed as being parabolic, and single-particle energy levels are obtained from highly degenerate harmonic oscillator solutions [2,3]. The Hamiltonian is given by

$$H = \frac{p^2}{2m^*} + V(x, y) \tag{9.5}$$

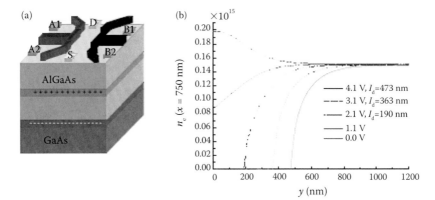

FIGURE 9.9 (a) Schematic drawing of a gate-defined Hall bar; the light- and dark-shaded gates depict the metal deposited on the surface and the white contacts are the injection (S) and collectors (D), together with the probe contacts (A1, A2, B1, and B2). (b) Actual electron density distribution considering different bias voltages. Interestingly, at zero bias, more electrons are populated beneath the gates. Since the surface is pinned to the Fermi level, the gates are relatively attractive.

where p is the momentum operator in 2D, m^* is the effective mass (equal to $0.067m_e$ in GaAs), and $V(x,y) = V_G(x,y) + V_D(x,y) + V_l(x,y) + V_H(x,y)$ is the mean-field potential composed of gates, donors, leads, and Hartree terms, respectively.

To get started, we first consider a parabolic confinement potential

$$V(x,y) = \frac{m^*\omega^2}{2}\left(x^2 + (1-\beta)y^2\right)$$ (9.6)

in describing a quantum dot (QD), with a deformation parameter β, similar to one considered by Hackenbroich and Weidenmueller [14] and Hackenbroich, Heiss, and Weidenmüller [15]. We show the energy spectrum of such a QD featuring a radius of $r = 7a_B^*$, where $a_B^* \approx 9.81$ nm is the effective Bohr radius, in Figure 9.10 for different deformation values. For the completely symmetric case ($\beta = 0$), all the states are (n_x, n_y)-fold degenerates except, of course, the ground state. The asymmetry of the parabolic confinement lifts this degeneracy; however, at certain β values, the single-particle eigenvalues bunch together ($\beta \sim 0.3; 0.36$), where some of the bunched levels present avoided crossings.

Even such a simple calculation shows that it is possible to obtain some thick levels, i.e., where the single-particle energy levels bunch together; therefore the resulting density of states (DOS) is large compared to the neighboring levels. It is useful to define the DOS to investigate the coupling between the dot and the leads, and also for further calculations, as follows:

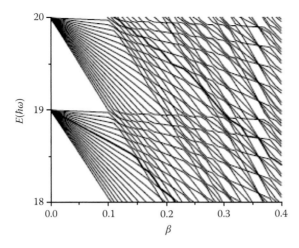

FIGURE 9.10   The energy spectrum of a parabolic QD with asymmetry in the y direction obtained by solving the Strum–Louisville differential equation in 1D. The resolution of the graph is not enough to present all the avoided crossings.

$$D_\Delta(E) = \sum_{n=0}^{N} F_n(E,\Delta) \int_{-r}^{r} |\psi_n(x,y)|^2 \, dxdy, \tag{9.7}$$

where the spectral function is defined as

$$F_n(E,\Delta) = \frac{\pi\Delta}{(E - E_n)^2 + (\pi\Delta)^2}, \tag{9.8}$$

with a broadening parameter Δ, and $\psi_n(x,y)$ is the corresponding eigenfunction.

In Figure 9.11, we show the DOS calculated for selected broadening parameters considering a wide range of asymmetry. It can be clearly seen that the wide DOS (levels) pop up once again at certain β values; however, the single-particle-level spacing is relatively small. The inset in Figure 9.11f depicts the $D(E)$ calculated for different $\Delta_a = 2|E_2 - E_1| \times 10^{-\alpha}$. If one considers a relatively large broadening Δ_2, the single-particle states are no longer visible, like the bottom-most solid (blue) line corresponding to $\Gamma \gg \delta$. Meanwhile, for Δ_3, the single-particle levels are almost visible again, at $\Gamma \sim \delta$. Certainly, introducing such a phenomenological broadening is due to the lack of leads. The level width, in fact, is determined by the lead states; therefore, it is common to describe the level width by considering the transmission rate from the pth lead state $\phi_p^*(x,y)$ to the nth dot state, where p is the eigenmode of the incoming wave (described by plane waves here), by

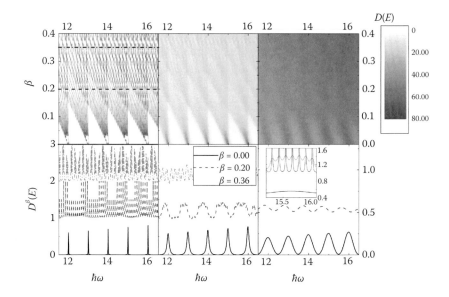

FIGURE 9.11 The calculated DOS for $\alpha = 2$, $\alpha = 2.5$, and $\alpha = 3$ (upper, left to right). Cuts are along three selected β values; peaks denote the single-particle energies. Inset: comparison of the DOS broadening at $\beta = 0.3$ for a fixed α.

$$\hbar\Gamma_{pm} = 4\kappa^2 W_n^2(a,b)(E)\left|\int dy\psi_n(x_b,y)\phi_p^*(x_b,y)\right|^2, \qquad (9.9)$$

where κ is the wave vector at the matching point, $W_n(a,b)$ is the Wentzel-Kramers-Brillouin (WKB) transmission coefficient calculated at the barrier along the classical turning points (a,b), via

$$W_n(a,b)(E) = \frac{e^{\xi(E)}}{1+\frac{1}{4}e^{\xi(E)}}, \xi(E) = -2\int_a^b dx\sqrt{\frac{2m}{h^2}(v(x)-E)}.$$

Such a description, obviously, cannot account for coupling asymmetries observed in real samples, in which we are interested.

In our toy model calculations, the leads are completely ignored; therefore, it is not clear how to relate the single-particle probability distributions $|\psi_n(x,y)|^2$ with the transmission amplitudes to the leads. Usually, the tails of the wave functions penetrating to the leads are the measure by which to define the level width, Γ. However, here we can only investigate the DOS broadening, which presents some wide states resulting from the level-bunching.

In the next step, we include phenomenological leads to our system, which break the rotational symmetry and, therefore, lift the n-fold degeneracy, even at $\beta = 0$.

For this purpose, we consider the effective potential to simulate the dot, $V_{\text{dot}}(x, y) = \frac{m^*\omega^2}{2}\left(x^2 + (1 - \beta)y^2\right)$, $x^2 + y^2 \leq r$, and the leads $V_L(x, y) = \frac{m^*\omega^2}{2}\left(r^2 + (1 - \beta)r^2\right) + V_0 y^2$, $x^2 + y^2 > r$. It is useful to investigate the $|\psi_n(x,y)|^2$ distribution to distinguish eigenstates, which are elongated along the current direction, to have an understanding of the coupling of the dot states to the leads. The spatial distribution of the wave functions considering the 23rd–30th eigenstates is shown in Figures 9.12a–9.12g, together with the potential profile, Figure 9.12h. At the bottom right, we present the energies of the eigenstates as a function of level number, n (Figure 12i).

Here, the horizontal line corresponds to a typical Fermi energy, E_F, of the QD system, namely, $E_F \sim 12$ meV $\sim 2R^*$, whereas the vertical line indicates the energy eigenstates plotted in Figures 9.12a through 9.12g. For the lowest n shown here, it is clearly seen that the electron wave function is well localized in the QD elongated in the symmetry direction. Note that we have set the asymmetry parameter, $\beta = 0{:}3$, in y. The next three eigenstates are mainly on the lead sides, however, at $n = 26$ the probability on the right lead side is enhanced.

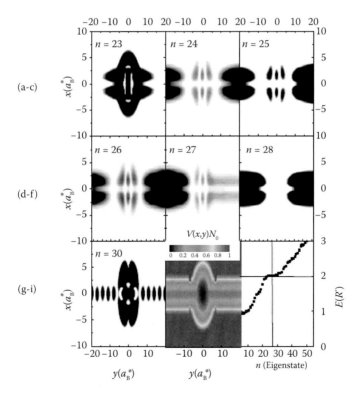

FIGURE 9.12 (a–g) The spatial distribution of the wave function considering neighboring states. (h) The toy potential profile, including open leads; red denotes high and blue denotes low potential. (i) Energy distribution as a function of the state number n, together with the horizontal line indicating the E_F.

The interesting feature occurs at $n = 27$, where the electron is mostly coupled to the left lead and is weakly coupled to the right lead. For the other states in this energy plateau, the wave functions are evenly coupled to both leads, or the electron is well-localized in the dot. So far, we have considered either a toy model of the dot and the leads or the demonstrative potential describing the system under investigation. We have shown that introducing an asymmetry to the parabolic dot lifts the n-fold degeneracy and, meanwhile, level-bunching occurs for certain β values. This level-bunching is interpreted as a broad level, taking into account the DOS distribution. Next, we included effective leads to our toy model, breaking the rotational symmetry. Together with the parabolic asymmetry, we have observed that the transmission amplitude depends almost linearly on the energy of the incoming state, assuming plane waves and within the WKB approximation. We proceed with our work by considering the real geometry and the potential profile calculated within the self-consistent Thomas–Fermi–Poisson (TFP) theory. In the following, we first discuss the limitations of such a mean-field approximation and compare our method with the existing calculation schemes in the high electron occupation regime, i.e., $N \gtrsim 100$. We show that the single-particle energy states and energies can be well-described in this regime by considering TFP theory.

The calculation of the electrostatic potential considering real sample geometries together with the electron–electron (e–e) interaction is a challenging issue. Because such a calculation cannot be done analytically in almost all cases, numerical techniques are usually deployed. It is clear that, for "more than a few" electron regimes ($N > 10$), exact diagonalization methods are either impossible or very costly in terms of computation.

It is favorable to use a mean-field approximation to describe the e–e interactions, which is questionable in the "less than a few" electron regime. The commonly used approach to determine the bare confinement potential generated by the gates is the "frozen charge" approximation [16], which takes into account the gate pattern and the effect of the spacer between the gates and the 2DES. Since it is not self-consistent, this approximation cannot account for the induced charges on the metallic gates defining the QD. The effects resulting from the induced charges and donor layer can be handled by solving the 3D PE self-consistently. Almost a decade ago, M. Stopa introduced a very effective numerical scheme to describe the electrostatics of such samples [17], including the e–e interactions, either using a full-Hartree, i.e., solving the PE and SE self-consistently, or considering the Thomas–Fermi approximation (TFA).

The exchange-correlation interaction was accounted for by a local density approximation (LDA) using the density functional theory (DFT). It was shown that the TFA is powerful enough to describe the electrostatic potential, even if the electrons are fully depleted in some regions of the sample [17].

Here, we stay with the TFA to calculate the electrostatic properties of the real sample geometry using the algorithm developed by A. Weichselbaum et al. [4,9], which implements an efficient grid relaxation technique to solve the 3D PE as mentioned. This approach was shown to be reliable in obtaining the potential profiles in the "more than a few" electron regime, considering QDs [11] and QPCs [10]. The next step in our calculation scheme is to obtain the single-particle energies and states, which perform as described in the previous section. Figure 9.13 presents the calculated potential profile for the sample geometry measured by Avinun-Kalish et al. [13]. We apply negative voltages to the gates shown in the inset. The upper and lower two gates (denoted by red areas) are kept at the same potential, V_2, while the center gate (left black) and the plunger gate (right black) are biased with a fixed voltage, V_1. Here, we consider a unit cell of 440×440 nm^2 spanned by a 128×128 mesh matrix to calculate the self-consistent potential. The surface potential is fixed to -0.75 V, pinning the Fermi energy to the mid-gap. The 2DES is some 100 nm below the surface, followed by a thick GaAs layer. To achieve numerical convergence and satisfy the open boundary conditions, three mesh points of dielectric material are assumed at all boundaries. In Figure 9.13, fixed voltages of $V_1 = -1:5$ V and $V_2 = -2:2$ V are applied, the bulk electron density is estimated to be 3×10^{11} cm^{-2}, corresponding to $E_F \approx 12.75$ meV; with the given density, the number of electrons in the dot N is close to 200. Figure 9.14 presents the calculated single-particle wave functions as a function of spatial coordinates, together with the potential counterplot and the corresponding eigenenergies versus the state number. We show the states residing at the energy plateau, which lies in the close vicinity of E_F (depicted by the horizontal solid line in Figure 9.14i). The states shown in the upper panel present a chaotic behavior, whereas the first two states of the mid-panel are the nonpropagating states. At $n = 165$, a resonant channel is observed; meanwhile, the highest state shown presents the chaotic behavior. These results show that, qualitatively, transport through state 165 is much more probable compared to the others residing at the same plateau. Although the single-particle energy eigenvalues are close to each other, a single channel is in

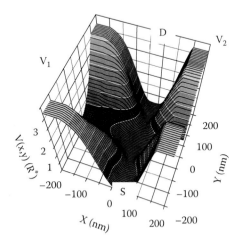

FIGURE 9.13 The self-consistent potential plotted in 3D as a function of the lateral coordinates. The lengths are in units of effective Bohr radius and energy is normalized with the effective Rydberg energy. The inset depicts the sample geometry, where S stands for source lead and D stands for the drain lead. The coupling of the QD to the leads is manipulated by changing the applied potential V_2. (From Mese, A. I., A. Bilekkaya, S. Arslan, S. Aktas, and A. Siddiki. 2010. *Journal of Physics A: Mathematical and Theoretical* 43:354017.)

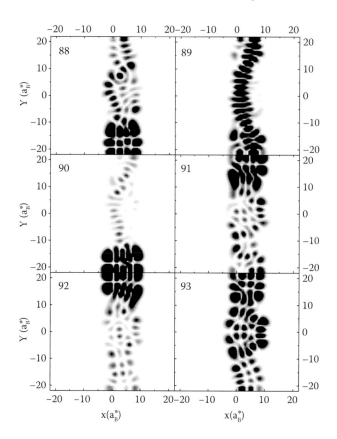

FIGURE 9.14 (a–g) Selected eigenfunctions residing at the $E = 1.96\ R^*$ plateau (indicated by the horizontal line in i) calculated for the color-plotted potential (h), together with the energy spectrum near E_F. The state = 160 presents a slight asymmetry in coupling to the source lead compared to the drain. The asymmetric potential distribution within the dot is visible. (From Mese, A. I., A. Bilekkaya, S. Arslan, S. Aktas, and A. Siddiki. 2010. *Journal of Physics A: Mathematical and Theoretical* 43:354017.)

charge of transport. At these gate voltages, the QD is loosely defined, as one can see that it is also possible to find an electron at the left side of the actual QD. This situation is changed by applying a higher negative potential to the central and plunger gates, $V_1 = -2{:}0$ V. However, the QD potential is not rotationally symmetric even if one neglects the gates, since the center gate is geometrically different from the plunger gate.

To summarize, we have calculated the self-consistent electrostatic potential exploiting the smooth variation of the bare potential within the TFA. Next, we obtained the single-particle eigenstates and energies considering a real sample geometry and crystal structure. We found, similar to the toy models, that some single-particle levels bunch and present an energy plateau while changing the state number. It was observed that, within these plateaus, not all the states contribute to the transport, since the overlap of the dot wave functions and lead wave functions simply vanish.

More interesting, we found that at intermediately high energies, the wave functions are coupled to at least one of the leads much more strongly than the ones in their close energy vicinity. This result, we believe, supports the phenomenological models, which attribute the abrupt change of the phase lapses to e–e interactions [18,19].

9.2.3 SOLVING POISSON AND SCHRÖDINGER EQUATIONS IN 1D

So far, we have only calculated the electrostatic quantities by solving the PE self-consistently. In Section 9.2.2.2, we have also shown wave functions that are calculated for a given effective potential. In these calculations the Schrödinger equation is solved only once; however, the energy eigenstates $\psi_a(\mathbf{r})$ and discrete eigenvalues E_α are not used to obtain the electron density via

$$\rho^-(\mathbf{r}) = n_{el}(\mathbf{r}) = \sum_\alpha \int dE \left| \psi_\alpha(\mathbf{r}) \right|^2 f(E, E_\alpha, T) \qquad (9.10)$$

The next step is to include quantum mechanics to our calculation scheme and obtain self-consistently the potential and density profile. Although one can write code to solve Equations 9.3 and 9.10 iteratively, we show our results obtained by Matlab® subroutine Aquilla, which is quite accurate and is also useful in visualization. An important advantage of Aquilla is that the Schrödinger equation is solved only in a predefined region, which reduces the computational effort considerably. Moreover, the calculation can be generalized to holes, and different stoichiometries are also accessible.

The first illustration in Figure 9.3 depicts a single interface; the self-consistently calculated conduction and valance bands at equilibrium are shown together with the wave functions of the two lowermost energy subbands. Because the number of dopant donors is sufficiently small, the Fermi energy lies in-between the ground state and the first excited state; therefore, a monoenergetic 2DES is formed. Due to the asymmetric confinement (triangular-shaped), the electron gas is mainly confined near the interface. The implications of such an inversion asymmetry will be discussed in the context of spin-orbit coupling in the following. Another important application of band engineering is to create multi-2DESs, placed parallel to each other. The simplest version is the so-called bilayer system, similar to Figures 9.15 and 9.16; however, interesting many-body physics takes place whenever these two electron layers are brought in close proximity. Once the intralayer direct Coulomb interactions become as important as the interlayer direct Coulomb interactions, the drag phenomena is observed [20,21] in the presence of a high perpendicular magnetic field. Moreover, at a high-mobility limit, tunneling between these layers is also observed [22]. Quite recently, a couple of groups reported an excitonic Bose–Einstein condensation featuring these bilayer systems [23–25]. We will briefly discuss these effects in later sections. Last but not least, by repeating the bilayer structures, one can obtain super-lattices and engineer artificial crystals. Since their discovery, such artificial crystals have also become a paradigm [26].

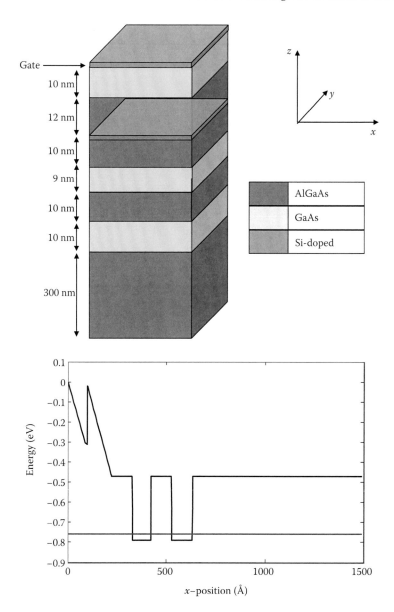

FIGURE 9.15 Illustration of the double layer structure (left) and calculated energy band diagram in z direction (right). Both layers are equally occupied and sufficiently separated such that no wave functions are overlapping the bilayer.

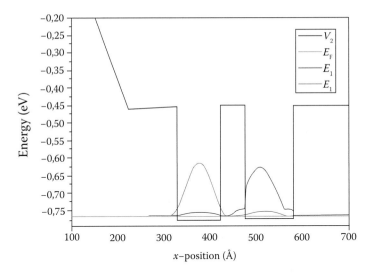

FIGURE 9.16 Two electron layers are brought into close vicinity, hence the electron wave functions overlap, yielding a finite tunneling probability under potential bias at the top gate. Note that the upper layer is less occupied, leading to a density mismatch.

We aim to discuss the transport properties of the single and bilayer systems, mainly in the presence of a high perpendicular magnetic field. In particular, the quantum Hall effects will be introduced, with a special emphasis on single-particle interactions.

9.2.4 SOLVING THE POISSON AND SCHRÖDINGER EQUATIONS IN 2D WITH A MAGNETIC FIELD

Here, the basic properties of a 2D electron gas will be introduced, first in the absence of a magnetic field, and then we will focus on the quantizing effect of the external field. A 2DES is a system where electrons are forced to move only on a plane at the interface of a semiconductor heterojunction, therefore, the system is described by the following single-particle Hamiltonian:

$$H_i = \frac{1}{2m^*}\left(p_{x_i}^2 + p_{y_i}^2 + p_{z_i}^2\right) + V\left(x_i + y_i + z_i\right), \tag{9.11}$$

where i indexes the particle. Because the Hamiltonian can be separated in z direction, one can write the energy eigenfunctions and energies of an unbounded, noninteracting (i.e., $V(x_i, y_i, z_i) = 0$) particle as

$$\psi_i(x, y, z) = \phi_i(x, y)\delta\left(z - z_i\right); \quad E_{i,n_x,n_y,n_z} = E_0 + \frac{\hbar^2\left(k_{x_i}^2 + k_{y_i}^2\right)}{2m^*}, \tag{9.12}$$

assuming mass isotropy and only the lowest subband is occupied in the growth direction. It immediately follows that the Fermi energy is given by $E_F = \dfrac{\hbar^2 k_F^2}{2m^*}$ and the number of electrons within the Fermi circle is $N = \dfrac{A k_F^2}{2m^*}$, in terms of Fermi momentum k_F. Hence, the density of states is constant, whereas the available states at the momentum space increase with the same speed of the area A, i.e., $D(E) = \dfrac{\hbar^2}{m^* \pi}$ and $E_F = \dfrac{\pi \hbar}{m^*} n_{el}$. The many-body Hamiltonian of the 2D reduced system is then

$$H = \sum_i H_i = \sum_i \left(\frac{1}{2m^*} \left(\mathbf{p}_i \right)^2 + V_{ext} \left(\mathbf{r}_i \right) + \sum_j V \left(r_i, r_j \right) \right), \qquad (9.13)$$

where the vectors are two dimensional, the second term is the external potential generated by the gates, donors, etc., and the last term prescribes the many-body interactions. To solve such a many-body Hamiltonian is usually very difficult; therefore, one makes simplifications while keeping an eye on the essential physics. The first approximation is the so-called mean-field approximation, where one replaces the many-body interaction term (the j sum) by an effective potential that represents the effect of other particles to the selected "ith" particle. This approach reduces the dimensionality of the problem drastically; however, it usually leads to the underestimation of many-body effects, such as exchange and correlation effects, that take into account the symmetry of the wave functions. In fact, the exchange effects can also be included at a mean-field level, which is then called the Hartree–Fock approximation. The replaced potential neglecting the exchange part is called the Hartree potential, which properly takes into account the direct Coulomb interaction and is given by

$$V_H(\mathbf{r}) = \int d\mathbf{r}' \rho^- \left(\mathbf{r}' \right) K \left(\mathbf{r}', \mathbf{r} \right), \qquad (9.14)$$

where particle density is given by

$$\rho^-(\mathbf{r}) = \sum_\alpha \int dE \left| \varphi_\alpha(\mathbf{r}) \right|^2 f \left(E - E_\alpha, T, E_F \right) \qquad (9.15)$$

This approximation works quite well for particles for which the spin properties can be neglected. The mean-field Hartree approximation is commonly accepted as a reasonable method to obtain electron density and potential profile. A further simplification can be made if the external potential varies smoothly on the length scale of the extend of the wave functions, namely the Thomas–Fermi approximation (TFA) [27]. Then, one can replace the actual single-particle wave function by a delta function, namely $|\phi_\alpha(r)|^2 \rightarrow \delta(r)$. This semiclassical approach takes into account quantum mechanics only by density of states and Fermi function; hence, charge density is given by [28]

$$\rho^-(\mathbf{r}) = \int dE D(E) f \frac{1}{e^{(E-E_F)/kT}+1} \qquad (9.16)$$

Considering a constant density of states, the self-consistent equation takes the form

$$\rho^-(x,y) = D_0 \int dE f\left(E, E_F, T\right), \qquad (9.17)$$

and

$$V(x,y) = V_{ext}(x,y) + \frac{2e^2}{k} \iint dx' dy' \rho\left(x',y'\right) K\left(x,x',y,y'\right) \qquad (9.18)$$

Here, the kernel $K(x,x',y,y')$ is the solution of the PE in 2D for the given boundary conditions and κ is the dielectric constant of the material. The second term describes the direct Coulomb interaction at the Hartree level, within a mean-field approximation. Note that in 2D and without a magnetic field, the chemical potential, μ, is equal to the Fermi energy. Usually it is a formidable task to find a closed form of the kernel to perform analytical calculations for general boundary conditions; therefore, one commonly does numeric calculations [10,17].

In principle, one has to repeat the calculation scheme described above to obtain the density and potential profiles, in the presence of a high perpendicular magnetic field. However, the energy dispersion is quantized due to the B field, since now the momentum operator has to be replaced by the generalized momentum, i.e., $\mathbf{p} \rightarrow \mathbf{p}^-$ $(e/c)\mathbf{A}$, hence, the single-particle Hamiltonian looks like

$$H = \frac{1}{2m^*}(\mathbf{p} - (e/c)\mathbf{A})^2 + V(x,y), \qquad (9.19)$$

where $V(x,y)$ encloses the confinement and interaction potentials, which we will investigate in the next section.

9.3 QUANTIZED HALL EFFECTS AND RELATED PHENOMENA

One of the most fascinating discoveries in semiconductor physics is the observation of the integer-quantized Hall effect (IQHE) [29], where a standard Hall experiment is performed at low temperatures and a high **B** field considering a 2DES. It was observed that the longitudinal resistance presents oscillations at low fields and, at the high field limit, the R_l vanishes at certain field intervals. The zero R_l is accompanied by constant Hall resistance, which is multiplied by a one-over-integer value, i.e., $R_H = \frac{1}{n}\frac{h}{e^2}$ (e.g., 1.34). Such a precise and reproducible universal behavior was crowned by a Nobel Prize in physics in 1985. Soon after, at higher mobility 2DESs, D. Tsui and his coworkers also observed other plateaus and zero-resistance states. However, the value of the prefactor is no longer an integer but is instead a fractional number [30]. The fractional Hall effect is commonly attributed to many-body inter-actions and is a very strongly correlated system. Moreover, this strongly correlated system presents amazing projections of the quasiparticle states. Due especially to the dimensionality, one expects to observe properties of particles that obey fractional statistics [31,32]. Another, intriguing aspect observed in these systems is a coherent state measured at a bilayer that is related to Bose–Einstein condensation-like behav-ior of excitonic particles [23].

Before proceeding with complicated and rich many-body physics, we would like to summarize the mainstream models of the quantized Hall effect in an almost chrono-logical order. We start with a single-particle Hamiltonian and include the effect of disorder induced by the remote impurities on the energy levels. The next step is to include the confinement potential calculated for a homogeneous background donor layer. The effect of interactions is investigated, starting from a classical descrip-tion, i.e., screening, and is generalized by the inclusion of many-body effects at a mean-field level. The spin degree of freedom is included in a couple of steps to our discussion as, the Zeeman, the exchange (and correlation) interaction, and spin-orbit coupling terms are investigated separately. At a final subsection we provide a brief summary.

9.3.1 BULK THEORIES

In general, bulk theories assume an infinite 2D electron gas, which can be described by the single-particle Hamiltonian given below, where all interactions are completely neglected. In the absence of a confinement potential and impurities the single-particle Hamiltonian is only composed of the kinetic part,

$$H_i = \frac{1}{2m^*}\left(\mathbf{p}_i - (e/c)\mathbf{A}\left(\mathbf{r}_i\right)\right)^2 , \qquad (9.20)$$

where the gauge invariant vector potential $\mathbf{A}(\mathbf{r}_i)$, can be described either in the sym-metric $\left(\mathbf{A}(\mathbf{r}) = \frac{\mathbf{B}}{2}(-y,x,0)\right)$, or in the Coulomb gauge $\left(\mathbf{A}(\mathbf{r}) = \frac{\mathbf{B}}{2}(x,0,0)\right)$, which is also known as the Landau gauge. In both gauges the energy eigenvalues are given by the same equation:

$$E_n = \hbar\omega_c\left(n+\frac{1}{2}\right), \quad n = 0,1,2,\cdots , \qquad (9.21)$$

where $\omega_c = e\mathbf{B}/m^*c$ is the cyclotron frequency. In this part, we will essentially con-fine ourselves to Hall bar geometry. Therefore, at this point, we give only the wave functions calculated at the Landau gauge, which are given by

$$\psi_{n,k_y}(x,y) = \frac{1}{N} e^{ik_y y} \times e^{\left(\frac{x-X_0}{2l}\right)^2} H_n\left(\frac{x-X_0}{l}\right), \tag{9.22}$$

up to a normalization factor, N, where k_y is a quasicontinuous wave number in y direction (a good quantum number) and $X_0 = -l^2 k_y$ is called the center coordinate, which defines the zero of the harmonic oscillator generated by the $\mathbf{B}$ field. As can be seen from Equation 9.21, the energy eigenvalues do not depend on the center coordinate; therefore, for each choice of ky (i.e., X_0) there is a single eigenvalue. This points to an (almost) infinite degeneracy that enriches physics. Such a description is useful in understanding the drastic effect of the $\mathbf{B}$ field on the density of states; however, it is not sufficient to investigate the transport properties. In transport, and also when discussing screening properties, one usually uses the dimensionless filling factor, ν, which is the fraction of number of electrons N_{el} to number of flux N_ϕ, in units of flux quanta $\phi_o = h/e$. The number of parabolas, N_P, shown in Figure 9.21b is given by $L_x / \Delta X_0$, where L_x is the sample dimension in x. With $\Delta X_0 = \dfrac{\hbar}{e\mathbf{B}} \dfrac{2\pi}{L_y}$ one obtains

$$N_p = \frac{L_x L_y \mathbf{B} e}{2\pi\hbar} = \frac{A\mathbf{B}e}{h} = \varphi/\varphi_0 = N_\varphi, \tag{9.23}$$

hence, the filling factor is given by

$$\upsilon = \frac{N_{el}}{N_\varphi} = 2E_F/\hbar\omega_c \tag{9.24}$$

Since the very early days of the charge-transport theory, collisions played an important role [1]. Without collisions, the charged particles would radiate when subjected to an external electric field, which is not observed experimentally in metals or semiconductors. The problem of radiation is resolved by assuming collisions of the charged particles with some impurities. Such a scattering-based definition of conduction also applies for the system at hand, i.e., a 2D electron gas subject to perpendicular magnetic field. The original Drude theory of transport assumes some scattering centers to provide finite conductivity and almost all the transport properties can be obtained from the mean-free path, l_{mfp}, the average distance where electrons can move without scattering [1]. Assuming that electrons have an average and constant drift velocity, V_D, the mean-free path is given by $l_{mfp} = V_D \tau$, where τ is the time between two scattering events, hence a quantity directly related with the material properties. Similarly, in quantum mechanics, one obtains the conductivities by investigating the effect of impurity potential on the DOS via Green's function. To obtain the single-particle DOS, one includes a single impurity potential to the bare Hamiltonian [33]

$$H_i = H_i^0 + V_{imp}, \tag{9.25}$$

and calculate the energy dependent Green's function via

$$G(E) = \sum_{\alpha} \overline{< \phi_{\alpha} \left| \frac{1}{E-H} \right| \phi_{\alpha} >} , \tag{9.26}$$

where α stands for the quantum numbers of the system, e.g., Landau index n and center coordinate X_0, and the bar denotes that an averaging is performed over all configurations of the impurity potential. Next one obtains the DOS from $D(E) = -\frac{1}{\pi} \Im m G(E)$. In the absence of any disorder, the DOS of a 2DES is given by

$$D(E) = \frac{1}{2\pi l^2} \sum_{n} \delta\left(E - E_n\right) \tag{9.27}$$

Meanwhile, the disorder broadens the DOS depending on the properties of the scattering centers. Figure 9.17 illustrates the DOS of a homogeneous 2DES (a) without a **B** field, (b) with a magnetic field in the absence of disorder, and (c) with a **B** field and disorder. We will discuss the origin and properties of disorder, while we investigate the effect of interactions. At this point, it is sufficient to underline the fact that if the 2DES is subject to a high perpendicular **B** field in the presence of impurities, the Landau levels are broadened. It is important to note that the dimensionless quantity filling factor can be visualized more easily when considering the DOS in the presence of a **B** field: At zero temperature, all the levels below the Fermi energy are completely full, therefore if the Fermi energy falls between two Landau levels, the filling factor is an integer equal to the number of states below. If the Fermi energy is equal to one of the Landau energies, then the top level is partially occupied and, hence, the filling factor is a noninteger. In principle, the disorder has two main effects on

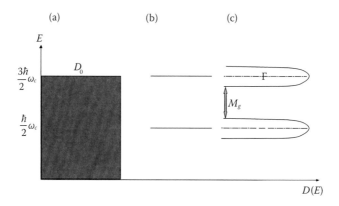

FIGURE 9.17   The density of states (a) without and (b) with **B**-field Landau levels. The LLs are broadened due to collision broadening, where the level width is given by Γ, determined by the impurities. In disordered samples, the energy gap $E_g = \hbar\omega_c$ is reduced to M_g, the so-called mobility gap.

the 2DES. The one we have briefly discussed is level-broadening, which is a purely quantum mechanical effect [34]. However, the total disorder potential also has a classical counterpart, the so-called classical localization, where electrons cannot contribute to the current because their energy is not sufficient [28]. It is important to note that, while calculating the DOS, one usually uses single impurity potentials $V_{imp}^k(r)$ and obtains quantum mechanical localization, whereas to obtain classical localization (or a percolation picture) one considers the total disorder potential, namely $V_{dis}(r) = \sum_k V_{imp}^k(r)$, where k indexes the single impurity. Next, we investigate the effect of disorder in a purely electrostatic manner.

9.3.1.1 Impurity Potential

The disorder potential resulting from the impurities experienced by the 2DES has quite complicated range dependencies. The potential generated by an impurity is (i) damped by the dielectric material between the impurity and the plane where the 2DES resides and (ii) is screened by the homogeneous 2DES depending on the density of states, which changes drastically with and without a magnetic field. Moreover, depending on the single impurity potential, the landscape of the total disorder potential profile varies considerably. It is common for theoreticians to calculate the conductivities from single impurity potentials, such as Gaussian [33], Lorentzian [35], or any other analytical functions [36,37]. However, as will be discussed, the landscape of potential fluctuations is also important to define the actual mobility of the sample at hand, particularly in the presence of an external magnetic field.

In this section, we discuss the different range dependencies of the Coulomb and Gaussian donors located at the center of a unit cell that presumes open boundary conditions. Next, the effect of the spacer thickness on the disorder potential is shown, namely the damping of the external (Coulomb) potential, and is compared with the Thomas–Fermi screening. The different damping/screening dependencies of the resulting potentials are discussed in terms of range. In the last part, we show the distinguishing aspects of the total screened disorder potential, such that the long-range part is suppressed but the short-range part is still effective.

9.3.1.2 Coulomb vs. Gaussian

We start with a textbook result: The electrostatic potential at (x_0,y_0,z_0), created by a single, positively charged particle (ionized donor) placed at x,y,z_D is given by

$$V\left(x_0,y_0,z_0\right) = \frac{e^2/\bar{\kappa}}{\sqrt{\left(x_0 - x\right)^2 + \left(y_0 - y\right)^2 + \left(z_0 - z_D\right)^2}}, \qquad (9.28)$$

where z_D and z_0 label the z position of the donor layer, and the electron gas, respectively, and $\bar{\kappa}$ is the average dielectric constant (~12.4 for GaAs).

Throughout this section, we assume that the 2DES resides on $z = z_0 = 0$ plane and the donors are placed at a finite distance (spacer thickness) $z_D > 0$; hence, the divergencies that may occur in the above equation are ruled out [38]. In principle, Equation 9.28 provides a correct description of the impurity potential generated by an ionized donor; unfortunately, such a description is not useful to define conductivities analytically [33]. Instead, one usually considers a Gaussian impurity with a potential amplitude V_{imp} generating a potential at the (x_0, y_0) plane

$$V\left(x_0, y_0, 0\right) = \frac{e^2 V_{imp}}{\kappa \left|z_D\right|} \exp\left[-\frac{\left(x_0 - x\right)^2 + \left(y_0 - y\right)^2}{2 z_D^2}\right] \tag{9.29}$$

These potentials are shown in Figure 9.18 for a unit cell of a square lattice with a relevant average dielectric constant, κ, considering a single donor residing at the center. Because the donor is at a finite distance from the plane where the electrostatic potential is calculated, no singularity is observed in the potential distribution. We should note that the electrostatic potential created by the donor is damped (we use the term *damped*, not to be confused with *screened*) by the dielectric material, which lays between the donor layer and the plane where we calculate the potential. The Coulomb potential contains a long-range component, which leads to long-range fluctuations due to overlapping if several donors are considered within the unit cell, whereas the Gaussian potential decays exponentially on the length scale comparable with the separation thickness. In Figure 9.19, we plot the potential generated by 10 donors distributed randomly at the $z = z_D \approx 30$-nm plane, both Coulomb type (Figure

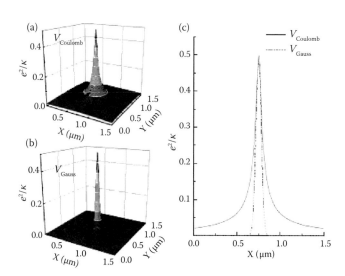

FIGURE 9.18 **(See color insert.)** (a) A single Coulomb and (b) a Gaussian impurity located at the center of a 1.5-μm × 1.5-μm unit cell, approximately 30 nm above the electron gas ($z = z0 = 0$). The short-range behaviors are similar, while the long-range parts are strongly different. Potential profiles projected through the center (x, $y = 0.75$ μm), for the Coulomb (solid black line) and Gaussian (broken red line) impurities.

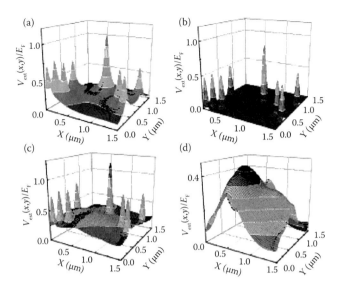

FIGURE 9.19 **(See color insert.)** External potential generated 30 nm below a plane containing 10 (a) Coulomb and (b) Gaussian donors. The range of the Gaussian potential is determined by the spacer thickness. (c) The long-range part of the Coulomb potential profile, where only the lowest two Fourier components are back-transformed to configuration space. (d) The Gaussian potential profile plus the long-range part of the Coulomb potential in order to compare the different potential landscapes.

9.19a) and Gaussian type (Figure 9.19b), together with the long-range part of the Coulomb potential (c). Because the Gaussian potential is relatively short-ranged, it is clearly seen that no overlapping of the single-donor potentials occur. Hence, the external potential experienced by the electrons can be approximated to a homogeneous potential fairly well on a length scale of $0.5 \gtrsim \mu$m. From the above observation one can conclude that approximating the impurity potential by Gaussian potentials is not sufficient to cover the long-range part of the disorder potential. Similar arguments are found also in the literature [39–41]. To overcome the difference observed at the long-range potential fluctuations between the Coulomb and the Gaussian impurities, the following procedure is applied: We perform a 2D Fourier transformation of the Coulomb potential and make a back transformation while keeping the first few momentum q components in each direction, hence, only the long-range part of the potential is left. Then, we add the long-range part of the Coulomb potential to the potential created by donors, i.e., the confinement potential. This is carried out to simulate the short-range part of the impurity potential by Gaussian impurities and calculate the Landau level-broadening and the conductivities described within the self-consistent Born approximation [33] (SCBA). The long-range part of the disorder potential is simulated by a (long-range) modulation potential and is added to the confining potential, as we describe in Section 9.3.3.

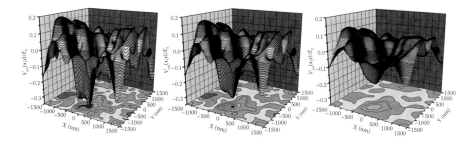

FIGURE 9.20   (See color insert.) Electrostatic potential profile generated by the random distribution of 120 ions at $z = 90$ nm (upper), $z = 120$ nm (middle), and $z = 200$ nm (lower). Similar to Figure 9.19, damping is due solely to the dielectric material in between 2DES and the donor layer.

9.3.1.3   Effect of Spacer Thickness

Next, we discuss the effect of the spacer thickness on the impurity potential experienced in the plane of 2DES. It is well-known from experimental [42] and theoretical [43] investigations that, if the distance between the electrons and donors is large, the mobility is relatively high and it is usually related to suppression of the short-range fluctuations of the disorder potential. In Figure 9.20, the external potential created by 120 donors placed 90 nm above the 2DES is shown. The potential axis is scaled with the Fermi energy, E_F, corresponding to an average electron density $n_{el} \approx 3 \times 10^{15}$ m^{-2}, $E_F \approx 10$ meV. The resulting potential shows considerable short-range fluctuations, with a variation amplitude $\sim E_F/2$ (without screening by the 2DES). If the distance between 2DES and the donors is increased to 120 nm, while keeping the donor distribution fixed, it is seen that the potential profile becomes smoother and the long-range part of the potential fluctuations is more pronounced, although the amplitude of the potential variation remains at the same order of magnitude. Increasing the spacer thickness further to 200 nm, it is observed that the short-range fluctuations are approximately smeared and a slowly varying potential landscape is left. These results agree with the experimental observations of high-mobility samples and are easy to understand from the z dependence of the Fourier expansion of the Coulomb potential,

$$V^{\vec{q}}(z) = \int d\vec{r}\, e^{-i\vec{q}\cdot\vec{r}} \sum_{j}^{N} \frac{e^2/\kappa}{\sqrt{\left(\vec{r}-\vec{r}_j\right)^2 + z^2}} = \frac{2\pi e^2}{\kappa q} e^{-|qz|} NS(\vec{q}) \tag{9.30}$$

where $S(\vec{q})$ contains all the information about the in-plane donor distribution, and N is the total number of the ionized donors. We observe that, if the spacer thickness is increased, the amplitude of the potential decreases rapidly. We also see that the short-range potential fluctuations, which correspond to higher order Fourier components, are suppressed more efficiently. Before proceeding to the screening properties, it is useful to note that for low-mobility samples, the range of the potential fluctuations generated by the donors is on the order of a few hundred nanometers and the amplitude of the variation is around 50% of the Fermi energy. For high-mobility samples, the range of the fluctuations is at the order of a few micrometers and the amplitude of the variation is only 20% of the Fermi energy. Although these quantitative estimations are only valid for our specific number of donors and their distribution, in general, they hold qualitatively, as we will show later. Therefore, we will consider the highly fluctuating confinement potential to be low mobility and the opposite to be high mobility, which then enables us to estimate mobility qualitatively by measuring the fluctuation (in fact, modulation) amplitude. We performed 3D self-consistent simulations to get appropriate estimates depending on the properties of the disorder. The screening treatment of the disorder potential is left to the interactions section, as an example of linearity of the problem.

The bulk picture has improved our understanding of the QHE since the early days; however, its reliability can be questioned if one starts to investigate narrow samples where edge effects may become important. Moreover, the most important features of the QHE are the high reproducibility and precession; these two properties are not well-covered by the bulk theory. First, at each disorder configuration, the level-broadening changes; this may change the constant Hall resistance value at every different sample, which is not the case. Second, at very high-mobility samples, the quantized Hall resistance should vanish; however, in such samples, one can still observe a well-developed IQHE. Consequently, since the mid-1980s people have investigated the effects arising from the finiteness of the system, namely the edge theories that we will discuss next.

9.3.2 Edge Theories

As can be understood by the name, edge theories consider the physical boundary conditions to be important in describing the QHE [44]. The main idea is to include the confinement potential to the single-particle Hamiltonian and study the transport properties. Figure 9.21a shows the Landau levels of an unbounded system, whereas Figure 9.21b demonstrates the effect of the boundaries when the Landau levels are

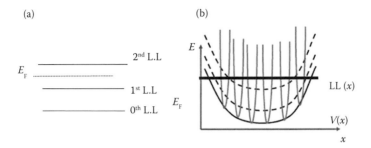

FIGURE 9.21 Landau levels of (a) an unbounded system and (b) a confined system. Landau levels are bent due to the confinement. The parabolas present the induced magnetic confinement.

bent at the edges due to the electrostatic confinement together with the magnetic confinement. The effective (single-particle) Hamiltonian is now

$$H_i = H_0 + V_{\text{conf}}(x) \tag{9.31}$$

Because we take the Landau gauge, it is convenient to assume a translational invariance in the y direction and drop the y coordinate. The actual shape of the confinement potential is determined by many parameters, such as how the crystal is grown and what edge-defining method is used (etching or gating). We have already discussed these issues, however, at the basic level we will consider here a slowly varying potential to be able to implement Thomas–Fermi approximation. Later, we also discuss hard wall-edge potential and its implications, such as a cleaved-edge over grown samples. The TFA essentially replaces the position-dependent confinement potential by a constant, which is determined by the local value of the real potential at X_0. To be explicit, we make the approximation $E_{N,X_0} \approx E_n + V(X_0)$. This is valid because k_y is quasicontinuous and, therefore, the distance between two center coordinates $\Delta X_0 = -l^2 \Delta k_y = -l^2 \dfrac{2\pi}{L_y}$ is infinitesimally small when $L_y \to \infty$. Hence, one can also bend the Landau levels following the confinement potential, as shown in Figure 9.21b. Such a smoothly varying potential is not always the case; it is useful to look at the solution of the Hamiltonian in the case of a linear potential. In this situation, the wave functions are shifted opposite to the slope of the potential and the shifting amount is given by [28]

$$\Delta X_e = -\frac{eE_x}{m\omega_c^2}, \tag{9.32}$$

where the electric field can be calculated by $\dfrac{1}{e}\dfrac{\partial V(x)}{\partial x}$. In the case of a hard wall, one has to solve the Schrödinger equation for a given center coordinate and the

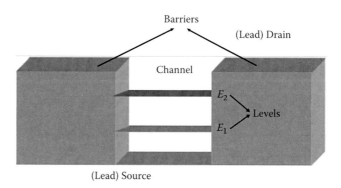

FIGURE 9.22 A schematic drawing of an ideal 1D quantum constriction, the first two eigenenergies are shown.

solutions are given by Weber functions of the second kind [45]. When including direct Coulomb interaction, such a rapidly varying potential affects edge states drastically. The important aspect of the edge theories is not defining the position-dependent Landau levels; it is determining transport of such systems by means of the above-mentioned edge states.

This transport formalism is known as the Landauer–Büttiker (LB) formalism, generalized by Büttiker [46] utilizing the arguments of Landauer [47] to describe ballistic transport within quantum mechanics in 1D. Figure 9.22 shows a schematic drawing of a double-barrier system, where a quasi-1D channel is induced perpendicular to the current direction. Note that there are an almost infinite number of states at the leads, whereas, due to the size quantization, there are finite and discrete energy levels within the barriers (at the channel).

Therefore, one obtains the conduction via tunneling probabilities from the nth state to m^{th} state T_{nm} by summing all transmission probabilities

$$G = \frac{2e^2}{h} \sum_{nm} |T_{nm}|^2 \qquad (9.33)$$

The prefactor 2 points the spin degeneracy of each channel (level). Consider a case such that the chemical potential of the system is larger than the ground state, but smaller than the first excited state. Then the electrons at the Fermi energy may tunnel through the barrier with a finite probability. If the tunneling probability is unity, two electrons will be transported from the injection lead to the drain lead, hence, conduction will be $2e^2/h$. The next step is to extend this formalism for a 2DES subject to a perpendicular magnetic field. Let's assume that the Fermi energy (i.e., the number of particles) is known, then one can estimate the crossing point of the E_F and Landau levels, where DOS is highly degenerate. Similar to the 1D case, transport takes place only at the channels that coincide with Fermi level. These channels are called the (Landauer–Büttiker) edge states (ES) and are used to explain many interesting aspects of the QHE. These edge states are nothing but quantized skipping orbits as demonstrated in Figure 9.24a. In equilibrium, the number of forward movers are equal to the number of backward movers (more physically, $I = n^+.k_y + n^-(-k_y)$ = 0), therefore the total current is zero. However, if one applies a potential difference between the source and drain, the electrochemical potential of forward movers are increased; hence, one measures a finite current. The edge states suppress backscattering; therefore, the longitudinal resistance vanishes. Meanwhile, the Hall resistance is quantized to the number of edge channels [3]. Although LB ES are useful, this formalism fails to explain the transition between the Hall plateaus and requires the utilization of localization (or, in general, bulk states). Another discrepancy of the LB-type edge picture arises from the electrostatics: As seen in Figure 9.24b, the corresponding electron density profile is far away from being stable due to its step-like shape. Therefore, it was proposed that interactions, and in particular direct Coulomb interactions, may become important in understanding the formation of the edge states [27,45,48,49]. In the following, we will discuss the effect of interactions starting from screening of a homogeneous "metal-like" 2D system.

9.3.3 INTERACTIONS

In the early days of QHE, the role of interactions was completely neglected since it was assumed that the kinetic energy of the system was dominating the Coulomb energy. One of the arguments was that the screening length $a_0 = a_B^*/2$ is an order of magnitude smaller that the mean electron distance (Fermi wavelength, $\lambda_F = 30$–40 nm); therefore, interactions should not be as important as the bulk or edge effects. However, in the late 1980s and early 1990s, A. M. Chang [50], R. R. Gerhardts [45], L. Glazman [51], and their coworkers phenomenologically, numerically, or analytically included interactions. It was conjectured by A. M. Chang that the 2DES splits into compressible and incompressible strips and current is carried by the incompressible strips. We will discuss these strips and their properties in greater detail in the following sections; however, it is sufficient for the moment to make the following statements: (i) a system is compressible whenever the chemical potential is a continuous function of the number of articles, which happens only if the energy dispersion is gapless; and (ii) the system is called *incompressible* if there is a gap in the energy spectrum, where chemical potential changes abruptly while changing the number of particles. Therefore, if the chemical potential equals one of the Landau levels, one can keep adding particles due to the highly degenerate DOS, and the system is compressible. In contrast, if the chemical potential is not equal to one of the two consequent Landau levels, one has to pay an amount of energy, $\hbar\omega_c$, to add another particle to the system; the system is then called *incompressible*.

In the following section, we will investigate the response of the electronic system to the external electric and magnetic fields, starting from screening. As can be guessed, screening properties of these strips are quite different and highly nonlinear, depending on the **B** field. The single-particle interactions will be discussed in the frame of a mean-field approximation and a special emphasis will be given to direct Coulomb interactions. Next, we will include the spin degree of freedom via the Zeeman effect and discuss the indirect interactions at the DFT level, namely the Thomas–Fermi–Dirac approximation, followed by a very brief discussion on spin–orbit coupling.

9.3.3.1 Screening

The response of a mobile charge medium to an applied external field is generally described by the dielectric function, ε. Here we assume that the external fields are constant in time, but smoothly vary on spatial coordinates. In such a case, the mobile charges try to cancel out the external field (or potential) by changing their positions, of course, if they are able to do so. The screened potential in the momentum q space is given by

$$V_{scr}(q) = V_{ext}(q)/\varepsilon(q) \qquad (9.34)$$

One should not use the relevant dielectric function to describe the response of the 2DES. As an example of linear screening we continue our discussion on the disorder potential.

9.3.3.2 Linear Screening of the Disorder Potential

Here we discuss electronic screening of the external potential created by the donors previously discussed. For a dielectric material, the relation between the external and the screened potentials are given by Equation 9.34, where $\varepsilon(q)$ is the dielectric function and is given by

$$\varepsilon(q) = 1 + \frac{2\pi e^2 D_0}{\bar{\kappa} |q|},\tag{9.35}$$

with the constant 2D density of states $D_0 = \dfrac{m}{\pi\hbar^2}$ in the absence of an external **B** field, and is known as the Thomas–Fermi (TF) function. When considering small length scales, it is convenient to rewrite the dielectric function in terms of effective Bohr radius, $a_B^* = \bar{\kappa}\hbar^2/(m^*c^2) \approx 9.81\,\text{nm}$. This simple linear relation, together with the TF dielectric function, essentially describes the electronic screening of the Coulomb potential given in Equation 9.30, if there are sufficient electrons [39] ($n_{el} > 0.1 \times 10^{15}\,\text{m}^{-2}$). Consider a case where the q component approaches zero. Then, the external (damped) potential is well-screened, hence, the long-range part of the disorder potential. The short-range part, in contrast, remains unaffected (i.e., high q Fourier components). Now, we turn our attention to the second type of impurities considered: the Gaussian ones. As is well-known, the Fourier transform of a Gaussian is also of the form of a Gaussian; therefore, similar arguments also hold for this kind of impurity. However, as seen in Figure 9.18, the long-range part of the single Gaussian impurity is already small compared to the Coulomb one; hence, the screened potential has almost the same shape as the external potential with a damping of its amplitude. To make quantitative estimations about the range of the potential fluctuations, we now numerically calculate the screened potential generated by the Coulomb donors, which are distributed randomly on the $z = 120$ nm plane. A 2D Fourier transformation is performed to obtain the potential screened by the 2DES at zero temperature and magnetic field. In the left panel of Figure 9.23, the external

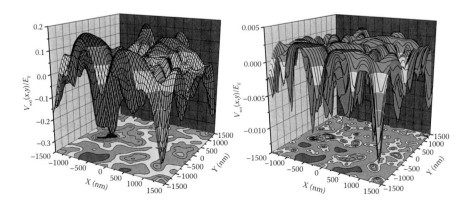

FIGURE 9.23 **(See color insert.)** The external potential generated by random distribution of 90 ions, residing 120 nm above the 2DES (left). Damping is due solely to the dielectric spacer. Screened potential by the 2DES at $T = 0$ and $B = 0$, for the given external potential, within the linear screening regime (right).

potential (damped by the background dielectric material) generated by such donors is depicted. As previously discussed, the amplitude of the variation is on the order of half of the Fermi energy, and the range of the fluctuations is around 200–300 nm, which will be considered as low mobility later. The electronically screened potential is shown in the right panel of Figure 9.23. The result is quite interesting: The long-range part of the external potential is well-screened, but the short-range fluctuations are still distinguishable, although the amplitude of the variation is now around 1% of the Fermi energy. This can be well understood if one considers the dielectric function (Equation 9.34), as q becomes smaller (i.e., for the long-range part), $\varepsilon(q)$ becomes larger, resulting in a small amplitude within the linear screening regime. One can, therefore, conclude that the screening is more effective for the small-q Fourier components, which correspond to the long range of the external potential. Here, we should emphasize once more the clear distinction between the effect of the spacer on the external potential and the screening by the 2DES, i.e., via $\varepsilon(q)$. The former depends on the Fourier transform of the Coulomb potential and the important effect is the different decays of the different Fourier components (see Equation 9.30), so that the short-range part of the disorder potential is well-dampened, whereas the latter depends on the relevant DOS of the 2DES and the screening is more effective for the long-range part.

We summarize our findings as follows: On one hand, the spacer thickness strongly reduces the amplitude of short-range fluctuations and hence increases the mobility, whereas the long-range fluctuations are less affected. On the other hand, screening of the 2DES strongly affects the long-range fluctuations and much less the short-range fluctuations. Therefore, it is reasonable to treat short-range and long-range potential fluctuations differently. The short-range part is included in our calculations via the SCBA while calculating the level-broadening and conductivities, and the long-range part is included at the self-consistent level. In the next section, we will discuss the long-range parts of the potential fluctuations while numerically investigating the Coulomb interaction of the 2DES.

9.3.3.3 Direct Coulomb

The effect of classical Coulomb interaction, i.e., the Hartree term, is included in our calculations by means of a single-particle mean-field approximation as described in Equation 9.15, where the density is given by Equation 9.17. The only difference between the zero magnetic field and finite magnetic field is due to the form of the DOS. That is, we replace the constant DOS (D_0) with the one given in Equation 9.26, if we are still in the TFA limits. Otherwise, one has to obtain the density starting from the calculated appropriate eigenfunctions. Before presenting numerical results, we would like to discuss the different screening properties of the 2DES, which arise from the quantization of the applied **B** field. As we have mentioned before, depending on the Fermi level, the system may become compressible or incompressible [50]. Previously, while considering an infinite system, the electron gas is either compressible or incompressible. However, when one also considers the edges, the Landau levels become position-dependent; hence, the 2DES splits into two different regions. At the compressible regions, the Fermi energy equals the Landau levels or, at the incompressible regions, E_F falls between to LLs. If one neglects direct Coulomb

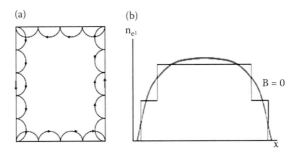

FIGURE 9.24 (a) Skipping orbits at filling factor one; quantization of the center coordinate leads to Landauer–Büttiker edge states. (b) The corresponding step-like density profile (black solid line) compared with zero magnetic field electron density (red line).

interactions, i.e., the Landauer–Büttiker formalism, only the edge channels are compressible and the regions between the channels are incompressible, where the electron density presumes a constant value [48]. These are the regions that essentially suppress backscattering and lead to IQHE. Introducing interaction(s) changes this picture slightly and enlarges the compressible regions. In short, interactions interpolate the density distribution between $\mathbf{B} = 0$ and the LB picture, which cures the electrostatic nonequilibrium arising from the step-like behavior (see Figure 9.24). The calculated results of the density and potential profiles are shown in Figure 9.25. (a) We see that the electron distribution varies if the system is compressible; the

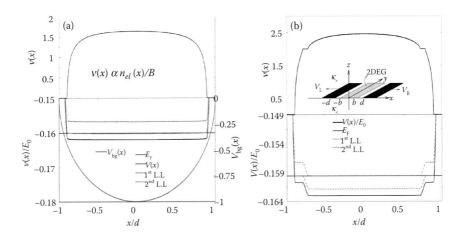

FIGURE 9.25 Electrostatic quantities calculated at two typical B values, where the system is (a) completely compressible and (b) two-edge incompressible strips are formed. (upper) Density changes monotonically at (a), where the screened potential is at the electron region 8 (lower). In contrast, (b) the density is constant at the incompressible strips and varies at the compressible regions. The screened potential varies at the incompressible strips due to poor screening. The green lines are the lowest two Landau levels that follow the total potential, whereas the Fermi level is constant throughout the sample (blue thick line).

screened potential is given in the left lower panel. (b) Such behavior points to a metal, whereas, at the incompressible regions, the density is constant and potential varies, similar to an insulator. Here, we would also like to focus on the different screening properties of these regions: At the compressible region, the DOS is high and screening is nearly perfect; in contrast, at the incompressible regions, there are no available states and screening is poor [52]. This behavior can be easily seen from Equation 9.35, where the DOS dependency of the dielectric function is given explicitly. Now, we proceed with our numerical results where the electron density and potential profiles are calculated self-consistently via solving Equations 9.10 and 9.18, where the single-particle Hamiltonian is given by

$$H_i = H_0 + V_{\text{conf}}(x) + V_H(x) \tag{9.36}$$

In our calculation scheme, we start with a given confinement potential generated by the donors (without magnetic field), similar to Section 9.2.3, then obtain the screened potential as an initial condition for the $\mathbf{B} \neq 0$ calculations. The explicit form of the kernel depends on the boundary conditions; however, we confine ourselves to the historical Chklovskii [51] geometry, where the donors, gates, and electrons reside on the same z plane depicted in the inset of Figure 9.25b. We assume that the negatively biased in-plane gates deplete the electrons from the edges, hence, the 2DES is formed in the interval $-b < x < b$, whereas the donor layer extends to the physical edges $(-d < x < d)$. The boundary conditions are then $V(-d) = V(d) = V_G$, and we set $V_G = 0$ without loss of generality. The electron density and potential distributions are shown in Figure 9.25, featuring two characteristic $\mathbf{B}$ values. In the left panel, the system is completely compressible, i.e., the lowest spin-degenerate Landau level is partially occupied; therefore, the electron distribution (in fact, the filling factor distribution) varies monotonously in the $x < |b|$ interval. Meanwhile, the potential is approximately flat at the compressible region due to nearly perfect screening. The right panel depicts the same quantities considering a slightly lower $\mathbf{B}$ field, where the center filling factor is larger than two. We observe that two incompressible (edge) strips are formed at two edges of the sample, where the Fermi energy falls between two consequent Landau levels. An interesting case occurs when the center of the sample becomes incompressible, namely $\nu(0) = 2k$. We call this case the *bulk incompressible region*, which will become very important when discussing the different regimes of the IQHE. The evolution of the incompressible strips as a function of $\mathbf{B}$ is shown in Figure 9.26, where the highlighted colors indicate the positions of the (even) integer filling factors. At high $\mathbf{B}$-field values, only the lowest LL is occupied and the system is compressible, similar to the case depicted in Figure 9.25a, and the density profile will look similar to Figure 9.25b if one cuts through the broken line in Figure 9.26. Decreasing the $\mathbf{B}$ further, one observes that the edge ISs become narrower and narrower until the next Landau levels start to be occupied at the bulk ($\mathbf{B} \sim 0.9$). Now, the question is whether our TFA is still valid when the IS becomes very narrow, which we address next. In the previous discussion, the TFA is utilized to obtain electrostatic quantities, which implicitly assumes that the total potential varies smoothly on the magnetic length. However, after the ISs become narrow, the total potential starts to vary rapidly on the length scales of the strip width; hence, TFA becomes

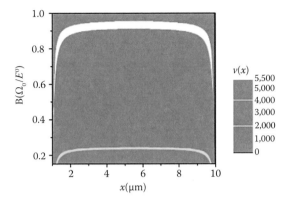

FIGURE 9.26 The evolution of the incompressible strips as a function of a magnetic field (*y* axis) and lateral coordinate (*x* axis); colored regions are compressible whereas the highlighted colors indicate the incompressible strips.

questionable. A better approximation is to use the unperturbed Landau wave functions to calculate the electron density, instead of using Dirac-delta functions, and relax the TFA to the so-called quasi-Hartree approximation [49]. In Figure 9.27, we show two cases where (a) the extent of the wave function is small compared to the width of the IS and (b) a case where the wave function exceeds the width, hence TFA fails. The quantum mechanical correction to the TFA yields "leaky" incompressible strips, explicitly saying the electrons now can scatter through the IS; therefore, the calculated longitudinal resistance is no longer zero. This is the very important outcome of the self-consistent Hartree calculations: The incompressible strips vanish if their width become comparable with the wave extend. The consequence of this statement is that, for a given magnetic field, there may exist only one edge state, different than the Landauer–Büttiker picture. Therefore, the useful method of counting edge states and obtaining the Hall resistance cannot be used anymore. This is somewhat disappointing for the transport calculations and one now has to develop a local conductivity model to obtain global resistances. We will introduce briefly such a local model in the following sections; however, there are other missing terms in our single-particle Hamiltonian, such as the spin-dependent many-body effects. Next, we discuss the bare spin-splitting and obtain odd-integer incompressible strips.

9.3.3.4 Zeeman Effect

The spin degree of freedom of the electrons has been neglected up to now because the effective Landé g^* (= −0.44) factor of the bulk AlGaAs material is relatively small when compared to the g factor in vacuum (~2) [28]. The single-particle Hamiltonian, which includes the Zeeman effect, then becomes

$$H_i = H_o + V_{conf}(x) + V_H(x) + \hat{\sigma} g^* \mu_B \mathbf{B}, \tag{9.37}$$

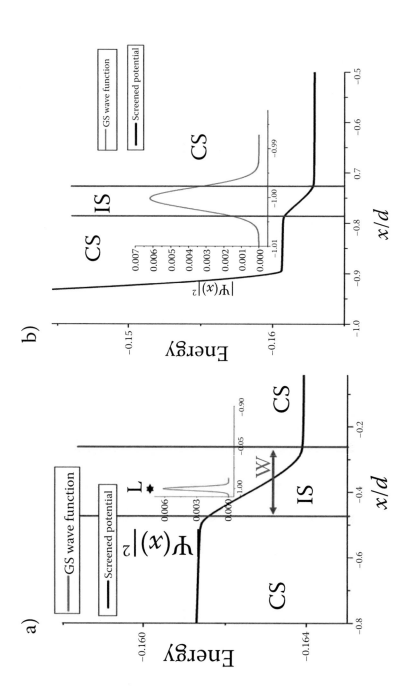

FIGURE 9.27 Wave extends of the ground state are compared with the widths of the incompressible strips. (a) The TFA is valid since potential varies smoothly on the scale of wave extend. (b) The TFA fails due to strong potential variation, resulting in leaky incompressible strips.

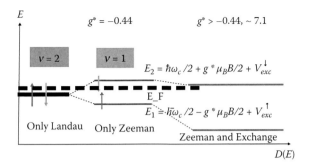

FIGURE 9.28 Schematic presentation of the density of states, without spin-splitting (left), with bare effective g^* (middle), and exchange enhanced split-Landau levels.

where $\hat{\sigma} = \pm\dfrac{1}{2}$ is the electron spin and μ_B is the Bohr magneton. As a rough estimate, the Zeeman energy is one-sixtieth of the cyclotron energy; therefore, one can easily neglect the spin-splitting of the Landau levels. However, as we will discuss in the next subsection, the effective g^* is enhanced due to many-body effects, specifically exchange interactions. A schematic presentation of the Landau-level splitting is shown in Figure 9.28, considering an infinite system, where we also show the effect of many-body interactions. In a first attempt to understand the spin-resolved edge states, we make a very simple approximation and insert an effective g^* taken from experimental findings. Figure 9.29 presents the filling factor widths at $\nu = 1$ considering different effective g^* factors. We see that, at the bare -0.44 value, a very narrow incompressible strip can be observed that changes by increasing the empirical value. The argument about the formation of incompressible strips depending on the wave extend also holds here; however, note that the Zeeman gap is relatively small compared to the Landau gap, therefore, the odd integer incompressible strips depend

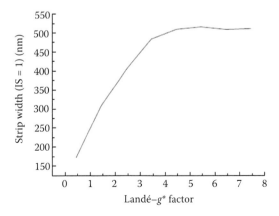

FIGURE 9.29 The variation of the $\nu = 1$ incompressible strip with the empirical g^* factor at a finite temperature ($T = 1$ K) and fixed $B = 7.8$ T.

more strongly on the external conditions [53,54] such as temperature, etc. The next step is to calculate exchange enhancement of the effective g^* factor in a mean-field approximation and compare the widths of the odd-integer incompressible strips with the ones calculated empirically.

9.3.3.5　Indirect Coulomb Interactions and Spin-Orbit Coupling

As mentioned in Section 9.1, it is a formidable task to perform numerical calculations considering many-body effects and, therefore, we employed a mean-field approximation to include interactions at a basic level. A reasonable way to deal with many-body interactions is to utilize DFT and obtain exchange and correlation potentials using the local density approximation (LDA). The literature is rich in the sense that one can choose an appropriate DFT [55] description of the exchange potential to determine the effective g^* factor. Literally, we will stay within the TFA and add an exchange potential to our single-particle Hamiltonian to mimic effects coming from many-body interactions; such an approach is known as the Thomas–Fermi–Dirac approximation and we utilize the Tanatar–Ceperly [56] parametrization to describe the exchange term,

$$V_{\text{exc}\uparrow} = \frac{\sqrt{2}}{4} \frac{e^2}{\varepsilon_0 \varepsilon_r \pi^{3/2}} \sqrt{n} \left[[1+\xi]^{3/2} + [1-\xi]^{3/2} - \frac{2n_\downarrow}{n} \left[[1+\xi]^{1/2} + [1-\xi]^{1/2} \right] \right] \tag{9.38}$$

and

$$V_{\text{exc}\downarrow} = \frac{\sqrt{2}}{4} \frac{e^2}{\varepsilon_0 \varepsilon_r \pi^{3/2}} \sqrt{n} \left[[1+\xi]^{3/2} + [1-\xi]^{3/2} - \frac{2n_\uparrow}{n} \left[[1+\xi]^{1/2} + [1-\xi]^{1/2} \right] \right], \tag{9.39}$$

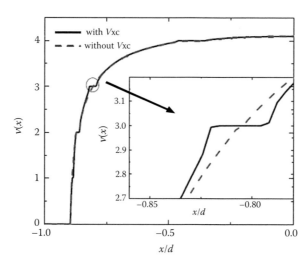

FIGURE 9.30　Spatial distribution of the filling factor, where $\nu(0) < 4$. The IS 3 is dramatically enhanced when compared without considering exchange potential.

where $\xi = \dfrac{n\uparrow - n\downarrow}{n}$ is the spin-polarization, and ε_0 and ε_r are the dielectric constants of vacuum and material. Figure 9.30 compares the widths of the incompressible strips when considering an effective $g^* = 5.2$ and the exchange enhancement. We clearly see that the exchange potential drastically widens the IS, whereas $\nu = 2$ strip is essentially unaffected due to the vanishing polarization. Another many-body effect is the correlation; however, such an effect arises from the wave function itself, which we neglected completely at this level. Therefore, correlation effects cannot be accounted for in our oversimplified approximation. The discrepancy from more realistic calculations [57,58] is mainly twofold: (1) We cannot investigate corrections to the kinetic part, and (2) the many-body gap induced by correlations cannot be included in our calculations, which essentially leads to fractional quantum Hall effect. The second topic will be touched on at the end of this chapter in a qualitative manner. To summarize, we have included the effects coming from many-body interactions within a mean-field approximation, where the exchange potential is taken from the Tanatar–Ceperly description. We have observed that the exchange potential acts differently on spin-up and spin-down electrons; hence, the Zeeman gap is enlarged effectively. Therefore, one observes larger odd-integer incompressible strips. The correlation part does not contribute to the widths (in opposite direction) because we have neglected the actual extends of the wave functions.

The next interaction that must be included in our single-particle Hamiltonian is the spin-orbit coupling (SOC), which is due to the interaction between the spin of the electron and the (cyclotron) orbit. We should note that the coupling strength, α_{SO}, is so small in energy that the actual energy dispersion is almost unchanged. However, the phase of the wave function is thought to be modified [59]. The spin-orbit coupling Hamiltonian is given by

$$\mathcal{H}_{SO} = \frac{\hbar}{(2m_e c^2)}\, \vec{\sigma} \cdot [\vec{\nabla}V \times (\vec{\mathbf{p}} + e\vec{\mathbf{A}})] \tag{9.40}$$

and can be generalized up to a third-order Dresselhaus term as

$$\mathcal{H}_D \propto \left\{\sigma_x \kappa_x, \kappa_y^2 - \kappa_z^2\right\} + \left\{\sigma_y \kappa_y, \kappa_z^2 - \kappa_x^2\right\} + \left\{\sigma_z \kappa_z, \kappa_x^2 - \kappa_y^2\right\} \tag{9.41}$$

where the curly brackets denote the anticommutator and K is the wave number. As an illustrative example, we show the calculated energy dispersion that also takes into account SOC in Figure 9.31. One can clearly see the splitting of the degenerate Landau levels due to strong SOC, proportional to the momentum in y direction and wire width W. We performed our calculations for electrons that are confined to a quantum wire because, as we will discuss in the next subsection, only the electrons within the incompressible strips contribute to current and will have an impact at the interference experiments. The energy scale is very small compared to the cyclotron

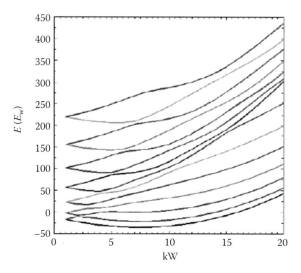

FIGURE 9.31 Level splitting of electrons confined to a wire due to spin-orbit coupling in the presence of an external magnetic field. Anticrossings are observed at large k_y values, i.e., center coordinates.

or Zeeman energy; however, the phase of the wave function changes because the real and the imaginary parts are differently affected for each spin direction [60]. Thus far, we have only calculated the potential and the density distributions, considering different interactions acting on the electron system. We have seen that the bulk theories include the effect of impurities and obtain broadened DOS together with the classical localization, neglecting the edge effects. Instead, the edge theories add the actual confinement potential and describe the IQHE by means of quantum mechanical conductivity models, where many edge states coexist. The direct Coulomb interaction is included in our Hamiltonian to cure the abrupt density profile in a self-consistent manner within a Thomas–Fermi mean-field approximation and we have shown that the 2DES is split to compressible and incompressible regions, with completely different screening properties. Once the potential varies rapidly on the quantum mechanical length scales, TFA breaks down and one has to do a better approximation—namely, a full Hartree calculation—which also takes into account the finite extend of the wave functions [49,61]. Then, we observed that if the incompressible strip becomes narrower than the extend of the wave function, the strip becomes leaky and back-scattering takes place. In the next step, we also included an exchange term and observed that the effective g^* factor is enhanced considerably, leading finite widths of the odd-integer ISs. Finally, our single-particle Hamiltonian also contained the spin-orbit interaction and we have seen that the spin-degenerate Landau levels are split. All of the above considerations do not give a prescription to obtain transport properties of a 2DES subject to a perpendicular **B** field. In the next section, we introduce a local-equilibrium transport model that successfully describes the IQHE, including interactions.

9.3.3.6 Calculating the Resistances

Once the density and the electric fields (electrostatic potentials) are calculated self-consistently, it is required to have a well-defined prescription to obtain the current distribution via Ohm's law [35,49,62]

$$\mathbf{j}(x, y) = \hat{\sigma}(x, y)\mathbf{E}(x, y), \tag{9.42}$$

where $\mathbf{j}(x,y)$ and $\mathbf{E}(x,y)$ are 2D vectors, together with the 2×2 conductivity tensor $\hat{\sigma}(x, y)$. Hence, if one can relate the local density (or filling factor) with the conductivity tensor elements, it is an easy task to obtain the current (density) distribution. We take the relation from the self-consistent Born approximation [33]; however, any other reasonable conductivity model can also be used [36,37,63]. By any other reasonable model, we mean that the longitudinal conductivity should vanish, i.e., σ_l ($\nu = k$) = 0 for the incompressible regions and the Hall conductivity assumes an integer at these regions. The essential outcome of such a conductivity calculation is shown in Figure 9.32, which satisfies these conditions. We should note that approximating the local conductivities to the conductivities calculated for an infinite system is valid only if TFA conditions are satisfied and the local conductivities have to be obtained from local potentials [37]. However, the essential features of the IQHE, such as quantized Hall plateaus and vanishing longitudinal resistance, are independent of the conductivity model considered. On the other hand, the transition regions strongly depend on the conductivity model; therefore, results for activation energy, localization length, etc. are affected. We also note that the local-probe experiments also point to the necessity of a local-transport model [64–67]. Now, let us relate the local conductivities with the global (measurable) quantities, such as resistances. Assuming a translational invariance and utilizing Maxwell equations, together with the equation of continuity, one obtains the longitudinal resistivity as

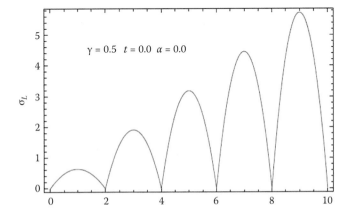

FIGURE 9.32 Longitudinal conductivities calculated within the self-consistent Born approximation depicting Gaussian impurities characterized by the level-broadening parameter and single impurity range $\alpha = R/l$, within the limit of delta potential.

$$R_l = 2E_y^0/I, \quad I = \int j_y(x)\,dx, \tag{9.43}$$

where E_y^0 is the constant electric field in y direction and is given by

$$E_y^0 = I\left[\int \frac{dx}{pl(x)}\right]^{-1} = \text{constant} \tag{9.44}$$

Note that the longitudinal resistivity and conductivity vanish simultaneously,

$$\rho_l(\upsilon = k) = \sigma_l^{-1}(\upsilon = k) = \frac{\sigma_l(\upsilon = k)}{\sigma_l^2(\upsilon = k) + \sigma_H^2(\upsilon = k)} = 0 \tag{9.45}$$

Hence, all the current is confined to the incompressible strips, which have integer filling factors. Meanwhile, the Hall conductivity assumes the quantized value; therefore, the Hall resistance is simply

$$R_H = \frac{V_H}{I} = E_y^0 \int \frac{\rho_H(x)}{\rho_l(x)} \tag{9.46}$$

The findings of the local experiments, which measure either the local potential, compressibility, or resistivity, coincide almost perfectly with our numerical findings, as discussed next. As an example, we demonstrate the filling factor and current density distribution in Figure 9.33, where we also abandoned the assumption of translation invariance, but implicitly imposed periodic boundary conditions while solving the

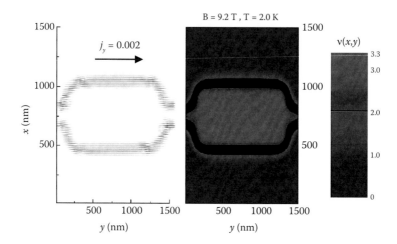

FIGURE 9.33 **(See color insert.)** Spatial distribution of the current density (left) and filling factor (right). One-to-one correspondence between the current and the incompressible strips (black contour) can be clearly seen. The higher current densities in the left panel point to the formation of hot-spots. System parameters are given at the figure labels, whereas current density is given in units of cyclotron energy.

PE. We modeled the current contacts by assuming metallic gates residing at the ends of the sample; such an approximation is crude, however useful, and is realistic up to some extent that can be tested experimentally [68]. Here, we see that the current is distributed all over the sample if there exists no incompressible strip. In contrast, in the presence of an incompressible strip (edge or bulk), all the current is confined to these regions. Hence, one observes the quantized Hall effect as shown in Figure 9.34, which depicts the incompressible strips (highlighted colors) and the calculated global resistances (lines) for an translational invariant system. Injecting current to the incompressible strips is an issue that must be answered. In principle, these states are completely filled and it is therefore not possible to add any other electrons. However, the current is injected to the incompressible strips near the contacts and it is reasonable to think that the current density near the hot-spots locally heats the electrons and the incompressible region is partially melted. Hence, at this region it is possible to inject electrons to the incompressible strip and, once the electrons are injected, the incompressible strip carries the current. The preliminary calculations [69], including Joule heating locally [70], support our microscopic injection picture. The calculation scheme briefly reported here gives a satisfactory description of the IQHE in the linear response regime, where the imposed current does not affect the electron density. In fact, the current itself induces the Hall potential, which has to be added to the self-consistent calculation and, therefore, modifies the electron and potential distribution, which is then the nonlinear transport regime. Similar calculations show that the induced potential imposes asymmetries in the resistances, which become important when considering unequally depleted samples [62,71]. The effects arising from the nonlinear transport regime itself are a very recent topic and still under investigation. Indeed, such current-induced asymmetries can also be seen in Figure 9.33, which are the so-called hot-spots. A full understanding of the formation

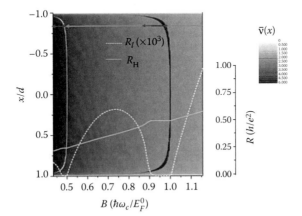

FIGURE 9.34 The interrelation between incompressible strips (highlighted background figure) and resistances (lines on top). If the system is completely compressible (indicated by a red arrow), the longitudinal resistance is finite while RH is not quantized. If there is an incompressible region within the sample (indicated by green arrows), a quantized Hall effect is observed.

of these regions are still under debate; however, one can obtain such spots within classical or semiclassical calculations [69,72,73]. The modeling of current contacts is another challenging problem; some very recent numerical approaches describe the contacts either without quantum mechanics or self-consistent interactions [72,73], which provide reasonable results at the injection and collection regions. These results are in accord with our results; however, our scheme has to be improved to give a systematic description.

9.3.3.7   Summary

In this section, we briefly mentioned the mainstream theories that describe different aspects of the IQHE, like the bulk and the noninteracting edge pictures. Next, we discussed the effects of single- and many-body interactions within a mean-field approximation at the Hartree or Hartree–Fock level. We discussed the conditions to have incompressible strips and their influence on the current distribution utilizing a local transport model and mentioned the agreement with the experiments. The many-body effects are included in our calculations at a basic level (LDA+DFT) and we have also obtained the odd integer plateaus. The contacts and the nonlinear transport regime are left unresolved; however, we pointed in the directions to cover the weak parts of our scheme. In the next section, we present some of the applications of the interaction theory of the IQHE, which is essentially a semiclassical microscopic model that takes into account, as much as possible, the real experimental conditions.

9.4   RECENT EXPERIMENTAL SYSTEMS AND THEIR MICROSCOPIC MODELING

As a first application, the ultimate precision and high reproducibility of the quantized Hall effect was used to obtain a resistance standard; however, the fascinating properties of the ballistic edge states that mimic quantum mechanical transport properties are also utilized to deepen our understanding of nanostructures. To investigate transport at a 2DEG, in the absence of edge states, the so-called Corbino geometry is used where gauge invariance is also tested. Moreover, coherent electronic states could also be inferred by double-layer systems and the statistics of real or quasiparticles are investigated via interference experiments. The mono-energetic edge states are used as laser beams and many of the optical interference experiments are done using electrons; however, the experimental outcomes from optics are not always as expected. Last but not least, momentum-resolved tunneling spectroscopy experiments performed on cleaved-edge overgrown samples enable us to investigate the extreme edge properties and, hence, to check the conditions of incompressible strip formation. All of these very interesting experiments are briefly discussed here, and we also present some of our recent results.

9.4.1   THE CORBINO GEOMETRY

In the early days of the quantized Hall effect, when single-particle theories were popular, it was common to investigate transport properties of a 2DES considering a Corbino geometry [74]. This is shown in Figure 9.35a and the incompressible strip

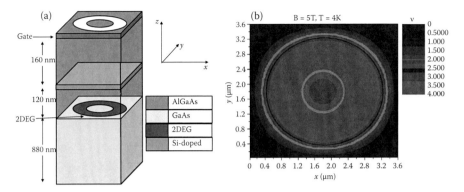

FIGURE 9.35 **(See color insert.)** Corbino disc is defined by inner and outer spherical gates or etching 280 nm below the surface. The spatial distribution of the local filling factors, where incompressible strips are highlighted by gray color. Note that strips carry equilibrium edge currents; however, there is no excess current imposed as in the Hall bar geometry.

distribution is depicted in Figure 9.35b. The essential difference from a Hall bar is that the edge states are perpendicular to the contacts; moreover, experimentally one measures conductances, not resistances. Ideally, a Corbino geometry is equal to a Hall bar: if one takes an infinitely long Hall bar and merges the ends of the bar, one obtains a Corbino geometry. However, due to the radial symmetry one usually solves the single-particle Hamiltonian in the symmetric gauge. As mentioned before, Corbino geometry is utilized to suppress edge effects and the IQHE is explained only by the bulk effects. The idea is, if the bulk of the Corbino geometry is compressible, the system behaves like a metal and conduction between the inner and the outer contact is finite or, if the bulk is incompressible, the Hall conductivity can only change by an integer multiple of e^2 / h because both the particle number and magnetic flux is quantized. In contrast, we show that, if there exists an incompressible strip along the edge of the constraint that decouples the two contacts, quantized Hall conditions are satisfied, although the bulk remains compressible. The recent experimental investigations, both local-probe-wise and transport-wise, strongly support our numerical results that the bulk of the Corbino sample is compressible at the low field edge of the conduction plateau. A systematic investigation of this geometry is still underway. The arrows in Figure 9.35b present the excess current direction, whereas the curved arrows show the direction of the equilibrium current. In a Hall bar, both of these currents are confined to the same incompressible region; however, at a Corbino geometry, they are perpendicular to each other. Here, we have shown that the Corbino and Hall bar geometries are not the same, either in their current distribution or measurable quantities. Moreover, we pointed out that the term "in the absence of edge states" is not correct. Edge states exist, however, they do not contribute to the excess current.

9.4.2 Quantum-Point Contacts

As we have discussed, one can induce narrow constrictions on the 2DES by depositing gates on the crystal surface [12]. One of the most commonly used quantum

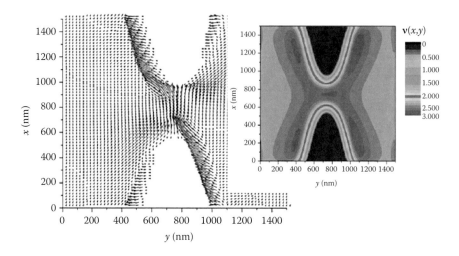

FIGURE 9.36 (See color insert.) The spatial distribution of the classical current in the close vicinity of a gate-defined QPC (right) and the local filling factor distribution, where two incompressible strips come close to each other. The conductance is not quantized; however, interference may occur due to partitioning between edge states.

constrictions is the so-called quantum-point contact, where two gates are placed in a close proximity to create a quasi-1D channel similar to Figure 9.36. In the presence of a perpendicular **B** field, as shown previously, Landau levels are formed and transport takes place only if a percolating incompressible strip exists. A similar case is also observed at the QPCs, i.e., if the Fermi level falls between two Landau levels, only a fixed number of electrons can be transported, leading to a conductance plateau. Otherwise, the conduction increases linearly with the number of particles (namely, the chemical potential) injected into the constraint. A typical QPC geometry and distribution of incompressible strips, together with the conduction versus particle number (or **B** field) illustration, is shown in the inset of Figure 9.36. Since the conduction plateau is strongly affected by the formation of the incompressible strips and incompressible strips are sensitive to the geometry of the sample, it is important to perform calculations considering realistic constrain geometries. This statement is quite different than the conventional assumption that all the QPCs are the same; in fact, the quantum transport is strongly related to the structure itself and a systematic numerical investigation shows that one should keep an eye on the real structural properties of the system [10,75]. Most important, screening (interactions, in general) properties of the material at hand have a strong impact on the transport, which is usually underestimated for simplicity.

9.4.3 DOUBLE-LAYER SYSTEMS SUBJECT TO ⊥ B-DIRECT COULOMB INTERACTION

Another interesting system where direct Coulomb interactions become important is the double layer system. In Section 9.2.3, we have shown that if two (or more) quantum wells are brought into close vicinity, quantum mechanical properties, like tunneling [22], may become important [21]. Here, we discuss such a system; however, we focus our attention on interlayer Coulomb interactions in the presence of a **B** field and look at the effects coming from the formation of the incompressible strips due to their screening properties. In Figure 9.37, we show the calculated density profiles considering different top gate potentials using our 3D routine introduced at the beginning of this chapter. We observe that the bottom layer is not affected by the high bias applied to the top gate until the top electron layer is depleted, since the top layer can screen the external potential almost perfectly. The case dramatically changes if subjected to a **B** field and considering a density mismatch. The top layer has a lower electron density, hence, the bulk incompressible region forms at a lower **B** field value and the screening becomes poor. Therefore, the bottom layer sees the modulation potential (i.e., the long-range disorder potential) with an enhanced amplitude. The case is similar with the edge incompressible strips; however, screening is better since the compressible region is much larger compared to the incompressible region(s) [76]. The change in the external field affects screening properties of the system locally, which can be directly measured by the second layer transport properties. The findings of the interaction theory agree very well with the experiments [76,77] that show hysteresis-like behavior, since the sweeping direction of the magnetic field becomes important: if one starts with low fields, first two narrow-edge incompressible strips are formed. This makes the system relax more easily to equilibrium when compared to a large bulk incompressible region.

Therefore, starting from high fields changes the equilibration processes and the incompressible bulk at the top layer dominates the longitudinal resistance of the

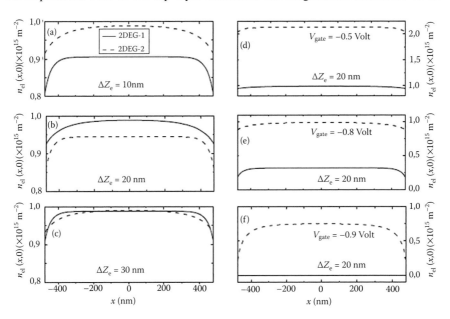

FIGURE 9.37 Electron number density in a bilayer system illustrating different separation thicknesses (left) and a fixed separation by applying different voltages to the top gate. The top layer (2DES-1) is completely depleted at the highest gate potential (−0.9 V).

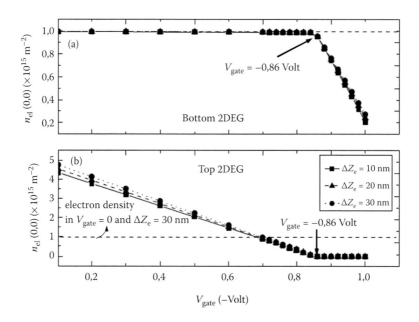

FIGURE 9.38 Bulk electron densities of the (a) bottom and (b) top layers as a function of the applied gate potential. The bottom layer is almost unaffected until the top layer is depleted.

bottom layer. A more interesting phenomena observed at the bilayer systems is the Coulomb drag, where the current is driven only from one (active) layer and the developed Hall field at this layer generates a Hall field at the other (passive) layer [78]. Most of the theories developed to explain this effect assume an infinite system and obtain the transverse resistances via (phonon induced) momentum transfer [78]. The only microscopic model that tackles Coulomb drag microscopically is the interaction theory [79]. However, a systematic and full self-consistent treatment is still missing. The results of a first calculation is shown in Fig. 9.38, where the electron density is changed by a top gate. The self-consistent calculation results in a highly nonlinear voltage-density relation, leading complicated magneto-transport phenomena, namely the negative Coulomb drag.

9.4.4 INTERFEROMETERS

At low temperatures, low-dimensional electron systems show numerous peculiar quantum transport properties [13,80–83]. Some of the most interesting transport systems are the particle interferometers. Interferometers like Mach–Zehnder [80] and Aharanov–Bohm [83] are induced in a 2DES by metallic gate electrodes and/or chemical etching. The strong magnetic field applied perpendicular to the plane of the interferometers quantizes the current carrying states and generates edge states where the current is carried without backscattering, i.e., ballistic transport. These quantized edge states replace the (monochromatic) light beams at the optical versions of the interferometer(s), therefore a coherent transport takes place. Various unexpected findings of recent experiments performed at these (quantum-Hall-based) interferometers indicate that an accurate treatment of interaction effects, taking full and realistic account of sample and geometry dependent details, is essential for a satisfactory understanding of the observed phenomena, such as the path length independent interference pattern. A promising theoretical candidate is the screening/ interaction theory of the integer Hall effect, which was able to explain microscopically both the vanishing longitudinal resistance and the exact quantization of the plateaus, together with the transition between the plateaus. The measurements under quantized Hall conditions ($T < 10$ K and $\mathbf{B} > 2$ Tesla) are not only performed considering integer charged particles, but are also performed using fractionally charged (quasi-) particles that present interference patterns as well [84], and the patterns cannot be explained within noninteracting single-particle theories or DFT-based interaction models [85]. The importance of application of the above mentioned screening theory to the fractional Hall effect and interferometers is promoted by the specialists of the community.

Here, we will briefly address these issues using the self-consistent solution schemes of the Schrödinger–Poisson equations concerning the particle interferometers in two different geometries (such as Mach–Zehnder and Aharonov–Bohm), within Hartree- and spin-dependent generalizations thereof. In doing so, we shall build on our previous work with this method, which was shown to describe successfully various subtle geometry-dependent effects observed in quantum Hall systems, and investigate different working regimes of the interferometers such as high or low temperatures and magnetic fields. A special case of the self-consistent calculation, i.e., Thomas–Fermi–Poisson approximation (SCTFPA), which has already been used to successfully explain the reproducibility of the highly accurate Hall plateaus and the local probe experiments.

We aim to generalize the solution techniques of the Schrödinger–Poisson equations, and investigate the transport properties under the influence of interactions and thereby give an explanation of the experimental findings observed at different interferometers. Finally, we aim to design our own interferometer under optimized working conditions and search for application possibilities.

In the past decade, high-mobility samples enabled the experimentalists to build structures in which one can infer quantum mechanical properties of the edge channels, like the phase of the particles that carry the current. By phase, we mean the Aharonov–Bohm phase and also the classical phase arising from different velocities of the particles on different paths. In Figure 9.39, we show a schematic drawing of the Mach–Zehnder and Aharonov–Bohm interferometers, together with the filling factor distributions of a typical $\mathbf{B}$ field value. The spatial positions of the incompressible edge states are also shown in Figure 9.40 at selected $\mathbf{B}$ field values, (a) where interference is expected since two-edge states come close enough to each other, (b) where no interference pattern may be observed while two states are far apart, or (c) where no partitioning takes place, i.e., two states overlap. As mentioned, the AB phase that the electrons (or quasiparticles) acquire is due to the loop that particles enclose $\Delta\phi_{AB} \propto \Phi/\Phi_0$ and if they move with different velocities on two paths of the arms (l) of the interferometer $\Delta\phi_{vel} \propto eV_D l/v\hbar$, hence, leading to a differential conductance [86]

(a)

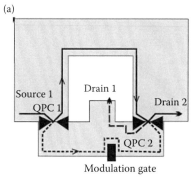

(b) Aharonov-Bohm Interferometer

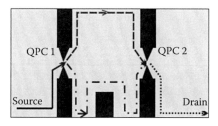

FIGURE 9.39 Schematic drawing of the (a) Aharonov–Bohm and (b) Mach–Zehnder interferometers. The edge states replace coherent light beams, whereas QPCs act as semitransparent mirrors. Path lengths of the interferometer arms are changed by the modulation gate, hence, the area enclosed by the coherent electrons, which affect the phase acquired.

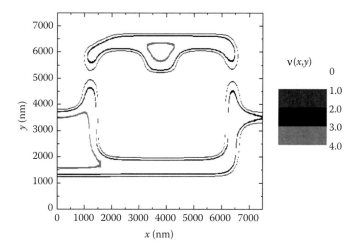

FIGURE 9.40 The incompressible edge state distribution at the real Mach–Zehnder interferometer calculated at absolute zero temperature within a semi-self-consistent scheme. Different integer filling factors are indicated by colors. At this B field value, no interference pattern is expected since the outermost edge states are far apart and the $\nu = 1$ channel breaks in some regions on its way from source to drain, hence phase coherence is lost.

$$\delta G_D(\Phi_{AB}, V_D) = \delta G_0 \cos(2\pi\Phi/\Phi_0)\cos(eV_D l/v\hbar), \tag{9.47}$$

where δG_0 is the conductance independent of the phase, and V_D is the drift velocity. The actual velocity of the edge states is defined by the widths of the incompressible strips [87], since $v = \dfrac{1}{\hbar}\dfrac{\partial E(k)}{\partial k}$, therefore it is important to take into account the real interferometer geometry and perform calculations self-consistently. Steps toward this have been put forward; however, the variety of sample structures and complexity of the geometries limit our efforts. In the near future, a more systematic investigation will be made to deepen our understanding of the interference patterns.

9.4.5 Cleaved-Edge Overgrown Samples

We have discussed the importance of the incompressible strips to investigate a variety of different experimental setups; moreover, we have also seen that the formation of these strips are limited by quantum mechanics as well as electrostatics.

An extreme sharp-edge limit is obtained by the so-called cleaved-edge overgrowth (CEO) method [88]. The idea is to grow two 2DES perpendicular to each other and create a very sharp edge at the CEO side [89], as shown in Figure 9.41. Hence, an incompressible strip will form on the normal edge; however, due to the strong variation at the total potential, the incompressible strip vanishes on the CEO edge. Such an asymmetry at the confinement potential has quite a strong effect on the transport properties, both in the linear and nonlinear response regimes. The side metallic gate allows us to manipulate the edge profile in a very precise manner; therefore, one can observe the effects of the edge on the nonlocal transport. We proposed a couple of unexpected effects that can be observed at such extreme sharp edges [90]. One of those is the rectification of the longitudinal resistance depending on the current direction at nonlinear response [71], which coincides with the preliminary experimental results performed at the Grayson lab at Northwestern University.

These experiments support the prediction of the edge-to-bulk transition of the IQHE, explicitly stating that one can infer the two distinct regimes of the IQHE,

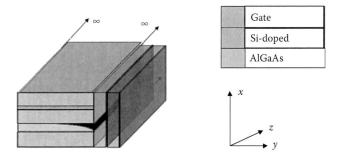

FIGURE 9.41 Illustration of a cleaved-edge overgrown structure, two 2DES are placed perpendicular to each other. The steepness of the confinement potential of the horizontal 2DES at the CEO edge is controlled by the perpendicular gate on the right-hand side.

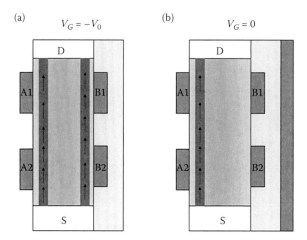

FIGURE 9.42 The CEO sample biased by a negative side gate (gray region), resulting in a "normal" sample with two incompressible strips (pink-colored) at both (a) edges and (b) the unbiased system where there exists only one incompressible strip at the soft edge. The arrows indicate the excess current, whereas the color grade (green) depicts the smooth change at the electron density. S denotes source, D denotes drain, and the yellow region is the insulator between the two 2DESs.

namely the bulk incompressible region, where bulk theories are applicable, and the edge regime, where all the physics is dominated by the edge states. Figure 9.42 depicts the two different regimes considering a CEO sample by showing the spatial distribution of the incompressible strips. The calculations assume that the CEO edge is extremely sharp; however, since we are at very high fields (i.e., the extent of the wave function or magnetic length is still small enough), the TFA provides reliable results. Moreover, in Figure 9.43, we show the (a) actual potential and (b) density distributions, together with the same quantities at the first-growth direction, near the CEO edge calculated using Aquilla.

Calculations performed considering the extreme sharp-edge limit [90] needs more attention to be accurate and reliable. The main difficulty is that one has to obtain self-consistent solutions together with interactions and narrow intervals (typically at the order of tenths of the magnetic length) such that a calculation can only be done utilizing finite element methods, which we are currently working on.

9.4.6 Curved Crystals

The 2DESs that we have discussed so far are always confined to a straight plane; however, it is also possible to create a 2DES on a surface of a curved geometry [91], as shown in Figure 9.44. If such a system is subject to a magnetic field in the z direction, the field component normal to the surface changes proportional to the curvature angle and an inhomogeneous **B** field is experienced by the 2DES. Such an inhomogeneous field induces different filling factors locally, assuming that the density is homogeneous. Therefore, one can model the current distribution using the local Ohm's law similar to a plane 2DES except the fact that one has to include

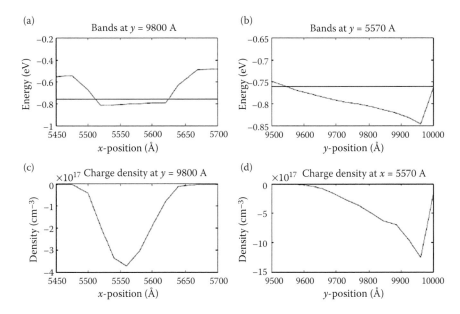

FIGURE 9.43 The valance band energy as a function of position in the (a) x and (b) y directions, focused to the quantum confinement region. Together with the calculated charge density distribution (lower). At the sharp edge, potential and density varies rapidly, leading to a narrow incompressible strip.

an additional effect—the so-called *static skin effect*—observed at curved metals. This effect forces current to be confined to a thin area at the edge of the metal and has been well-studied [92]. A similar effect also takes place at the curved 2DES, but only if the bulk of the sample becomes metallic, as we have shown. Hence, if the bulk of the sample becomes metallic, the current is confined only to one edge;

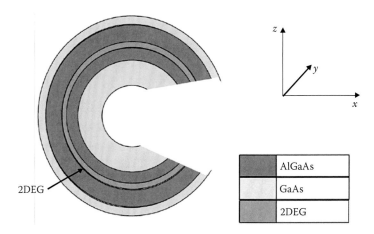

FIGURE 9.44 Schematic presentation of cylindrical 2DES, obtained by removing some amount of the substrate. The strain caused by removal of the crystal curves the 2DES.

therefore, when measuring longitudinal resistances, one observes dramatic asymmetries depending on the field direction. These experiments are performed at the Paul–Drude institute by K.-J. Friedland and results coincide with our interaction theory [93]. Further numerical and experimental investigations are necessary to understand the electrochemical potential induced by the curvature; however, making quantitative estimations is a formidable task.

9.5 RELATED INTERESTING SYSTEMS

The topics that are covered up to now are mainly modeling of 2D devices usually within a single-particle mean-field approximation; however, under strong magnetic fields the 2D systems also present many-body (MB) effects.

These effects are generally very complicated and are hard to handle numerically considering real experimental conditions; hence, analytical methods are basically used for idealized (and sometimes unrealistic) systems. The numerical deficiency arises from the large Hilbert space and, at most, 15–20 particles can be considered. Therefore, it is desired to have a seminumerical microscopic model to describe some of the interesting effects. We start with the fractional quantum Hall effect, where particles are strongly correlated and many-body interactions essentially dominate the physics. The second interesting experimental finding is observed at bilayer systems at total filling $\nu_T = 1$, where both layers have partial filling factors near 1/2. It was reported by several groups that both the longitudinal and Hall resistances vanish due to a strongly correlated coherent many-body state. The zero resistance states were attributed to Bose–Einstein condensation of excitonic particles, which was unexpected to be observed. Recent calculations considering ultracold Fermionic atoms present Landau-level-like discrete energy states and compressible/incompressible rings were obtained. Here, the rotation of the trap replaces the magnetic field and it is possible to simulate QHE with atoms. The last subsection is devoted to a brief discussion of these atomic systems.

9.5.1 FRACTIONAL QUANTUM HALL EFFECT

Historically, IQHE was attributed to the noninteracting single particles; however, the fractional quantum Hall effect, where resistance plateaus shown in Figure 9.34 are also seen at fractional values (see Figure 9.45), was explained by many-body interactions [32]. In this chapter, we have shown that the IQHE is also due to interaction; however, this interaction is a single-particle-direct Coulomb interaction. If one now also considers a many-body Hamiltonian

$$H = \sum_i H_i = \sum_i \left(\frac{1}{2m^*} \left(p_i - \frac{e}{e} A(r_i) \right)^2 + V_{\text{ext}}(r_i) + \sum_j V(r_i,r_j) \right), \quad (9.48)$$

where $\sum_j V(\mathbf{r}_i,\mathbf{r}_j)$ describes the many-body interactions, the solution becomes fairly complicated. As we have discussed, the simplest way is to replace the MB term by

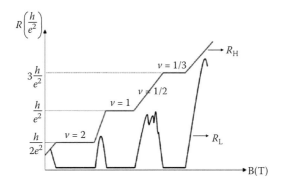

FIGURE 9.45 Illustration of the fractional quantum Hall effect: (black solid line) longitudinal and (red solid line) Hall resistances as a function of external magnetic field. The longitudinal resistance vanishes at $\nu = 1/3$, meanwhile Hall resistance is quantized to $3h/e^2$. Note the local minimum near $\nu = 1/2$, mimicking a zero effective magnetic field experienced by composite Fermions.

a mean-field approximated potential. In the first attempts, the single-particle term was neglected and the entire system was described by the interaction term because the correlation energy between the electrons dominated the single-particle energy. Then, the system was transformed to a single-particle Hamiltonian by assuming a Chern–Simons vector potential, which describes properly the statistical interactions. Hence, a many-body system can be solved as a single-particle system. The new quasiparticles are called the composite Fermions [30], where $2p$ flux quanta is attached to each electron and the noninteracting single-particle integer quantum Hall effect is obtained for composite Fermions by an effective magnetic field $\mathbf{B}^*$, hence $\nu = 1/2$ is mapped to $\mathbf{B} = 0$. Although, this statement is almost correct, the physics is much more complicated and is far beyond the scope of the present chapter. We believe that it is sufficient to note that such a strongly correlated MB system reconstructs the partially occupied lowest Landau level ($\nu < 1$) such that a new incompressible state is manifested due to interactions. Hence, one may repeat the above considerations on the incompressible strips to the FQHE [48,50]. For sure, such a treatment is limited by the discussed conditions on the MB wave extend, etc., and a better approximation to calculate real sample geometries requires both computational and analytical improvements.

9.5.2 EXCITONIC BOSE–EINSTEIN CONDENSATION

Another very exciting observation is reported in bilayer systems, where both of the layers are around $\nu = 1/2$, i.e., half-filled and brought within close proximity comparable with the wave extend. We have seen such a case in Figure 9.16, where the electron wave functions have finite probability in both wells. At $\nu = 1/2$ situation, one of the layers can be interpreted as half-filled electron gas and the other layer to be a half-filled hole gas, as shown in Figure 9.46. Hence, one can imagine that an exciton is formed with integer spin, i.e., obey Bose statistics [32]. If these quasiparticles form a macroscopic many-body wave function at low temperatures, one can consider

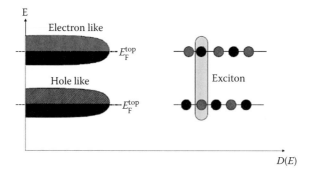

FIGURE 9.46 The density of states depicting a bilayer system, where both layers are half-filled. One can interpret the occupation as half-filled electrons or half-filled holes. An electron-hole pair, i.e., an exciton, is shown in the inset. This excitonic state forms a many-body coherent ground state under BEC conditions, i.e., at low temperatures and strong interactions.

this state to be a Bose–Einstein condensate; hence, a superfluid [23]. Therefore, this state is charge neutral and insensitive to applied potentials and the resistances vanish. This picture makes sense; however, it is incomplete in explaining the observed activation energies and also to explain similar experiments at bilayer-Corbino geometries [25]. A microscopic calculation that also takes into account direct and indirect interactions is desirable, though it is also a formidable task. Our interaction theory is a promising candidate, which has to be extended considering strong correlations of a exciton system. We are currently working on such a system, utilizing finite element methods.

9.5.3 ROTATING ULTRACOLD ATOMS

An analogue of a 2D electron gas subject to a high perpendicular magnetic field are the rapidly rotating, trapped ultracold atoms [94]. To give a mathematical intuition, we consider a system of rapidly rotating Fermi gas of ultracold atoms trapped in a 2D symmetric harmonic trap [95]

$$
\begin{aligned}
H &= \frac{p^2}{2M} + \frac{1}{2}M\omega^2 r^2 - \Omega L_z \\
&= \frac{1}{2}(\mathbf{p} - M\Omega \times \mathbf{r})^2 + \frac{1}{2}\left(\omega^2 + \Omega^2\right)r^2
\end{aligned}
\tag{9.49}
$$

where M is the mass of the atoms, is the rotation frequency, ω is the harmonic trap potential, and L_z is the angular momentum of the system in z direction. Here, Ω determines the rotation frequency. One can immediately see that the vector potential $A(\mathbf{r})$ is replaced by the rotation frequency crossproduct position vector and the confinement potential is replaced by $\frac{1}{2}\left(\omega^2 - \Omega^2\right)r^2$. Hence, our derivation of the

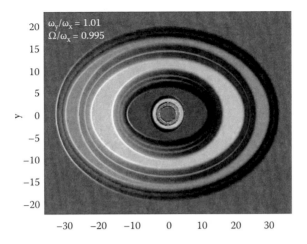

FIGURE 9.47   The atomic density distribution of a rapidly rotating Fermionic ultracold gas, calculated as in reference Ghazanfari and Oktel. The fully occupied Landau levels lead to formation of incompressible rings, whereas thin, light lines represent Friedel oscillations. (From Ghazanfari, N., and M. Ö. Oktel. 2010. *The European Physical Journal D* 59(3):435–441.)

Landau levels also applies here; one can then calculate the particle density at finite temperatures by

$$N_{\text{atom}}(r) = \sum_{\alpha} |\psi_{\alpha}(r)|^2 f\left(E, E_{\alpha}, \mu, kT\right) \tag{9.50}$$

An illustration is shown in Figure 9.47 for a typical rotation frequency and trapping parameters. We see the well-developed incompressible rings, together with Friedel oscillations (thin, equidensity lines) [96]. It is interesting to look at some effects observed at 2D—quantized Hall systems and geometries like the Corbino geometry, bilayer systems, etc.—moreover, the role of interactions may become important if one would like to investigate the fractional Hall states. Such calculations are on their way and our preliminary results agree well with the Corbino geometry discussed above. The next step is to include dipolar interactions (instead of direct Coulomb interaction) within a mean-field approximation similar to our calculations considering 2DES [97].

9.5.4   GRAPHENE

In the very last subsection of this chapter, we would like to touch a hot topic of the field, which utilizes 2D mass-less Dirac fermions to explain the quantized Hall effect observed at graphene monolayers [98]. Similar to our previous subsection, a Hamiltonian analogy is established between the Dirac Hamiltonian, and the Hamiltonian describes a graphene crystal formed of honeycomb lattices. Because the energy dispersion is parabolic at the Fermi energy, which is zero without doping, the particles can be described by a relativistic Hamiltonian. Moreover, the (Landau) energy gap of the discrete levels is large compared to normal semiconductors. Hence, quantum effects can be observed also at very high (essentially room) temperatures.

A huge amount of scientific activity is observed considering many aspects of QHE at graphene. The QHE also becomes a great laboratory for high-energy physicists and attracts attention from very different fields of the physics community. The subject is vast and our numerical and/or analytical approaches presented here are not sufficient to attack such a problem. However, in the field of QHE, many open questions remain other than the graphene, even in the theory, of standard QHE.

9.6 CONCLUSION

This chapter provides a brief report on the numerical modeling of 2D charge systems, where real sample structures, device geometries, and, more important, single-particle interactions are taken into account close to the experimental conditions. The crystal growth parameters are incorporated to solve the PE and SE self-consistently in one, two, and three dimensions. One of the most powerful tools that we have at hand is a home-improved numerical routine that can solve PE considering open boundary conditions for many interesting device geometries, the EST3D. The quantum mechanical calculations focusing on electron and potential distribution utilize the Matlab routine Aquilla and enable us to obtain the actual wave functions and energy band diagrams of materials we are interested in. Our (numerical-wise) last tool is an almost homemade FORTRAN routine, where we can obtain the filling factor, potential, and current density distribution under experimental conditions. Here, the Thomas–Fermi approximation is used at the Hartree level and we have simulated many interesting devices like Hall bars, Corbino disks, QPCs, and contacts.

Our numerical investigations point to a couple of interesting results: (1) The actual sample growth parameters and defining method (etching/gating/oxidation) of the geometric patterns strongly influence the electrostatic quantities. (2) The inclusion of single-particle interactions into the Hamiltonian of a 2DES subject to perpendicular **B** field leads to formation of compressible incompressible strips for which we have calculated, in a self-consistent manner, the widths and positions at finite temperatures and shown that the incompressible strips become leaky if their widths are comparable with quantum mechanical length scales. Hence, only one edge state is left at a given quantized Hall plateau that carries all the imposed excess current. (3) We applied our method to recent experimental systems most interesting to the interference experiments that utilize mono energetic edge states. The visibility oscillations and the unexpected behaviors are to be investigated with our tools.

We mentioned the related quantum Hall systems and their vast variety of applications. The many-body interactions based experiments are distant to our approach at the moment; however, reasonable directions are discussed by which to study these effects.

We have also reported on our very preliminary results that consider ultracold rapidly rotating atoms. Further investigation is under way.

ACKNOWLEDGMENTS

The results presented here are a collection of the author and his collaborators. Hence, I would like to thank all of them for their kind help in preparing the manuscript. I would like to thank Professor Rolf R. Gerhardts, who introduced most of the concepts discussed here during my Ph.D. work, especially the idea of including interactions in a self-consistent way and utilizing the local Ohm's law. Much of the experimental knowledge and intuition were gained during my stay at the Max Planck institute Stuttgart in Professor Klaus von Klitzing's group. In particular I would like to thank Professor W. Dietsche, Dr. S. Lok, Dr. M. C. Giesler, Dr. Omar Stern, PD J. Weis, and Dr. E. Ahlswede. As a "probe" postdoctoral fellow, I learned a lot from PD Stefan Ludwig, Dr. Dawid Kupidura, and our MS student Jose Horas in the group of Professor J. Kotthaus, who I also would like to thank for their kindness. At my real postdoctoral work, I had the privilege of discussions with Professor Florian Marquardt, Dr. V. Golovach, and Professor J. von Delft, which helped me to improve my understanding of the interference phenomena. The members of the Heiblum group, and especially Prof. M. Heiblum and Nissim Ofek (Mach–Zehnder and Aharanov–Bohm interferometers); also Dr. K. J. Friedland (cylindrical QHE) contributed much to my vision of the QHE-related systems. I also very much appreciate the comments, lectures, and discussions of Associate Professor Matthew Grayson. He enlightened me on many issues about the quantized Hall effect, and especially on CEO crystals. Also, thanks to Assistant Professor O. Oktel for introducing me to the interesting cold atoms.

Our nanoelectronics group in Turkey performed calculations and prepared a considerable amount of the data presented here; moreover, they also helped me in preparing the illustrations and schematic drawings (especially I. Pekyilmaz, S. Mirioglu, H. Atci, and N. Ghazanfari). I would like to thank (for assistance with figures and data) A. Salman, S. Aktas, A. I. Mese, E. Cicek, D. Eksi, S. E. Gulebaglan, G. B. Akyuz, S. Sarikurt, I. Sokmen, A. Yildiz, E. Koymen, U. Erkarslan, G. Oylumluoglu, and N. Ghazanferi for their contributions. Finally I also acknowledge A. U. Siddiki for her critical reading of the manuscript.

These works were partially supported by DPG, DFG, MPG, DIP, and TUBiTAK (TBAG:109T083), the Feza Gursey Institute, I.T.A.P Marmaris, IU-BAP:6970, and the municipality of Akyaka.

REFERENCES

1. Neil W. Ashcroft and N. David Mermin. 1976. *Solid state physics*. Pacific Grove, CA: Brooks Cole.
2. J. H. Davies. 1998. In *The physics of low-dimensional semiconductors*, New York: Cambridge University Press.
3. S. Datta. 1995. In *Electronic transport in mesoscopic systems*. Cambridge: Cambridge University Press.
4. A. Weichselbaum and S. E. Ulloa. 2003. Potential landscapes and induced charges near metallic islands in three dimensions. *Physical Review E* 68(5):056707.

5. P. M. Morse and H. Feshbach. 1953. *Methods of theoretical physics*, vol. II, 1240. New York: McGraw-Hill.

6. S. Arslan. 2008. *Image processed modelling of qpc's*. M.Sc. Thesis, Technical University of Munich.

7. M. Huber, M. Grayson, M. Rother, W. Biberacher, W. Wegscheider, and G. Abstreiter. 2005. Structure of a single sharp quantum Hall edge probed by momentum-resolved tunneling. *Physical Review Letters* 94:016805.

8. M. Heiblum. 2007. Private communication.

9. A. Weichselbaum and S. E. Ulloa. 2006. Tunability of qubit Coulomb interaction: Numerical analysis of top-gate depletion in two-dimensional electron systems. *Physical Review B* 74(8):085318.

10. S. Arslan, E. Cicek, D. Eksi, S. Aktas, A. Weichselbaum, and A. Siddiki. 2008. Modeling of quantum point contacts in high magnetic fields and 77 with current bias outside the linear response regime. *Physical Review B* 78(12):125423.

11. E. Cicek, A. I. Mese, M. Ulas, and A. Siddiki. 2009. *Spatial distribution of the incompressible strips at Aharonov-Bohm interferometer*. ArXiv Condensed Matter e-prints. September, 2009.

12. B. J. van Wees, H. van Houten, C. W. J. Beenakker, J. G. Williamson, L. P. Kouwenhoven, D. van der Marel, and C. T. Foxon. 1988. Quantized conductance of point contacts in a two-dimensional electron gas. *Physical Review Letters* 60:848.

13. M. Avinun-Kalish, M. Heiblum, O. Zarchin, D. Mahalu, and V. Umansky. 2005. Crossover from 'mesoscopic' to 'universal' phase for electron transmission in quantum dots. *Nature* 436:529–533.

14. G. Hackenbroich and H. A.Weidenmueller. 1995. *Independent-electron model for the phase of the transmission amplitude in quantum dots*. ArXiv Condensed Matter e-prints /9502033, February, 1995.

15. G. Hackenbroich, W. D. Heiss, and H. A. Weidenmüller. 1997. Deformation of quantum dots in the Coulomb blockade regime. *Physical Review Letters* 79:127–130.

16. J. H. Davies, I. A. Larkin, and E. V. Sukhorukov. 1995. Modeling the patterned two-dimensional electron gas: Electrostatics. *Journal of Applied Physics* 77(9):4504.

17. M. Stopa. 1996. Quantum dot self-consistent electronic structure and the Coulomb blockade. *Physical Review B* 54:13767–13783.

18. P. G. Silvestrov and Y. Imry. 2007. Level-occupation switching of the quantum dot, and phase anomalies in mesoscopic interferometry. *New Journal of Physics* 9:125.

19. C. Karrasch, T. Hecht, A. Weichselbaum, J. von Delft, Y. Oreg, and V. Meden. 2007. Phase lapses in transmission through interacting two-level quantum dots. *New Journal of Physics* 9:123.

20. P. J. Price. 1983. Hot electron effects in heterolayers. *Physica* 117B:750.

21. T. J. Gramila, J. P. Eisenstein, A. H. MacDonald, L. N. Pfeiffer, and K. W. West. 1991. Quantum magnetotransport calculation for 2d electron systems with weak 1d modulation and anisotropic scattering. *Physical Review Letters* 66:1216.

22. J. P. Eisenstein, I. B. Spielman, L. N. Pfeiffer, and K. W. West. 2002. Tunneling in a quantum Hall excitonic condensate. *Int. J. Mod. Phys. B* 16:2923.

23. M. Kellogg, J. P. Eisenstein, L. N. Pfeiffer, and K. W. West. 2004. Vanishing Hall resistance at high magnetic field in a double-layer two-dimensional electron system. *Physical Review Letters* 93(3):036801.

24. E. Tutuc, R. Pillarisetty, S. Melinte, E. P. DePoortere, and M. Shayegan. 2003. Layer-charge instability in unbalanced bilayer systems in the quantum Hall regime. *Physical Review B* 68:201308.

25. L. Tiemann, J. G. S. Lok, W. Dietsche, K. von Klitzing, K. Muraki, D. Schuh, and W. Wegscheider. 2008. Exciton condensate at a total filling factor of one in Corbino two-dimensional electron bilayers. *Physical Review B* 77(3):033306.

26. R. Tsu and L. Esaki. 1973. Tunneling in a finite superlattice. *Applied Physics Letters* 22:562–564.

27. K. Lier and R. R. Gerhardts. 1994. Self-consistent calculation of edge channels in laterally confined two-dimensional electron systems. *Physical Review B* 50:7757.

28. R. R. Gerhardts. In *Low-dimensional electron systems in semiconductors*. In preparation, 2010.

29. K. v. Klitzing, G. Dorda, and M. Pepper. 1980. New method for high-accuracy determination of the fine-structure constant based on quantized Hall resistance. *Physical Review Letters* 45:494.

30. D. C. Tsui, H. L. Stormer, and A. C. Gossard. 1982. Electronic processes at the breakdown of the quantum Hall effect. *Physical Review Letters* 48:1559.

31. S. Das Sarma, M. Freedman, and C. Nayak. 2005. Topologically protected qubits from a possible non-abelian fractional quantum Hall state. *Physical Review Letters* 94(16):166802.

32. Zyun F. Ezawa. 2000. *Quantum Hall effects: Field theoretical approaches and related topics*. Singapore: World Scientific.

33. T. Ando, A. B. Fowler, and F. Stern. 1982. Electronic properties of two-dimensional systems. *Reviews of Modern Physics* 54:437.

34. B. Kramer, S. Kettemann, and T. Ohtsuki. 2003. Localization in the quantum Hall regime. *Physica E* 20:172.

35. K. Güven and R. R. Gerhardts. 2003. Self-consistent local-equilibrium model for density profile and distribution of dissipative currents in a Hall bar under strong magnetic fields. *Physical Review B* 67:115327.

36. T. Champel, S. Florens, and L. Canet. 2008. Microscopics of disordered two-dimensional electron gases under high magnetic fields: Equilibrium properties and dissipation in the hydrodynamic regime. *Physical Review B* 78(12):125302.

37. T. Kramer. 2006. A heuristic quantum theory of the integer quantum Hall Effect. *International Journal of Modern Physics B* 20:1243–1260.

38. A. Siddiki. 2005. *Model calculations of current and density distributions in dissipative Hall bars*. PhD thesis, Julius-Maximilians-University Würzburg.

39. A. L. Efros. 1988. Non-linear screening and the background density of 2deg states in magnetic field. *Solid State Communications* 67:1019.

40. A. Siddiki and R. R. Gerhardts. 2007. Range-dependent disorder effects on the plateau-widths calculated within the screening theory of the IQHE. *Int. J. of Mod. Phys. B* 21:1362.

41. J. A. Nixon and J. H. Davies. 1990. Potential fluctuations in heterostructure devices. *Physical Review B* 41:7929–7932.

42. J. Horas, A. Siddiki, J. Moser, W. Wegscheider, and S. Ludwig. 2008. Investigations on unconventional aspects in the quantum Hall regime of narrow gate defined channels. *Physica E* 40:1130–1132.

43. W. Cai and C. S. Ting. 1986. Screening effect on the landau-level broadening for electrons in gaas-gaalas heterostructures. *Physical Review B* 33:3967.

44. B. I. Halperin. 1982. Self-consistent local-equilibrium model for density profile and distribution of dissipative currents in a Hall bar under strong magnetic fields. *Physical Review B*, 25:2185.

45. U. Wulf, V. Gudmundsson, and R. R. Gerhardts. 1988. Screening properties of the two-dimensional electron gas in the quantum Hall regime. *Physical Review B* 38:4218.

46. M. Büttiker. 1986. Four-terminal phase-coherent conductance. *Physical Review Letters* 57:1761.

47. R. Landauer. 1981. Can a length of perfect conductor have a resistance? *Phys. Lett.* 85A:91.

48. D. B. Chklovskii, B. I. Shklovskii, and L. I. Glazman. 1992. Electrostatics of edge states. *Physical Review B* 46:4026.

49. A. Siddiki and R. R. Gerhardts. 2004. Incompressible strips in dissipative Hall bars as origin of quantized Hall plateaus. *Physical Review B* 70:195335.

50. A. M. Chang. 1990. A unified transport theory for the integral and fractional quantum Hall effects: Phase boundaries, edge currents, and transmission rejection probabilities. *Solid State Communications* 74:871.

51. D. B. Chklovskii, K. A. Matveev, and B. I. Shklovskii. 1993. Ballistic conductance of interacting electrons in the quantum Hall regime. *Physical Review B* 47:12605.

52. A. Siddiki and R. R. Gerhardts. 2003. Thomas–Fermi–Poisson theory of screening for laterally confined and unconfined two-dimensional electron systems in strong magnetic fields. *Physical Review B* 68:125315.

53. A. Siddiki. 2008. The spin-split incompressible edge states within empirical Hartree approximation at intermediately large Hall samples. *Physica E* 40:1124–1126.

54. G. Bilgeç, H. Üstünel Toffoli, A. Siddiki, and I. Sokmen. 2010. The self-consistent calculation of exchange enhanced odd integer quantized Hall plateaus within Thomas-Fermi-Dirac approximation. *Physica E: Low-dimensional Systems and Nanostructures* 42(4):1058–1061.

55. W. Kohn and L. Sham. 1965. Spatial spin polarization and suppression of compressible edge channels in the integer quantum Hall regime. *Physical Review* 140:A1133.

56. B. Tanatar and D. M. Ceperley. 1989. Quantized Hall conductance in a two-dimensional periodic potential. *Physical Review B* 39:5005.

57. C. Sohrmann and R. A. Römer. 2006. Compressibility in the integer Quantum Hall Effect within Hartree–Fock approximation. *Physica Status Solidi C* 3:313–316, February 2006.

58. S. Ihnatsenka and I. V. Zozoulenko. *"0.7 anomaly" and magnetic impurity formation in quantum point contacts.* ArXiv Condensed Matter e-prints, January 2007.

59. T. Hakioglu. *Electronic Mach–Zehnder interferometry under spin-orbit coupling.* Unpublished, 2007.

60. S. Sarikurt, S. Sakiroglu, A. Siddiki, T. Hakioglu, and I. Sokmen. 2010 *Electronic Mach-Zehnder interferometry under spin-orbit coupling.* Unpublished.

61. T. Suzuki and T. Ando. 1993. Transport properties between quantum Hall plateaus. *Journal of the Physical Society of Japan,* 62:2986.

62. A. Siddiki. 2009. Current-direction induced rectification effect on (integer) quantized Hall plateaus. *EPL* 87:17008–17014.

63. R. R. Gerhardts. 1975. Path-integral approach to the two-dimensional magneto-conductivity problem ii application. *Zeitschrift für Physik B* 21:285.

64. E. Ahlswede, P. Weitz, J. Weis, K. von Klitzing, and K. Eberl. 2001. Hall potential profiles in the quantum Hall regime measured by a scanning force microscope. *Physica B* 298:562.

65. E. Ahlswede, J. Weis, K. von Klitzing, and K. Eberl. 2002. Hall potential distribution in the quantum Hall regime in the vicinity of a potential probe contact. *Physica E* 12:165.

66. S. Ilani, J. Martin, E. Teitelbaum, J. H. Smet, D. Mahalu, V. Umansky, and A. Yacoby. 2004. The microscopic nature of localization in the quantum Hall effect. *Nature* 427:328.

67. G. A. Steele, R. C. Ashoori, L. N. Pfeiffer, and K. W. West. 2005. Imaging transport resonances in the quantum Hall effect. *Physical Review Letters* 95(13):136804, September 2005.

68. O. Goektas. 2009. *Small alloyed ohmic contacts to 2DES and submicron scale Corbino devices in strong magnetic fields: Observation of a zero bias anomaly and single-electron charging.* PhD thesis, Stuttgart University.

69. D. Eski, O. Kilicoglu, and A. Siddiki. 2010. *Screening Model of Metallic Non-Ideal Contacts at Integer Quantized Hall Regime.* eprint arXiv:1003.5963.

70. S. Kanamaru, H. Suzuuara, and H. Akera. Spatial distributions of electron temperature in quantum Hall systems with compressible and incompressible strips. *Journal of the Physical Society of Japan* 75(6):064701.1–064701.10.

71. A. Siddiki, J. Horas, D. Kupidura, W. Wegscheider, and S. Ludwig. 2009. *Asymmetric non-linear response of the IQHE*. ArXiv Condensed Matter e-prints, November 2009.

72. C. Uiberacker, C. Stecher, and J. Oswald. 2009. Systematic study of nonideal contacts in integer quantum Hall systems. *Physical Review B* 80(23):235331.

73. T. Kramer, V. Krueckl, E. J. Heller, and R. E. Parrott. 2009. *On the self-consistent calculation of electric potentials in Hall devices*. ArXiv Condensed Matter e-prints, November 2009.

74. V. T. Dolgopolov, A. A. Shashkin, N. B. Zhitenev, S. I. Dorozhkin, and K. v. Klitzing. 1992. *Physical Review B* 46:12560.

75. A. Siddiki and F. Marquardt. Self-consistent calculation of the electron distribution near a quantum-point contact in the integer quantum Hall effect. *Physical Review B*, 75:045325, 2007.

76. A. Siddiki. 2007. Self-consistent Coulomb picture of an electron-electron bilayer system. *Physical Review B* 75:155311.

77. A. Siddiki, S. Kraus, and R. R. Gerhardts. 2006. Screening model of magnetotransport hysteresis observed in bilayer quantum Hall systems. *Physica E* 34:136.

78. F. von Oppen, S. H. Simon, and A. Stern. 2001. Oscillating sign of drag in high Landau levels. *Physical Review Letters* 87(10):106803.

79. K. Güven, A. Siddiki, P. M. Krishna, and T. Hakioglu. 2008. A self-consistent microscopic model of Coulomb interaction in a bilayer system as an origin of drag effect phenomenon. *Physica E* 40:1169–1171.

80. Y. Ji, Y. Chung, D. Sprinzak, M. Heiblum, D. Mahalu, and H. Shtrikman. 2003. An electronic Mach-Zehnder interferometer. *Nature* 422:415.

81. L. V. Litvin, H.-P. Tranitz, W. Wegscheider, and C. Strunk. 2007. Decoherence and single electron charging in an electronic Mach-Zehnder interferometer. *Physical Review B* 75(3):033315.

82. P. Roulleau, F. Portier, D. C. Glattli, P. Roche, A. Cavanna, G. Faini, U. Gennser, and D. Mailly. 2007. *Direct measurement of the coherence length of edge states in the integer quantum Hall regime*. ArXiv Condensed Matter e-prints, 710, October 2007.

83. F. E. Camino, W. Zhou, and V. J. Goldman. 2005. Aharonov-bohm electron interferometer in the IGH regime. *Physical Review B* 72:155313.

84. F. E. Camino, W. Zhou, and V. J. Goldman. 2007. *e/3* Laughlin quasiparticle primary-filling $v = 1/3$ interferometer. *Physical Review Letters* 98(7):076805, February 2007.

85. S. Ihnatsenka and I. V. Zozoulenko. 2008. *Interacting electrons in the Aharonov–Bohm interferometer*. ArXiv Condensed Matter e-prints, 803, March 2008.

86. D. T. McClure, Y. Zhang, B. Rosenow, E. M. Levenson-Falk, C. M. Marcus, L. N. Pfeiffer, and K. W. West. 2009. Edge-state velocity and coherence in a quantum Hall Fabry-Pérot interferometer. *Physical Review Letters* 103(20):206806, November 2009.

87. D. Eksi, E. Cicek, A. I. Mese, S. Aktas, A. Siddiki, and T. Hakioglu. 2007. Theoretical investigation of the effect of sample properties on the electron velocity in quantum Hall bars. *Physical Review B* 76:075334.

88. L. Pfeiffer, K. W. West, H. L. Stormer, J. P. Eisenstein, K. W. Baldwin, D. Gershoni, and J. Spector. 1990. Formation of a high quality two-dimensional electron gas on cleaved GaAs. *Reviews of Modern Physics* 56:1697–1699.

89. M. Huber, M. Grayson, M. Rother, R. A. Deutschmann, W. Biberacher, W. Wegscheider, M. Bichler, and G. Abstreiter. 2002. Tunneling in the quantum Hall regime between orthogonal quantum wells. *Physica E* 12:125–128.

90. U. Erkarslan, G. Oylumluoglu, and A. Siddiki. 2009. *Edge-to-bulk transition of the IQHE at cleaved edge overgrown samples: A screening theory based experimental proposal*. ArXiv Condensed Matter e-prints /0906.3796, June 2009.

91. K.-J. Friedland, R. Hey, H. Kostial, A. Riedel, and K. H. Ploog. 2007. Measurements of ballistic transport at nonuniform magnetic fields in cross junctions of a curved two-dimensional electron gas. *Physical Review B* 75(4):045347, January 2007.

92. A. V. Chaplik. 2000. Some exact solutions for the classical Hall effect in an inhomogeneous magnetic field. *Soviet Journal of Experimental and Theoretical Physics Letters* 72:503–505.

93. K.-J. Friedland, A. Siddiki, R. Hey, H. Kostial, A. Riedel, and D. K. Maude. 2009. Quantum Hall effect in a high-mobility two-dimensional electron gas on the surface of a cylinder. *Physical Review B* 79(12):125320.

94. M. H. Anderson, J. R. Ensher, M. R. Matthews, C. E. Wieman, and E. A. Cornell. 1995. Observation of Bose–Einstein condensation in a dilute atomic vapor. *Science* 269:198–201.

95. A. L. Fetter. 2009. Rotating trapped Bose–Einstein condensates. *Reviews of Modern Physics* 81:647–691.

96. N. Ghazanfari and M. Ö. Oktel. 2010. Rapidly rotatting fermions in an anisotropic trap. *The European Physical Journal D* 59(3):435–441.

97. N. Ghazanfari, A Siddiki, and M. O. Oktel. 2010. *Self-consistent modeling of current contacts*. Unpublished.

98. A. K. Geim. 2009. Graphene: Status and prospects. *Science* 324:1530.

99. A. I. Mese, A. Bilekkaya, S. Arslan, S. Aktas, and A. Siddiki. 2010. Investigation of the coupling asymmetries at double-slit interference experiments. *Journal of Physics A: Mathematical and Theoretical* 43:354017.

Index

Printed and bound by CPI Group (UK) Ltd, Croydon, CR0 4YY

21/10/2024

01777112-0004